Organic Photochemistry and Photophysics

MOLECULAR AND SUPRAMOLECULAR PHOTOCHEMISTRY

Series Editors

V. Ramamurthy
Professor
Department of Chemistry
University of Miami
Miami, Florida

Kirk S. Schanze
Professor
Department of Chemistry
University of Florida
Gainesville, Florida

Organic Photochemistry and Photophysics

edited by
Vaidhyanathan Ramamurthy
University of Miami
Miami, Florida

Kirk Schanze
University of Florida
Gainesville, Florida

CRC Press
Taylor & Francis Group
Boca Raton London New York

CRC Press is an imprint of the
Taylor & Francis Group, an **informa** business
A TAYLOR & FRANCIS BOOK

Published 2006 by CRC Press
Taylor & Francis Group
6000 Broken Sound Parkway NW, Suite 300
Boca Raton, FL 33487-2742

First issued in paperback 2019

No claim to original U.S. Government works

ISBN 13: 978-0-367-45396-1 (pbk)
ISBN 13: 978-0-8493-7608-5 (hbk)

This book contains information obtained from authentic and highly regarded sources. Reasonable efforts have been made to publish reliable data and information, but the author and publisher cannot assume responsibility for the validity of all materials or the consequences of their use. The authors and publishers have attempted to trace the copyright holders of all material reproduced in this publication and apologize to copyright holders if permission to publish in this form has not been obtained. If any copyright material has not been acknowledged please write and let us know so we may rectify in any future reprint.

Visit the Taylor & Francis Web site at
http://www.taylorandfrancis.com

and the CRC Press Web site at
http://www.crcpress.com

Library of Congress Cataloging-in-Publication Data

Catalog record is available from the Library of Congress

Preface

Photochemistry is vital to the survival of life on Earth. Light-triggered events serve primary functions in various biological events. The role of light in materials science needs no emphasis, and the value of solar energy has never been more evident than now. Photochemistry is no longer a discipline practiced by a few who are interested in understanding the excited state photochemistry and photophysics of selected molecules. The concepts of photochemistry have pervaded various disciplines, and research in this area is pursued not only by photochemists but also by physical organic chemists, chemical physicists, biologists, materials scientists, and industrial chemists.

This volume contains seven chapters written by international experts on the photochemistry and photophysics of organic, inorganic, and biological molecules. The first chapter by Steer and coworkers summarizes the current status of upper excited state physics of organic and inorganic molecules. Almost five decades ago it was believed that no measurable processes are likely from S_2, S_3, T_2, and T_3 states; however, Chapter 1 makes it evident that this belief is no longer valid.

Excited state proton transfer has been known and investigated for almost four decades. Recent developments in pico and femto second spectroscopy have revived interest on this topic. In Chapter 2, Shizuka and Tobita summarize recent advances in excited state proton transfer.

Organic photochemistry deals with exploring and generalizing the excited state behavior of various chromophores. Excited state behavior of most other chromophores is understood on the basis of carbonyls, olefins, enones, and aromatics, which have played leading roles in this process. In Chapter 3, Nau and Pischel highlight the excited state behavior of azoakanes by comparing them directly with alkanones.

Photosubstitution reactions of aromatics could become very useful synthetic reactions provided all we know about them is understood in terms of simple models, as discussed by Fagnoni and Albini in Chapter 4.

Two of the most fundamental processes in chemistry are electron and proton transfer. Electron transfer has received considerable attention during the last two decades. In Chapters 5 and 6, Yoon, Yasuda, and coworkers, who have made fundamental contributions to electron transfer photochemistry, summarize their work on photoamination of olefins and photocyclization of phthalimides, two of the most synthetically useful reactions.

In Chapter 7, Armitage discusses a fascinating area of current research in photochemical sciences, namely, aggregation of cynanine and porphyrin dyes. In this chapter, factors that lead to controlled assembly of dyes on DNA surfaces are presented.

The volumes in this series serve the needs of chemists who have interest in using light to control the excited state chemistry of organic, inorganic, and biological molecules. It is believed that the chapters in this volume will serve not only experts but also as supplementary reading material in graduate courses.

Contributors

Angelo Albini
Department of Organic Chemistry
University of Pavia
Pavia, Italy

Bruce A. Armitage
Department of Chemistry
Carnegie Mellon University
Pittsburgh, Pennsylvania

G. Burdzinski
Quantum Electronics Laboratory
Faculty of Physics
Adam Mickiewicz University
Poznan, Poland

Maurizio Fagnoni
Department of Organic Chemistry
University of Pavia
Pavia, Italy

J. Kubicki
Quantum Electronics Laboratory
Faculty of Physics
Adam Mickiewicz University
Poznan, Poland

A. Maciejewski
Photochemistry Laboratory
Faculty of Chemistry
and
Center for Ultrafast Laser Spectroscopy
Adam Mickiewicz University
Poznan, Poland

Patrick S. Mariano
Department of Chemistry
University of New Mexico
Albuquerque, New Mexico

Jin Matsumoto
Department of Chemical Science and Engineering
Miyakonojo National College of Technology
Miyakonojo, Miyazaki, Japan

Werner M. Nau
School of Engineering and Science
International University of Bremen
Bremen, Germany

Uwe Pischel
Department of Chemistry
Faculty of Sciences
University of Porto
Porto, Portugal

Kensuke Shima
Department of Chemical Science and Engineering
Miyakonojo National College of Technology
Miyakonojo, Miyazaki, Japan

Tsutomu Shiragami
Department of Chemical Science and Engineering
Miyakonojo National College of Technology
Miyakonojo, Miyazaki, Japan

Haruo Shizuka
Gunma University
Kiryu, Japan

R.P. Steer
Department of Chemistry
University of Saskatchewan
Saskatoon, Saskatchewan, Canada

Seiji Tobita
Gunma University
Kiryu, Japan

S. Velate
Department of Chemistry
University of Saskatchewan
Saskatoon, Saskatchewan, Canada

Toshiaki Yamashita
Department of Chemical Science and Engineering
Miyakonojo National College of Technology
Miyakonojo, Miyazaki, Japan

Masahide Yasuda
Department of Applied Chemistry
Faculty of Engineering
University of Miyazaki
Gakuen-Kibanadai, Miyazaki, Japan

E.K.L. Yeow
Department of Chemistry
University of Saskatchewan
Saskatoon, Saskatchewan, Canada

Ung Chan Yoon
Department of Chemistry
College of Natural Sciences
Pusan National University
Pusan, Korea

Contents

1 Photochemistry and Photophysics of Highly Excited Valence States of Polyatomic Molecules: Nonalternant Aromatics, Thioketones, and Metalloporphyrins

G. Burdzinski, J. Kubicki, A. Maciejewski,
R.P. Steer, S. Velate, and E.K.L. Yeow

CONTENTS

1.1 INTRODUCTION

Almost a full century of research in the post-quantum-physics era has resulted in a vast literature on the photochemistry and photophysics of polyatomic molecules. For most compounds with closed shell ground states (S_0), only the lowest triplet state (T_1) and the lowest excited singlet state (S_1) need be considered to obtain a rather complete description of the behaviour of the molecule subsequent to one-photon excitation anywhere in the uv-visible region of the spectrum. The phenomenological rationalization for this observation rests with the proposition that excitation of such molecules in their higher energy singlet-singlet absorption systems, $S_n(n \geq 2) \leftarrow S_0$, produces excited states that undergo very rapid radiationless relaxation to S_1 and T_1. In most closed shell molecules, these latter states are much longer-lived, if bound, and thus exhibit intramolecular radiative and nonradiative photophysical processes and intermolecular photochemistry that can be readily measured.

The most convenient direct spectroscopic means of obtaining the rates of photophysical relaxation processes involves measuring the temporal decays and quantum yields of fluorescence and phosphorescence of radiative excited states. Kasha summarized the relevant empirical data that existed in the field of luminescence spectroscopy about a half century ago:

> The emitting level of a given multiplicity is the lowest excited level of that multiplicity [1].

This statement came to be known as Kasha's rule and lead to extensions, variously known as Vavilov's law and the Kasha–Vavilov rule, that the emission spectrum and photochemical quantum yield are independent of excitation wavelength.

Even 50 years ago, the spectroscopic behaviour of azulene was known (or suspected) to contravene Kasha's rule; it emitted relatively strongly from a higher excited state, S_2, but hardly at all from S_1 or T_1 [2]. More recently, with improvements in the sensitivity of fluorescence detection systems and the time resolution of pulsed light sources, it has become clear that Kasha's rule is not a "rule" at all. Rather, the current state of knowledge would be better summarized in a restatement of the definition of the quantum yield of emission:

> A bound electronic state of a polyatomic molecule emits with a probability equal to the ratio of its radiative to total (radiative + nonradiative) decay rates.

Whether or not one can observe emission from any bound excited electronic state is therefore determined only by the oscillator strength of the radiative transition, the rates of competing radiationless transitions and the detection limit of the observation system.

Many polyatomic molecules are now known to decay radiatively with measurable quantum yields from highly excited electronic states. Most are organic or organometallic compounds with closed shell ground states and exhibit $S_n(n \geq 2) \rightarrow S_0$, fluorescence quantum yields in the $\phi_f < 10^{-5}$ range and have $S_n(n \geq 2)$

lifetimes, $\tau_{Sn} < 1$ ps. However, some chromophores are clearly exceptional in that molecules containing them often exhibit much larger upper state quantum yields and much longer fluorescence lifetimes. Most prominent among systems investigated to date are azulene and other nonalternant aromatic compounds, molecules containing the thiocarbonyl chromophore, and some porphyrins and metalloporphyrins. This review focuses on recent advances in our understanding of the photochemistry and photophysics of these exceptional molecules when excited to states higher than S_1 or T_1 in their respective singlet and triplet electronic manifolds.

These exceptional molecular systems all have some spectroscopic and structural elements in common. First, their radiative upper states are bound, eliminating potential competing fast intramolecular photochemical relaxation processes. Second, in each case $S_n(n \geq 2)$ is coupled to the ground state by an electric dipole allowed transition of relatively large oscillator strength, so that the radiative decay constant in the numerator of the fluorescence quantum yield expression is relatively large. Third, the spacings between the radiative state (usually S_2) and the next lowest excited singlet state (usually S_1) are relatively large (typically, $\Delta E(S_2 - S_1) \geq 1$ eV). This disposition of electronic states results in a slowing of the rate of $S_2 - S_1$ internal conversion compared with molecules in which $\Delta E(S_2 - S_1)$ is smaller (as is the usual case), and allows radiative decay to occur with a higher probability. In most of these exceptional molecules, no additional electronic states that are strongly coupled to the radiative upper state, S_n, are interposed between S_n and S_{n-1}.

The recent interest in the photophysics and photochemistry of molecules with relatively long-lived highly excited electronic states has been prompted by advances in optoelectronics and by the development of molecular logic devices [3,4]. Photochemically stable molecules with relatively long-lived fluorescent $S_n(n \geq 2)$ states could act as excitation wavelength-sensitive electron or electronic energy donors/acceptors when incorporated into larger supramolecular or polymeric systems [5,6]. The potential value of incorporating such species into molecular devices increases with increased stability and excited state lifetime. Thus, there has been considerable interest recently in quantifying the factors that control the lifetimes (inverse rates of radiationless decay) of these highly excited states and in identifying additional chromophores with these properties. There is a clear need for further research in this field.

In this chapter we begin by briefly describing the theories of radiative and radiationless transitions. This material underpins interpretations of the data derived from measurements of the relaxation rates of highly excited valence states of polyatomic molecules and is drawn upon in later sections of the chapter. We then proceed to review recent progress in cataloguing and understanding the photochemical and photophysical behaviour of the more highly excited valence electronic states of azulene and other nonalternant aromatic molecules, of the thiocarbonyls, and of selected porphyrins and metalloporphyrins. We do not include research on the highly excited states of small molecules except when such work illustrates phenomena that are more generally applicable in larger systems. We also restrict our consideration to valence shell excited states, therefore specifically excluding long-lived highly excited Rydberg states such as those accessed in ZEKE spectroscopy. Our treatment builds on previous reviews and is complementary to them.

1.1.1 Radiative Transitions

The theory of radiative transitions is well established, and there is no need to reproduce a comprehensive treatment of it here [7]. For our present purposes it is sufficient to note that one can calculate an approximate value of the rate constant, k_r, for the radiative decay of any excited electronic state (to the ground state) from the integrated intensity of the corresponding absorption band system. Such calculations provide data that are useful in assessing the accuracy of fluorescence quantum yield and excited state lifetime data from which experimental values of $k_r = \phi_f/\tau$ are determined.

Consider the integrated molar absorptivity for an $S_n \leftarrow S_0$, absorption, $\int\varepsilon(\bar{\upsilon})d\bar{\upsilon}$, obtained from the appropriate segment of the uv-visible absorption spectrum (using unpolarized light and isotropic samples) and plotted as the molar decadic extinction coefficient (absorptivity), $\varepsilon(\bar{\upsilon})$ in $dm^3mol^{-1}cm^{-1}$, as a function of $\bar{\upsilon}$ in wavenumbers. An approximate value of the radiative rate constant for the corresponding $S_n \rightarrow S_0$ transition may be obtained directly from:

$$k_r = 8 \times 2303\pi c\bar{\upsilon}_0^2 n^2 \mathcal{N}^{-1}G\int\varepsilon(\bar{\upsilon})d\bar{\upsilon} \approx 3 \times 10^{-9}\bar{\upsilon}_0^2\varepsilon_{max}\Delta\bar{\upsilon}_{1/2} \qquad (1.1)$$

where $\bar{\upsilon}_0$ is the wavenumber of the absorption at $(\lambda_{max}, \varepsilon_{max})$, n is the refractive index of the medium, $\mathcal{N}$ is Avogadro's constant, $G = g_\ell/g_u$ (the ratio of the lower to upper electronic state degeneracies), and $\Delta\bar{\upsilon}_{1/2}$ is the full width of the electronic band system at half maximum, expressed in wavenumbers. Equation (1.1) is strictly applicable only to atomic transitions in which the absorption and emission lines are coincident. For molecules, in which there is a significant Stokes shift between the absorption and emission spectrum, Strickler and Berg [8] derived a more precise relationship:

$$k_r = 2.88 \times 10^{-9}<\bar{\upsilon}_f^{-3}>_{av}^{-1}n^2G\int\varepsilon(\bar{\upsilon})d\ln\bar{\upsilon} \qquad (1.2)$$

where $<\bar{\upsilon}_f^{-3}>_{av}^{-1}$ is the reciprocal of the mean value of $\bar{\upsilon}^{-3}$ in the fluorescence spectrum. Equation (1.2) strictly applies only to strongly allowed transitions, like those for which upper state emission can be most readily observed. Equation (1.2) also applies strictly only to those transitions in which the geometric distortion and/or displacement of the potential surface in the excited state is not very different from that of the ground state, *i.e.*, when the Stokes shift is not too large. In the three series of compounds reviewed here, the latter condition is met for the porphyrins and metalloporphyrins excited in their strong Soret bands, but is not met in the thiocarbonyls where $\pi \rightarrow \pi^*$ excitation ($S_2 \leftarrow S_0$) results in a rather substantial elongation of the C=S bond in the excited state.

1.1.2 RADIATIONLESS TRANSITIONS

When the fluorescence quantum yield of a highly excited valence state is small, as is usual, the lifetime of the excited state is determined primarily by the rates of its radiationless transitions. The theory of radiationless transitions in the excited electronic states of polyatomic molecules was developed several decades ago and is not reviewed in detail. Readers are referred to several excellent comprehensive reviews for further information [9,10]. Here we provide only those materials needed to understand the photophysics of the highly excited electronic states of those polyatomic molecules that exhibit unusually long lifetimes.

The course of the radiationless decay of any bound excited electronic state of a polyatomic molecule is determined by the relative disposition of its potential energy surfaces and the interactions between them. Strong interactions between zero order electronic states of the same symmetry result in excited state potential hypersurfaces that can contain several local minima in a large molecule. On the other hand conical intersections between more weakly interacting states of different symmetry (avoided crossings) are characterized by "holes" in an excited state surface and can provide a path for radiationless relaxation over or through a finite potential energy barrier.

No such surface intersections occur in the weak coupling case, in which the potential energy surfaces are "nested." Here, radiationless transitions between an initial zero order state, $|i>$, and a set of weakly interacting final states, $|f>$, may be classified according to the relative magnitudes of the density of final states, ρ_f, and the average interaction matrix element, $\upsilon = <i|H'|f>$, where H' is the perturbation operator. If $\upsilon >> \rho_f$, the system conforms to the small molecule limit where accidental coincidences between $|i>$ and equiergic levels in $|f>$ result in spectroscopically observable perturbations but no radiationless transition. If $\upsilon << \rho_f$, the system conforms to the large molecule or statistical limit where the spacings between final states equiergic with $|i>$ are small compared with the width of $|i>$ itself. (In the isolated molecule case, the width of $|i>$ is determined by its lifetime; in condensed media this width is determined in large measure by its interactions with the surrounding medium.) In this case, $|i>$ interacts with a pseudo-continuum of final states and the radiationless transition from $|i>$ to $|f>$ is statistical in nature and irreversible. Between these two limits is the intermediate case in which coherent excitation of $|i>$ may sometimes result in observable quantum beats superimposed on the fluorescence decay due to recurrences in the system. We need usually consider only the statistical limit in describing the radiationless decay of exceptionally long-lived highly excited valence states of polyatomic molecules. The molecules we shall describe are large ($3N - 6 \geq 27$) and the spacings between their adjacent electronic states are also large $\Delta E(S_2 - S_1) \geq 1$ eV), at least in the singlet manifold. Thus for $S_2 - S_1$ internal conversion in these anomalously fluorescent compounds, the density of final states is large and the condition $\upsilon << \rho_f$ easily met.

In the statistical limit, the probability (s^{-1}) of a radiationless transition from vibronic state $|i>$, with wavefunction Ψ_i and electronic energy E_i, to vibronic state $|f>$

with wavefunction Ψ_f and electronic energy E_f is given by the Golden Rule of time-dependent perturbation theory:

$$k_{nr} = 2\pi\hbar^{-1}\upsilon^2\rho_f \qquad (1.3)$$

The matrix element, υ, may be partitioned into electronic and vibrational terms using the Born-Oppenheimer approximation, $\Psi(q,Q) = \psi(q,Q)\chi(Q)$, where the vibrational wavefunction is $\chi(Q) = \Pi\chi_n(Q_n)$, and the electronic wavefunction, $\psi(q,Q)$, depends only parametrically on the nuclear coordinates, Q. Thus,

$$\upsilon^2 = <\Psi_i(q,Q)|H'|\Psi_f(q,Q)>^2 = \upsilon_{if}^2F \qquad (1.4)$$

and

$$k_{nr} = 2\pi\hbar^{-1}\upsilon_{if}^2F\rho_f \qquad (1.5)$$

where the electronic coupling matrix element (energy) is $\upsilon_{if} = <\psi_i(q,Q)|H'|\psi_f(q,Q)>$, and $F = <\chi_i(Q)|\chi_f(Q)>^2$ is an average squared vibrational overlap or Franck–Condon factor. To first order, the perturbation coupling the two electronic states is just T_N, the nuclear kinetic energy operator, if the interaction is vibronic (*i.e.*, for the process of internal conversion between two states of the same electron spin multiplicity). To first order $H' = H_{SO}$, the spin-orbit coupling Hamiltonian, if the radiationless transition involves states of different electron spin multiplicity (*i.e.*, for intersystem crossing).

Equation (1.5) is the simplest of all Golden Rule expressions because both the electronic matrix element and the Franck–Condon factor are taken as averages over all interacting vibronic states. A better model employs a Franck–Condon weighted density of states, $\rho_f(F)$, in order to account for the fact that not all states in the dense manifold couple with $|i>$ with the same probability. In any case, the variation in the Franck–Condon factors with electronic energy gap, $\Delta E = E_i - E_f$, determines the relative rates of radiationless transitions in compounds that contain the same chromophores and hence exhibit similar values of υ_{if}. The relative magnitudes of the Franck–Condon factors for different vibrational modes also determines the nature of the accepting modes populated preferentially by the radiationless transition.

Recent work by Lim and coworkers [11,12] on small thiones of C_s and C_{2v} excited state symmetry has clearly demonstrated the symmetry requirements of vibronically induced radiationless transitions. In the low pressure gas phase, C_{2v} molecules such as thioformaldehyde, H_2CS, exhibit both strong $S_1 - S_0$ fluorescence and strong $T_1 - S_0$ phosphorescence. These molecules therefore exhibit rates of $S_1 - S_0$ internal conversion and $T_1 - S_0$ intersystem crossing that are small in comparison with the excited state radiative decay rates, despite the fact that the density of states in S_0 is sufficiently large that the molecule's radiationless transitions should fall into the statistical limit case. On the other hand similar thiones of C_s excited state symmetry exhibit small quantum yields of $T_1 - S_0$ phosphorescence and may (*e.g.* thiophosgene, Cl_2CS) or may not (*e.g.* thiocyclobutanone) exhibit strong $S_1 - S_0$ fluorescence. The differences are due to the symmetry requirements of the promoting mode in the radiationless transition and the extent of non-planarity in the excited state. In the tetra-atomic molecules of C_{2v} ground state symmetry, S_1 of A_2 electronic symmetry can only couple with S_0 of A_1 symmetry via a vibration of a_2 symmetry. However, tetra-atomic molecules of C_{2v} symmetry have no a_2 modes, so these molecules are "vibrationally deficient," *i.e.* there is no vibration of the symmetry needed for the matrix elements υ in Equation (1.3) to be non-zero and the rates of radiationless transition are therefore small. Similar considerations should apply to $S_2 - S_1$ internal conversion in the tetra-atomic thiones, but such vibrational deficiencies will not affect the larger polyatomic molecules under consideration here.

The Franck–Condon factors for radiationless transitions depend upon a number of parameters, including the number of quanta of vibrational energy of a given mode needed to bridge the electronic energy gap, the wavenumber of the vibration, and the displacement of the vibrational coordinate(s) in |i> compared with |f>. In all cases, for a sufficiently large electronic energy gap, the Franck–Condon factors decrease approximately exponentially with increasing number of vibrational quanta, as shown in Figure 1.1 [13]. This relationship describes the energy gap law of radiationless transition theory, by which the rate of a radiationless transition varies approximately as the inverse exponential of the electronic energy gap, $\Delta E = E_i - E_f$, between the two coupled states [14]. Moreover, the Franck–Condon factors are largest for a given large ΔE for those vibrations of highest frequency (and which therefore require the fewest quanta to bridge a given electronic energy gap). Thus C–H stretching vibrations are the best accepting modes for radiationless transitions in which ΔE is large compared with $\overline{\upsilon}_{vib,f}$. In such cases, an appreciable H/D isotope effect on the rate of radiationless decay is also to be expected for similar reasons.

1.2 AZULENE AND OTHER NONALTERNANT AROMATIC HYDROCARBONS AND THEIR DERIVATIVES

1.2.1 AZULENE AND ITS SIMPLE DERIVATIVES

Azulene is the best-known exception to Kasha's rule and serves as a model for other nonalternant aromatic compounds, which also exhibit "anomalous" fluorescence from their second excited singlet states. This anomalous emission of the S_2 state was first observed unambiguously by Beer and Longuet-Higgins [2] and has been confirmed many times in more recent studies [15,16]. The second excited singlet state of azulene has a lifetime of *ca.* 1 to 2 ns in both the gas phase and in solution, and exhibits dual emission, decaying radiatively to S_1 with a minute quantum yield ($< 10^{-6}$) [17,18] and to S_0 with a quantum yield most recently determined to be 0.041 (in ethanol at room temperature) [16]. Earlier studies placed the $S_2 - S_0$ fluorescence quantum yield near 0.03 [19,20]. Small *et al.* also recently measured the quantum yield of azulene's $S_2 - S_0$ nonradiative decay using a completely independent

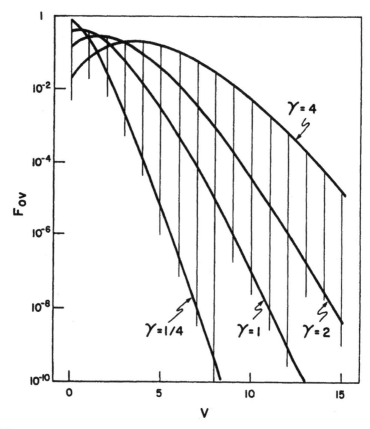

FIGURE 1.1 Semilogarithmic plot of the Franck–Condon factor, F, as a function of lower state vibrational quantum number, v, for several values of the displacement parameter, γ. The displacement parameter is related to the frequency of the vibration, ω, and the displacement, Q, of a harmonic oscillator by $\gamma = \omega Q^2 \hbar / 2$. (From W. Siebrand in "Modern Theoretical Chemistry" Vol. 1, Plenum Press (Kluwer), 1976. With permission.)

photoacoustic calorimetry method and obtained a value of 0.96 ± 0.01 [21]. The balance of a population of thermally equilibrated S_2 molecules relaxes radiatively to S_0 with a quantum yield of 0.04 ± 0.01.

The S_1 state of azulene is much shorter-lived ($\tau \approx 1$ ps, almost independent of temperature) in both the gas phase and in condensed media. Consequently $S_1 - S_0$ fluorescence is very weak (the radiative rate is modest, yielding $\phi_f \approx 10^{-6}$) and pump-probe methods with ps or fs temporal resolution are needed to measure this state's relaxation rates. The reason for S_1's short lifetime has now been well-established through high level *ab initio* calculations of the molecule's potential surfaces by Robb and coworkers [22]. The S_1 and S_0 surfaces exhibit a conical intersection at a geometry in which the transannular bond length in S_1 is significantly smaller than in the ground state and for which the S_1 potential energy minimum lies near the S_0 surface. Consequently the S_1 state decays mainly by surface hopping from a shallow minimum in S_1 to the steep wall of the S_0 potential surface. In doing so the molecule generates $ca.$ 14,000 cm^{-1} of vibrational energy that must be dissipated to the surroundings. The dynamics of the subsequent vibrational relaxation processes have also been studied by pump-probe methods.

Hunt and coworkers [23,24] first observed that azulene vapour exhibited substantially broader linewidths in its absorption $S_1 - S_0$ spectrum (~10 to 20 cm^{-1}) than in its $S_2 - S_0$ spectrum (~1 cm^{-1}). Recently, the vibrational energy dependence of the lifetimes of the excited S_1 state in jet-cooled (isolated) azulene were estimated from linewidths in the $S_1 - S_0$ cavity ring-down absorption spectrum. Under similar conditions, a fs time domain measurement by Zewail and coworkers established a dephasing time of less than 100 fs and an $S_1 - S_0$ internal conversion time constant of 900 ± 100 fs at a vibrational energy of ~2000 cm^{-1} [25].

Building on early work by Heilbronner $et\ al.$ [26,27], Liu and coworkers have recently shown that a degree of "orbital control" can be imposed on the relative energies of azulene's HOMO, LUMO, and LUMO+1 orbitals by substituting the molecule with simple electron withdrawing and/or donating groups (-F, -CHO, -CN, -CH$_3$) ($cf.$ a recent review [28]) [29]. Not only can the colour of the compound be controlled in this way, but it is also possible to increase the LUMO to LUMO+1 spacing significantly by substituting resonantly electron-donating groups such as fluorine at the 1-, 3-, 5- and 7-positions. Consequently, in these fluorine-substituted derivatives the S_2-S_1 electronic energy gap, $\Delta E(S_2 - S_1)$, is large (almost 2 eV) and greater than the corresponding $S_1 - S_0$ electronic energy gap.

Most of the photophysical data for azulene and its simple derivatives have come from measurements of the quantum yields of $S_2 - S_0$ fluorescence, ϕ_{S2}, and time-domain measurements of the lifetimes of the fluorescent S_2 state, τ_{S2}, in fluid solution. The first order rate constants for the parallel radiative and nonradiative processes that depopulate S_2 are then determined from $k_r = \phi_{S2}/\tau_{S2}$ and $\Sigma k_{nr} = (1 - \phi_{S2})/\tau_{S2}$. Whereas $S_2 - S_1$ internal conversion is clearly the major, and perhaps the exclusive, pathway for S_2's nonradiative decay in azulene itself, other nonradiative processes can be expected to occur in some substituted azulenes.

Data from experiments designed to quantify the effects of structure and environment on azulene's excited state dynamics are tabulated in Table 1.1. The following trends are clear:

- The S_2 radiationless transition rate is sensitive to both solvatochromic and substituent-induced shifts of the $S_2 - S_1$ energy gap. Figure 1.2 shows that an almost linear dependence of the logarithm of the rate constant of radiationless decay on the $S_2 - S_1$ energy gap is observed for a series of 1- and 1,3-fluorine-substituted azulenes in a range of solvents, as expected if the energy gap law of radiationless transition theory applies and internal conversion is the sole S_2 decay process. 1,3-Difluoroazulene in ethanol exhibits the largest S_2-S_0 fluorescence quantum yield ($\phi_f = 0.20$) and the longest-lived S_2 state ($\tau_{S2} = 9.5$ ns) measured to date for a polyatomic molecule in solution at room temperature [30].

- Intersystem crossing from S_2 to the triplet manifold is significant in heavy atom-substituted azulenes. Eber *et al.* first reported an internal heavy atom effect in chlorine- and bromine-substituted azulene derivatives and concluded that $S_2 - T_1$ intersystem crossing cannot be neglected in these molecules [30,31]. That conclusion has been verified by Steer and coworkers [16].

- Substitution of alkyl groups increases the nonradiative decay rates by increasing the effective number of coupled states while the electronic coupling matrix element remains constant (*cf.* Figure 1.2) [16].

- The effectiveness of a heavy atom in inducing intersystem crossing is dependent upon the position of substitution. Kim *et al.* measured the radiationless decay rate constants for the 2-haloazulenes in cyclohexane and concluded that these molecules did not exhibit a heavy atom effect since the radiationless decay rates yielded a linear energy gap law plot similar to Figure 1.2 [32]. However, the slope of their plot, which is a measure of the interstate coupling energy, is almost a factor of two larger than that of Figure 1.2, indicating a mix of energy gap and intersystem-crossing control of the radiationless decay rates.

- Substitution of a group, such as formyl (-CHO), can result in a much faster rate of S_2 radiationless decay than would be predicted solely on the energy gap law correlations shown in Figure 1.2. However, as Liu *et al.* have suggested [28,29], these formylazulenes contain n,π^* excited electronic states of energy similar to that of the fluorescent S_2 state, thus creating an alternate or a greatly perturbed S_2 radiationless decay path.

The effects on the $S_2 - S_1$ energy gap of substitution with various groups at different positions on the azulene rings have been explained very nicely by Liu and coworkers using calculations of the electron distributions of the molecular orbitals of azulene [28,29,33]. The HOMO and LUMO+1, the largest MO contributors to S_0 and S_2, respectively, have very similar electron density distributions, with nodes at positions 2 and 6. However, the LUMO of azulene is very different, and has high electron densities at the even numbered positions and very low electron densities at the odd-numbered positions in contrast to HOMO and LUMO+1, which have high electron densities at the odd-numbered positions. Thus electron donating substituents at the 1-, 3-, 5-, or 7-positions raise the HOMO and LUMO+1 energy level without significantly affecting the LUMO energy (*i.e.*, increasing $\Delta E(S_2 - S_1)$) whereas electron-withdrawing substituents at the 1-, 3-, 5-, or 7-positions have the opposite effect and decrease $\Delta E(S_2 - S_1)$. When electron-donating groups are substituted at the 2-, 4-, 6-, or 8-positions, HOMO and LUMO+1 are effected little, but LUMO increases in energy due to increased HOMO-LUMO repulsion and thus $\Delta E(S_2 - S_1)$

TABLE 1.1
Photophysical Data for the Radiative and Nonradiative Decay of the S_2 State of Azulene and its Derivatives in Several Solvents

Compound	Solvent	$E(S_2)/10^3$ (cm^{-1})	$\Delta E(S_2 - S_1)/10^3$ (cm^{-1})	ϕ_f	τ_2/ps	$k_r/10^7 s^{-1}$	$\Sigma k_{nr}/10^8 s^{-1}$
Azulene[a]	CH	28.38	14.01	0.0460	1330	3.50	7.17
	EtOH	28.51	13.99	0.0410	1310	3.10	7.32
FAZ[b]	H	28.01	14.97	0.0567	2430	2.33	3.88
	EtOH	28.14	14.94	0.0579	2460	2.35	3.83
DFAZ[b]	H	28.01	16.13	0.1470	6860	2.14	1.24
	EtOH	28.12	16.03	0.1970	9450	2.08	0.85
Azulene-d_8[a]	CH	28.45	14.05	0.0720	1710	4.20	5.43
	EtOH	28.58	14.02	0.0600	1660	3.60	5.66
IAZ[b]	H	29.07	14.41	0.0131	510	2.57	19.40
FIAZ[b]	H	28.82	15.39	0.0191	940	2.03	10.40
DFIAZ[b]	H	28.82	16.56	0.0508	2690	1.89	3.53
GAZ[a]	CH	27.20	13.50	0.0180	330	5.40	29.80
	EtOH	27.40	13.60	0.0170	330	5.10	29.80
FGAZ[b]	H	27.40	14.80	0.0914	1910	4.79	4.76
DCAZ[a]	CH	27.14	14.35	0.0730	990	7.40	9.36
	EtOH	27.40	14.30	0.0620	880	7.00	10.70
DBAZ[a]	CH	27.03	13.87	0.0160	200	8.00	49.20
TMAZ[a]	CH	27.82	12.29	0.0005	15	3.30	670.00
	EtOH	27.78	12.00	0.0005	14	3.20	710.00
2-Cl-azulene[c]	CH	27.90	12.56		217		
2-Br-azulene[c]	CH	27.76	12.47		198		
2-I-azulene[c]	CH	27.13	11.93		75		

Note: H = *n*-hexane; CH = cyclohexane; EtOH = ethanol; FAZ = 1-fluoroazulene;
DFAZ = 1,3-difluoroazulene; IAZ = 6-isopropylazulene; FIAZ = 1-fluoro-6-isopropylazulene;
DFIAZ = 1,3-difluoro-6-isopropylazulene; DCAZ = 1,3-dichloroazulene; DBAZ = 1,3-dibromoazulene;
TMAZ = 4,6,8-trimethylazulene; GAZ = guaiazulene.
[a] From Ref. [16]; [b] from Ref. [30]; [c] from Ref. [32].

decreases. The opposite occurs when electron-withdrawing groups are substituted at the even positions and hence $\Delta E(S_2 - S_1)$ increases. The largest $S_2 - S_1$ energy gaps, and hence the longest S_2 lifetimes, are therefore expected to be found in azulene substituted at the odd numbered positions by strongly π-donating groups such as fluorine and at the even numbered positions by electron withdrawing groups such as –CN. Eliminating all of the high frequency C–H vibrations in the molecule by full substitution would also result in smaller Franck–Condon factors for the accepting modes in the $S_2 \rightarrow S_1$ internal conversion process.

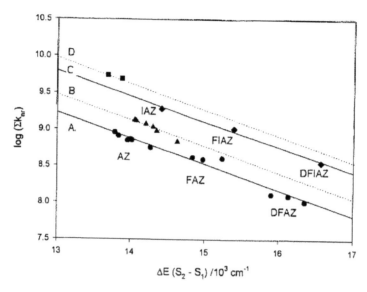

FIGURE 1.2 Log-linear energy gap law plots for azulene and their derivatives. (A) azulene (AZ), 1-fluoroazulene (FAZ), and 1,3-difluoroazulene (DFAZ) in several solvents; (B) 1,3-dichloroazulene; (C) 6-isopropylazulene (IAZ), 1-fluoro-6-isopropylazulene (FIAZ), and 1,3-difluoro-6-isopropylazulene (DFIAZ) in *n*-hexane; (D) 1,3-dibromoazulene. (From R. Steer et al., *J Phys Chem, A,* ACS, 1999. With permission.)

The effect of the molecular environment on the photophysics of the S_2 state of azulene and several of its simple derivatives of C_{2v} symmetry has been investigated by measuring the excitation and emission spectra and excited state fluorescence lifetimes of their Van der Waals complexes in a supersonic expansion [34–36]. A red shift in the $S_2 - S_0$ origin band, of magnitude directly proportional to the polarizability of the bound atom, is observed in the fluorescence excitation spectrum of all azulene-rare gas complexes, as expected if binding is dominated by dispersive interactions. Complexation with only one rare gas atom is sufficient to destroy the vibrational coherence effects observed earlier by Demmer and coworkers [37]. Lifetime shortening is observed for the 1:1 complexes and 1:2 complexes of azulene with Xe, and is attributed to an enhancement, via the external heavy atom effect, of the rate of $S_2 - T_1$ intersystem crossing.

In comparison with the S_1 and S_2 states, the photophysical properties of still higher electronic states of azulene have received little attention. Excitation to higher excited states (S_3 and S_4) produces $S_2 - S_0$ fluorescence spectra almost identical to that obtained by direct excitation to the S_2 state, suggesting that fast internal conversion occurs from S_3 and S_4. However, quantum yields of emission have not been measured as a function of excitation wavelength in this region, so these data do not eliminate the possibility that a dark radiationless decay path may occur at these higher energies [38]. Some additional information has been provided by Foggi *et al.*

[39] using femtosecond pump-probe spectroscopy. In the $S_n - S_1$ transient absorption spectrum of the short-lived S_1 state, absorptions were assigned to $S_3(3A_1)$, which is also active in ground state spectrum, and two close lying excited states $S_6(5A_1)$ and $S_7(6A_1)$, which do not contribute significantly to the ground state spectrum. The transient absorption of the S_2 state contains only one transition with sufficiently high oscillator strength to be observed. It was assigned by means of calculation to the $S_{11} - S_2$ absorption.

1.2.2 OTHER NONALTERNANT AROMATIC HYDROCARBONS

Research has also been undertaken to identify nonalternant aromatic hydrocarbons other than azulene that exhibit similar excited state photophysics. Cyclo-penta[c,d] pyrene (CPP) has an $S_2 - S_1$ energy gap of 7000 cm^{-1}, and anomalous S_2 emission has been observed in both steady state and time resolved fluorescence spectroscopy experiments. Plummer and Al-Saigh observed emission from CPP at 25000 cm^{-1} with a quantum yield of about 10^{-3} and a lifetime of 3 ns [40], and assigned the emitting state to S_2 by matching the steady state excitation spectrum with the absorption spectrum in the $S_2 - S_0$ region. This assignment has been supported by PPP/SCF calculation and by higher resolution fluorescence spectroscopy of CPP in a Shpol'skii matrix [41].

Fujimori and coworkers [42–44] studied several azulene derivatives, including azulene[1,2-b] furan, azulenopyridines, and azulenofuran, all of which exhibit anomalous fluorescence from the second excited singlet state. The emission quantum yields observed for these compounds were quite small, in the 10^{-3} to 10^{-5} range.

There are several reports in the literature in which emission from a highly fluorescent impurity has been mistaken for weak anomalous emission. For example the emission of fluoranthene had been the subject of a long-standing controversy. Weak emission of fluoranthene at ca. 353 nm was initially interpreted as anomalous $S_2 - S_0$ emission [45,46], but this observation was not routinely repeatable [47,48]. Later, Hofstraat et al. investigated fluoranthene using high resolution Shpol'skii spectroscopy [49], and identified the near uv emission as fluorescence from a substituted phenanthrene impurity present in commercial fluoranthene samples.

1.3 THIOKETONES

Molecules containing the thiocarbonyl (> C=S) functional group were first found to exhibit relatively intense $S_2 - S_0$ fluorescence and to undergo photochemical reactions directly from the S_2 state some 30 years ago. Since that time the subject of thioketone photochemistry and photophysics has received considerable attention, and has been reviewed several times [50–57]. The 1993 critical review of Maciejewski and Steer is used as a starting point for this section of the chapter, which focuses primarily on the work done since that review was written [50].

The thioketones have several features in common with other groups of compounds that have relatively long-lived S_2 states and exhibit "anomalous" fluorescence [50,58–60]. The $S_2 - S_0$ transition is electric-dipole allowed ($^1A_1 - {}^1A_1$ in C_{2v} symmetry), the $S_2 - S_1$ electronic energy gap is relatively large (typically in the 7,000 to 11,000 cm^{-1} range for aromatic thiones), and often no other electronic states of the bound molecule are interposed between S_2 and S_1 [58,60]. Thus, the rates of radiationless decay of the S_2 states can be relatively slow and the quantum yields of anomalous fluorescence quite large (> 10^{-3}) [50,58,60]. In other respects molecules containing the thiocarbonyl group are quite unique. They exhibit strong $T_1 - S_0$ phosphorescence in degassed fluid solution at room temperature [50,61], a measurable quantity that is very useful in following the course of relaxation of higher states [62]. These molecules are also quite reactive in their S_2 ($^1(\pi,\pi^*)$) states, being easily deactivated by most common solvents and readily undergoing both intramolecular and intermolecular photochemical H-abstraction [50,58,59]. Fortunately, perfluoroalkanes are relatively inert in these respects, and their use as solvents (despite problems of limited solubility) has permitted investigations of the intramolecular relaxation paths of the excited thione with a minimum of interference from the solvent [63].

1.3.1 INTRAMOLECULAR RELAXATION PROCESSES

In the absence of excited state quenching, radiative $S_2 - S_0$ decay of molecules containing the thiocarbonyl group typically occurs with a quantum yield of 10^{-3} to 10^{-1} [58,60,64–70]. (Vibrationally deficient tetra-atomic thiones are exceptional and have a fluorescence quantum yield near unity in the low-pressure gas phase.) A large fraction of the population of thiones in their S_2 states therefore decays radiationlessly; much of the recent research on these molecules has been aimed at elucidating the detailed pathways of this process. Initially, it was thought that the S_2 states of the aromatic thiones relaxed nonradiatively exclusively by a photophysical $S_2 - S_1$ internal conversion process brought about by vibronic coupling of the S_2 and S_1 states [50,60]. However, the observation of a position-sensitive deuterium-substitution effect on the radiationless decay rates of the S_2 states of xanthione-d_4, -d_6, and -d_8, together with subsequent experiments and potential energy surface calculations, have shown that a parallel pseudo-photochemical relaxation channel is also operative [61,62,68].

Data illustrating the intramolecular relaxation processes of the S_2 states of thiones (Scheme 1.1) in perfluoroalkane solvents are given in Table 1.2. The purely photophysical channel for S_2 relaxation is vibronically induced $S_2 - S_1$ internal conversion. Direct $S_2 - S_0$ internal conversion can be neglected due to its much larger electronic energy gap ($\Delta E(S_2 - S_0) \approx 20,000$ to $25,000$ cm^{-1}) compared with that for $S_2 - S_1$ (ca. 7,000 to 11,000 cm^{-1}) [50,58,60]. The photochemical channel contains two reversible H-abstraction processes, one of which leads to the S_1 state via an S_2' transient and a second one that leads directly to the ground state via a second

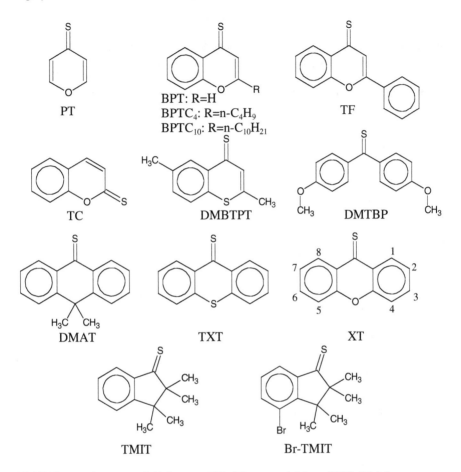

SCHEME 1.1 Structure of thioketones. PT: 4H-pyrane-4-thione; BPT: 4H-1-benzopyrane-4-thione; TF: thioflavone; TC: thiocoumarin; DMBTPT: 2,6-dimethyl-4H-1-benzopyrane-4-thione; DMTBP: p,p′-dimethoxythiobenzophenone; DMAT: 10,10-dimethylanthrathione; TXT: thioxanthione; XT: xanthione (XT-d_0 = XT-h_8, XT-d_4 = XT-1,4,5,8, XT-d_6 = XT-1,3,4,5,6,8, XT-d_8 = XT-1,2,3,4,5,6,7,8); TMIT: 2,2,3,3-tetramethylindanthione; Br-TMIT: 4-bromo-2,2,3,3-tetramethylindanthione.

transient, S_2'' [72]. The nature and relative rates of these photochemical pathways have been deduced from the following considerations:

- The effects of position-sensitive deuteriation on the lifetime, τ_{S2}, fluorescence quantum yield, ϕ_f, and quantum efficiency, $\eta(S_2 - S_1)$, of the $S_2 - S_1$ radiationless transition in xanthione [62,68]
- The effect of thione structure and rigidity on the excited state photophysical and photochemical decay constants [60,64,67,73]

TABLE 1.2
Photophysical and Spectroscopic Parameters for the S_2 State of Thioketones in Perfluorocarbon (PF) Solution at Room Temperature

Thioketone	PF Solvent	τ_{S2} [ps]	ϕ_f	k_r [× 10^7 s^{-1}]	k_{nr} [× 10^9 s^{-1}]	$E_{S2}^{0,0}$ [cm^{-1}]	$\Delta E(S_2 - S_1)$ [cm^{-1}]
PT	PFMCH	<20[a]				27490	9280
	PFH	4[b]	1 × 10^{-4} [a]	2.5	250.0		
BPT	PFDMCH	178.9[c]	0.023[a]	12.9	45.5	25540	8640
	PFH	210.0[a]					
	PFH	183.5[c]					
	PFH	187.2[c*]					
	PFMD	177.8[c]					
	PFTDHF	177.0[c]					
BPTC$_4$	PFH	102.1[c]	0.0126[c]	12.3	9.7	25330	8080
	PFH	103.4[c*]					
BPTC$_{10}$	PFH	96.6[c]	0.0118[c]	12.2	10.2	25330	8080
	PFH	97.5[c*]					
TF	PFDMCH	34.3[c]	0.0028[d]	8.2	29.1	23500	6400
		41.0[d]					
TC	PFDMCH	<3.0[e]	6.5 × 10^{-5} [e]	>2.2	>333	23500	5650
XT-d$_0$	PFDMCH	155[f]				23990	7970
	PFH	162[g]	0.014[h]	8.6	6.1		
	PFH	160[f]					
	PFMCH	165[g]					
	PFH	175[h]					
	PFMD	150[f]					
	PFTDHF	148[f]					
XT-d$_4$	PFDMCH	571[i]				23980	7960
	PFDMCH	502[f]	0.038[i]	7.6	1.9		
XT-d$_6$	PFDMCH	530[f]	0.042[i]	7.9	1.8	23980	7960
	PFDMCH	602[i]					
	PFH	540[f]					
	PFMD	527[f]					
XT-d$_8$	PFDMCH	626[i]				23980	7960
	PFDMCH	550[f]	0.038[i]	6.9	1.7		
TMIT	PFDMCH	880[i*]				28480	11030
	PFDMCH	780[i*]	0.14[j]	17.9	1.1		
		750[f]					
Br-TMIT	PFDMCH	30[f]	6 × 10^{-3} [k]	20.0	33.3	28570	11180
	PFH	35[k]					
DMBTPT	PFDMCH	101[a]				23470	7360
		68[f]	3.8 × 10^{-3} [a]	5.6	14.7		
TXT	PFDMCH	47[f]	2.3 × 10^{-3} [a]	4.9	21.2	21980	6800
		64[a]					
DMAT	PFH	410[a]	0.017[a]	4.1	2.4	26800	9100
DMTBP	PFH	35[a]	5.6 × 10^{-4} [a]	1.6	28.6	27800	8100

Note: PFH: perfluoro-*n*-hexane; PFMCH: perfluoromethylcyclohexane; PFDMCH: perfluoro-1,3-dimethylcyclohexane; PFMD: perfluoro-1-methyldecalin; PFTDHF: perfluorotetradecahydrophenanthrene.
* Measured in deoxygenated solution. $E_{S2}^{0,0}$ is determined from the absorption spectra.
[a] From Ref. [60]; [b] from Ref. [79]; [c] from Ref. [67]; [d] from Ref. [64]; [e] from Ref. [80]; [f] from Ref. [85]; [g] from Ref. [59]; [h] from Ref. [58]; [i] from Ref. [68]; [j] from Ref. [70]; [k] from Ref. [101].

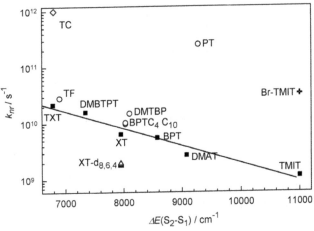

FIGURE 1.3 Semilogarithmic plot of the rate constant for the nonradiative decay processes of S_2 state k_{nr} vs. $S_2 - S_1$ energy gap for different thioketones (squares = rigid thioketones, triangles = deuterated thioketones, circles = nonrigid thioketones, diamond = TC, cross = Br-TMIT) in PF solution at room temperature. (From R. Steer et al., *Chem Phys*, Elsevier, 1984. With permission.)

- The effect of excess vibrational energy and molecular solvation (in supersonic expansions) on the excited state decay processes [74–78]. The details of the method used to extract rate constants for each of the three parallel radiationless decay processes may be found in References [62] and [72]. The reversibility of the photochemical reactions that promote S_2 relaxation has been established by comparing the overall quantum yield of photochemical thioketone consumption ($\phi_d < 10^{-3}$) with the net quantum efficiency of the photochemically induced nonradiative S_2 decay processes, which have a dominant role in $\eta(S_2 - S_1)$, a number that is much larger [58,62,73].

A typical energy gap law plot for the nonradiative relaxation of the S_2 states of aromatic molecules containing the thiocarbonyl group is shown in Figure 1.3. This plot can now be interpreted in the following way [72]:

1. The rate constant, k_{nr}, for the nonradiative decay processes of those rigid undeuteriated aromatic thiones that fall on the trend line is a composite of three rate constants, as indicated above. Internal conversion dominates for thiones with $\Delta E(S_2 - S_1) < 7,000$ cm^{-1}. The two reversible photochemical pathways dominate when the thione contains one or more H atoms in close proximity to the C=S group, *e.g.*, in the *peri*-position in aromatic thiones, and when $\Delta E(S_2 - S_1) > 8,000$ cm^{-1} for $S_n(n \geq 2)$.
2. The rates of the photochemically assisted pathways are substantially reduced when the thione contains deuterium in the *peri*-position due

either to a primary kinetic isotope effect or the presence of a barrier that is traversed by tunnelling [68].

3. Flexible thiones such as PT decay more rapidly than the rigid thiones because in such species an additional doubly excited (two-excited electron configuration) state lies below the S_2 state and provides an additional fast relaxation path [60,79].

4. The apparently anomalous behaviour of TC is attributed to a larger than usual excited state distortion, leading to a conical intersection of the S_2 and S_1 surfaces [80].

5. The much higher rate of radiationless relaxation of Br-TMIT relative to the unhalogenated species is attributed either to a heavy-atom enhanced intersystem crossing pathway or to reversible photochemical C–Br dissociation in the S_2 state of the former [72,81].

The photophysical properties of several thioketones (XT-d_0, XT-d_6, BPT) in their S_3 excited states ($^1(\pi,\pi^*)$, $^1A_1$ in C_{2v} symmetry) have also been investigated [82]. From the absence of measurable fluorescence and the $S_3 - S_0$ absorption oscillator strength, the excited state lifetimes have been estimated to be shorter than 1 ps. Although the excitation energies are relatively large, photochemical decomposition following excitation to the S_3 state occurs with a very small quantum yield. The efficiencies of the various S_3 radiationless relaxation processes have been obtained by quantitatively comparing their absorption spectra with their corrected fluorescence and phosphorescence excitation spectra [64,67,83]. Deactivation of the S_3 state occurs by three parallel nonradiative processes; $S_3 \rightarrow S_2$, $S_3 \rightarrow S_1$, and $S_3 \rightarrow S_0$, contrary to expectations based on Vavilov's law (*i.e.*, sequential deactivation) and the energy gap law. The extent of the contribution from reversible photochemical pathways has not been assessed, but should be taken into account. Available data for S_3's relaxation pathways are given in Table 1.3.

TABLE 1.3

Spectral Properties and Quantum Efficiencies of Nonradiative Transitions from the S_3 State for Three Thioketones in Perfluoro-1,3-dimethylcyclohexane at Room Temperature [82]

Property	XT-d_6	XT-d_0	BPT
E_{S3} [cm^{-1}]	28900	28900	30200
$\Delta E(S_3 - S_1)$ [cm^{-1}]	12900	12900	13400
$\Delta E(S_3\text{-}S_2)$ [cm^{-1}]	4900	5000	4800
$\Phi(S_3 \rightarrow S_2)$	0.90	0.72	0.74
$\Phi(S_3 \rightarrow S_1)$	0.00	0.04	0.11
$\Phi(S_3 \rightarrow S_0)$	0.10	0.24	0.15

1.3.2 INTERMOLECULAR RELAXATION PROCESSES

The nature of the intermolecular processes by which excited states relax vibrationally and electronically is a subject of substantial current interest. Thus, the dramatic effect of the solvent's molecular composition and structure on the rates of electronic relaxation of thiones in their S_2 states has been the subject of renewed recent investigation. The lifetimes of the S_2 states of many thiones make them ideal candidates for such investigations.

1. For thiones the processes of intramolecular (subpicosecond) and intermolecular (a few ps) vibrational relaxation occur on a time scale that is rather shorter than τ_{S_2} in at least some solvents (water [84] is exceptional). Thus the relaxation dynamics of the vibrationally equilibrated excited state can be observed in different solvents [54,58,59,63,67,85–87].

2. The thiones undergo minimal net photochemical decomposition, and the photolysis that is observed occurs largely from the longer-lived triplet state, rather than from the short-lived S_2 state [73,88–90].

3. Perfluoroalkane solvents act as very nearly ideal heat baths for these excited thiones, as demonstrated by the fact that the values of τ_{S_2} are similar in fluid perfluoroalkane solution at room temperature and in a supersonic expansion [63,74–78]. This observation permits the perfluoroalkanes to be used as the solvent of choice in establishing the kinetic parameters of the "unquenched" S_2 states of these molecules in fluid solution.

The mechanism of the intermolecular process by which the solvent efficiently deactivates the S_2 states of the thiones has been the subject of recent renewed interest. The overall outcome of the interaction leading to quenching is known because the quantum yield of triplet thione is almost the same when the thione is excited in an inert perfluoroalkane solvent or in a more reactive "quenching" medium [62]. That is, the quenching interaction increases the net rate of production of S_1 (and hence T_1 since the quantum efficiency of intersystem crossing from S_1 to the triplet manifold is unity [91,92]) via an alternate or enhanced $S_2 - S_1$ internal conversion process, rather than by providing a new path for S_2 to decay directly to the ground state, bypassing the triplet manifold [62,89]. After factoring out the change in nonradiative rate due to the solvatochromic effect on the energy gap, $\Delta E(S_2 - S_1)$ (cf. Figure 1.3), the value of the pseudo-first order rate constant for quenching by the solvent, k_q, can be obtained from the difference between the values of the total nonradiative decay rates in perfluoroalkane solution and in some other "reactive" solvent under otherwise identical conditions. The "efficiency" of excited state intermolecular quenching in a given solvent can then be obtained as the ratio $k_q/(k_r + k_{nr} + k_q)$, a number that represents the extent to which the solvent has enhanced the net rate of relaxation from S_2 to S_1. Such data are given in Table 1.4 for the solvent quenching of the S_2 state of the aromatic thione, BPT. Note that nonpolar hydrocarbon solvents and

TABLE 1.4

Photophysical Properties of the S_2 State of BPT in Different Solvents

Solvent	PFTDHP[a,b]	n-Hexane[c]	Acetonitrile[d]	CCl$_4$[e]	Methanol[f]
τ_{S2} [ps]	177.0 ± 2.0	20.0 ± 1.0	14.9 ± 1.0	30.0 ± 2.0	7.0 ± 1.0
Φ_F [× 10^{-3}]	25.0	2.2	1.7	3.4	0.62
k_r [× 10^8 s^{-1}]	1.5	1.1	1.1	1.2	0.9
$\Delta E(S_2-S_1)$ [cm^{-1}]	8590	8330	7150	7620	6370
k_{nr} [× 10^{10} s^{-1}]	0.54	0.64	1.5	1.1	2.6
k_q [× 10^{10} s^{-1}]	<0.001	4.4	5.2	2.2	12.0
Quenching efficiency [%]	0	87	78	68	82

[a] PFTDHF: perfluorotetradecahydrophenanthrene. [b] From Ref. [67]; [c] from Ref. [92]; [d] from Ref. [93]; [e] from Ref. [72]; [f] from Ref. [66].

acetonitrile, usually considered to be chemically unreactive, efficiently quench this thioketone's S_2 state [92,93].

Both earlier steady state photochemical and photophysical measurements and more recent fs transient absorption measurements suggest that a reversible photochemical interaction is operative in the quenching of the S_2 states of thiones by solvents that contain either C–H or O–H bonds [58,59,66,92–94]. Evidence for quenching via the formation of exciplexes has also been obtained for other solvents (e.g., acetonitrile) [86,93]. The following evidence is available from photochemical as well as steady state and fluorescence decay experiments. First, the early work by de Mayo and coworkers showed that thiones in their S_2 states are ubiquitous abstractors of H atoms, and that the process is efficient when compounds containing readily abstractable H atoms are present [95-97]. Second, the values of k_q for BPT, which has a higher S_2 energy ($E_{S2} \sim$ 72.8 kcal/mol) are consistently larger than those for XT ($E_{S2} \sim$ 67 kcal/mol) in the same solvent [92,94]. Third, a significant deuterium isotope effect is found when k_q is measured in fully or partially deuterated solvents [66,92,93], except those such as benzene [59] that have higher energy C–H bonds. These data all suggest that the mechanism of S_2 excited state quenching in many solvents involves an interaction that lies on the path to C–H bond dissociation [92-94] (and additional O–H dissociation in alcohols [66]). However the facts that the quantum yield of net thione photolysis is small and the yield of triplet remains large (> 70%) [61,62,92] suggest that this process must be reversible and occurs largely on the excited state potential surface.

Recent transient absorption experiments have provided more detailed evidence about the quenching process. For BPT in n-hexane, the first order rate of S_2 transient decay (τ = 19.9 ps), is identical to the rate of triplet transient appearance (τ = 19.4 ps), as required if the decay path is $S_2 - S_1 - T_1$ and the lifetime of S_1 is very short [92].

The same deuterium isotope effect is seen in the transient kinetics as is found in the steady state experiments, and the lifetime-lengthening effect of using a perfluorinated solvent has also been confirmed. Most important, however, no transient attributable to transient free radical products was observed [92], in keeping with the requirement that the net photochemical decomposition yield is small [88,98]. Two possible mechanisms might account for these observations: either (1) free radicals are formed by H-abstraction but the transferred H atom is returned from the singlet spin state of the radical pair to the thione on the S_1 excited state surface [95,96], or (2) the deactivation process involves an aborted H-atom transfer perhaps via a conical intersection between S_2 and S_1 along the C–H displacement coordinate [99,100]. The latter mechanism is preferred [58,67,92,94] and is illustrated graphically in Figure 1.4.

Acetonitrile appears to induce S_2 relaxation by two parallel mechanisms, one of which involves the same aborted C–H abstraction as seen in the aliphatic hydrocarbon solvents, and a second involving the tentative formation of an exciplex stabilized by charge-transfer, dispersion, and dipolar forces [86,93]. The alcohols appear to quench by a mechanism involving aborted hydrogen abstraction from C–H and O–H

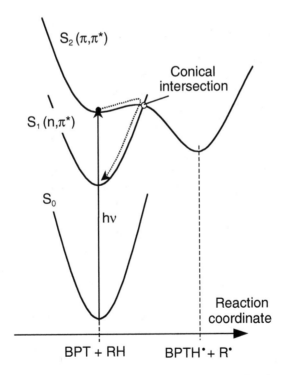

FIGURE 1.4 Schematic potential energy surface diagram for the aborted hydrogen abstraction process leading to quenching of BPT in the S_2 state.

bonds in a H-bonded complex with the thione [66]. Benzene is another interesting solvent because it exhibits only a very small H/D isotope effect on the quenching rate constant [59] but is nevertheless an efficient S_2 state quencher [59,101]. Here, aborted abstraction is not favoured due to the larger C–H bond energy, and a pure exciplex-mediated quenching process has been proposed [58,59] on the basis of a spectroscopically distinct species observed in the transient absorption spectrum of XT in benzene [102]. S_2 deactivation in water is exceptional, since it occurs in the hydrogen-bonded BPT–water complex. The lifetime of the complex in the S_2 state is found to be as short as 1.0 ± 0.1 ps in H_2O, and 2.2 ± 0.1 in D_2O [84].

Recently Vauthey and coworkers [103], using a fast transient grating technique, have examined the intermolecular electron transfer quenching of the S_2 state of XT in acetonitrile by high concentrations of weak electron donating additives such as the methoxybenzenes. Excited state quenching is observed, concomitant with $XT^{\bullet-}/D^{\bullet+}$ geminate ion pair formation, but the net yield of triplet thione is not significantly reduced. Fast spin-allowed charge recombination on the S_1 potential surface has been suggested to explain these observations. Note the interesting parallel between this process and the aborted H-abstraction quenching mechanism in H-containing solvents.

1.4   PORPHYRINS AND METALLOPORPHYRINS

The porphyrins and metalloporphyrins, which are ubiquitous in nature, have been the subject of intense spectroscopic, photophysical, and photochemical interest for many decades. Our basic understanding of the spectroscopy of these compounds was established in the early 1960s by Gouterman (cf. the excellent review in the series of monographs edited by Dolphin) [104]. We shall not re-review the spectroscopy here except to note that the S_2 and S_1 states are accessed via one-photon absorptions in the B (Soret) and Q (visible) bands respectively in their electronic spectra, and that the latter may be split into as many as four vibronic features. More recent reviews of the photochemistry and photophysics of these compounds have provided comprehensive compilations of information about their low-lying excited states [104–106]. These reviews, however, provide relatively little information about the unique excited state processes initiated by excitation of these molecules in their strong Soret and higher energy absorption bands in the violet and uv regions. Recent interest in higher excited state processes has been prompted by the experimental observation, using ps and fs pump-probe techniques, of energy and electron transfer involving the short-lived S_2 states of these molecules [107]. Levine and coworkers [4,108] have identified a general need for further information about such processes in higher electronic states to assist in the design of molecular logic systems. The metalloporphyrins, which offer excellent photochemical stability, are good potential candidates for use in molecular electronic and photonic devices.

We begin with a general *caveat*. Much of the older literature dealing with the spectroscopy and photophysics of the higher excited electronic states of the porphyrins and metalloporphyrins must be read and interpreted with caution. First, many of the earlier (and some more recent) experiments have been done at relatively large porphyrin concentrations (*i.e.*, > 10^{-5} M), where significant solute aggregation can occur in at least some solvents. Since such aggregates can contribute substantially to absorption in the Soret region, but generally radiate with much lower quantum yields than the excited monomer, the presence of a significant fraction of solute in aggregate form strongly affect quantitative measurements of spectra, quantum yields, and kinetics. Second, accurate quantitative measurements of metalloporphyrin $S_2 - S_0$ fluorescence spectra and quantum yields are difficult even in dilute solution, owing to the large oscillator strength of the Soret absorption and fluorescence reabsorption caused by the strong overlap of the $S_2 - S_0$ emission and absorption spectra. These two factors have led to an unusually large scatter in the reported quantum yields of S_2 fluorescence of the metalloporphyrins, and to errors in this state's radiative and nonradiative decay constants obtained from quantum yield measurements and calculated radiative rate constants (Equation 1.2).

1.4.1 INTRAMOLECULAR PROCESSES

Systematic investigations of the photophysics of the S_2 states of the porphyrins and metalloporphyrins started about 20 years ago. Kurabayashi *et al.* [109] measured the emission spectra of the S_2 states of a large number of metalloporphyrins and estimated their corresponding radiative and nonradiative rate constants by measuring their S_2 fluorescence quantum yields and using the Strickler-Berg formalism, Equation 1.2, to obtain the radiative rates. The estimated nonradiative rate constants for the S_2 states of ZnTPP (TPP = tetraphenylporphyrin), CdTPP and AlClTPP in acetonitrile, were observed to follow the log-linear relationship expected of the energy gap law [109]. Although the S_2 fluorescence of many compounds, including the diacids (H_4P^{2+}), were observed, no S_2 emission was detected for the free-base porphyrins (*e.g.*, H_2TPP, H_2P, and H_2OEP: P = porphine; OEP = 2,3,7,8,12,13,17, 18-octaethylporphine) [110,111]. The fast deactivation of the free-base porphyrin S_2 state has been attributed to strong coupling between the Q_y and B states, arising from an enhanced Q_y absorption band and a small $S_2 - S_1$ energy gap [110]. The high frequency N-H vibrational modes in free-base porphyrins were found not to be important in promoting the $S_2 - S_1$ internal conversion process. However, the values of the $S_2 - S_0$ emission quantum yields (ϕ_{S2}) for several Zn porphyrins were found not to correlate well with the $S_2 - S_1$ energy gap, even though their radiative rate constants are quite similar. In particular, the absence of significant S_2 emission from ZnOEP, although attributed to a larger Franck–Condon overlap between the B and Q states and hence an enhanced $S_2 - S_1$ internal conversion rate, was a troubling exception to the prediction of the energy gap law [110,111].

Direct experimental measurement of the lifetimes of the S_2 states (τ_{S2}) of the porphyrins and metalloporphyrins requires the use of ps or fs laser techniques; transient pump-probe and fluorescence upconversion methods have been most widely employed. The lifetimes so obtained range from a few ps to tens of fs. An early measurement established a lower limit of 1 ps for the S_2 fluorescence lifetime of isolated ZnTPP seeded in supersonic expansion from the line broadening of its fluorescence excitation spectrum [112], and this lifetime has been re-measured many times over the past 20 years. On the short time limit side, Zhong *et al.* have determined the S_2–S_1 internal conversion time for H_2TPP to be tens of femtoseconds using a time-resolved fluorescence depletion method [113].

The excited state behaviour of metalloporphyrins containing heavier metal atoms is more complex, particularly for those compounds that are paramagnetic. Fluorescence from the S_2 states of the lanthanide tetra-*p*-tolylporphyrins (TTP) was shown to be dependent on the nature of the metal ions (Sm^{3+}, Eu^{3+}, Gd^{3+}, Tb^{3+}, Yb^{3+}, and Lu^{3+}) [114,115]. In particular, the S_2 fluorescence quantum yield for GdTTP in ethanol is 0.001 whereas ϕ_{S2} values for YbTTP and EuTTP are < 0.0001. The results suggest that quenching of the S_2 states of YbTTP and EuTTP occur *via* a rapid transition from the upper excited state to a charge-transfer (CT) state whose energy is lower than that of the S_2 state, whereas in GdTTP the CT state is much higher. Direct measurement of τ_{S2} for LuTBP (TBP: tetrabenzoporphyrin), using a time-resolving electron-optical image converter, yielded a value of 4.5 ps [116]. In another study involving paramagnetic metalloporphyrins, Kaizu *et al.* showed that Er(dpm)TPP (dpm: 2,2,6,6-tetramethyl-3,5-heptanedione) exists as two conformers in methanol/ethanol glass at 77 K. One of the conformers has an Er^{3+} ion displaced out of the porphyrin plane and does not fluoresce from the S_2 state, while the other conformer has an Er^{3+} ion in the porphyrin plane and displays S_2 emission [117].

With the advent of femtosecond lasers, it is now routine to measure with great precision the ultrafast relaxation dynamics of metalloporphyrins excited to upper electronic states. Eom *et al.* carried out a fs pump-probe transient absorption spectroscopy study on the vibrational relaxation processes of NiTPP and NiOEP in toluene after photoexcitation to an unspecified highly excited state $S_n(\pi,\pi^*)$ [118]. A fast decay process of *ca.* 1 ps, corresponding to intramolecular vibrational relaxation (IVR) within the vibrationally hot ^{3}(d,d) electronic state of the Ni porphyrins, takes place in both NiTPP and NiOEP. A longer transient absorbance decay of *ca.* 10–20 ps was also detected for NiOEP, and this was attributed to subsequent intermolecular vibrational relaxation (EVR). Relaxation between the planar and ruffled conformers of NiOEP was thought also to contribute to the 10–20 ps decay process.

Zinc tetraphenylporphyrin (ZnTPP) has been used recently as a model compound for S_2 spectroscopic and dynamics studies, and has been the subject of intensive study. Some interesting questions regarding the actual relaxation mechanism still remain, however. Gurzadyan *et al.* recorded the S_2 fluorescence decay and the S_1 fluorescence rise profiles of ZnTPP in ethanol solution by fluorescence upconversion at an excitation wavelength (λ_{ex}) of 394 nm [119]. In their experiment, the S_2 state decayed exponentially with a 2.35 ps lifetime, identical to the risetime of its S_1 fluorescence. A 60–90 fs rise component in the S_2 fluorescence profile was ascribed to IVR in the S_2 state. Gurzadyan and coworkers also showed that τ_{S2} for ZnTPP varies significantly with solvent, and can be as low as 1.58 ps in toluene [120]. The S_2 fluorescence lifetime of MgTPP in ethanol was 3.25 ps, qualitatively consistent with the energy-gap law for these tetraphenyl-metalloporphyrins.

In a separate fluorescence up-conversion study, Mataga *et al.* showed that the S_2 fluorescence decay of ZnTPP in ethanol (λ_{ex} = 405 nm) at 430 nm and the S_1 fluorescence rise at 600 nm are described by a single-exponential function with the same time constant of 2.3 ps, in accordance with Gurzadyan's study [121]. When the emission was monitored in the region between the stationary S_2 and S_1 emission bands (*i.e.*, 490–580 nm), fluorescence from both unrelaxed vibronic states populated immediately after $S_2 - S_1$ internal conversion and higher vibronic states in S_1 are needed to adequately rationalize the fluorescence rise and decay profiles. A salient feature in Mataga's model is that the rate for IVR is reduced as the excess vibrational energy decreases. Similar behaviour was also observed for Zn-5,15-di-(3,5-di-t-butylphenyl)porphyrin in THF and toluene with λ_{ex} = 400 nm, and λ_{em} = 530-600 nm [122]. From these experiments, it was noted that the vibrationally hot S_1 states have measurable fluorescence lifetimes, and could in principle undergo energy and electron transfer to suitable acceptors [123]. It should be noted, however, that both these fluorescence upconversion studies were carried out on metalloporphyrin solutions whose concentrations were large enough to produce substantial solute aggregation and fluorescence reabsorption, the effects of which were not assessed.

Very recently, Yu *et al.* conducted a detailed and systematic investigation of the fluorescence dynamics of ZnTPP in benzene, mostly at concentrations low enough to avoid complications due to aggregation [124]. Upon direct one-photon excitation in the Soret band at 397 nm, the blue fluorescence of ZnTPP decays in 1.45 ps (measured by fluorescence upconversion), while the S_1 fluorescence profile is described by a rise time of 1.15 ps followed by a decay of 12 ps (EVR) and a decay of the S_1 population on a ns time scale. A short lifetime component of 240 fs, in addition to a longer lifetime component of 1.1 ps, was observed when the sample is excited by two-photons of 550 nm light. The discrepancy between the one- and two-photon absorption experiments was plausibly assigned to the presence of two distinct electronic states (S_2 and a higher energy S_2') each of which contribute to the Soret absorption band. In the event of a two-photon absorption, S_2' is preferentially formed from highly excited S_n (n>2) state and subsequently undergoes a rapid relaxation to the S_1 state in 240 fs, perhaps *via* a conical intersection. The S_1 fluorescence rise time (1.15 ps) is then attributed to a convolution of contributions from both the S_2 (lifetime = 1.45 ps) and S_2' states, while the 12 ps decay time is attributed to cooling of the higher vibrational Q states to the surrounding solvent (*i.e.*, EVR). A recent time-dependent density functional theory study has shown that the Soret absorption band of ZnP comprises not only the $a_{1u} \rightarrow e_g$ and $a_{2u} \rightarrow e_g$ transitions, as predicted by Gouterman's four-orbital model [104], but also contains a contribution from the $b_{2u} \rightarrow e_g$ transition [125]. The S_2' state could be assigned to either the latter symmetry-forbidden transition or a charge transfer state, however more experimental evidence is needed to confirm this.

According to Yu *et al.*, the temporal profile of the transient absorption of ZnTPP (λ_{ex} = 397 nm and probe wavelength λ_{probe} ranges from 571 to 702 nm) displays at most two decay times (1.5 ps and 12 ps) in benzene [124]. This differs markedly from Enescu and coworkers' measurement of the transient absorption decay of ZnTPP in ethanol (λ_{ex} = 404 nm and λ_{probe} = 430 and 470 nm), which exhibits an ultrafast component of 150 fs and a slower component of 2 ps [126]. The former was assigned to IVR within the S_2 state and the latter to fluorescence from the relaxed Soret state. The contribution of the S_2' state to the 150 fs time-constant requires further investigation.

Most recently, accurate measurements on dilute solutions containing negligible dimer showed (i) that the $S_2 - S_0$ and $S_1 - S_0$ fluorescence quantum yields of ZnTPP in ethanol decrease for excitation on both the blue and red wings of the Soret band, and (ii) that S_1 is not populated quantitatively when exciting in the uv-violet [127]. These results have been interpreted as providing support for the presence of an S_2' state that absorbs throughout the Soret region, but relatively more significantly on the red and blue edges of the narrow main Soret band. The depopulation of the S_2' state was suggested to follow two parallel pathways: $S_2' - S_1$ internal conversion and a second relaxation route that bypasses the S_1 state altogether (*e.g.*, $S_2' - S_0$ or $S_2' - T_n - S_0$).

Another class of porphyrins, substituted by 3,5-di-*t*-butylphenyl groups at the α and γ meso positions, has recently been studied using the fs fluorescence up-conversion technique [128]. Following excitation into the S_2 state, the zinc metalloporphyrin species undergoes an S_2–S_1 internal conversion with a time constant of 150 fs before intramolecular vibrational relaxation takes place in the Q state with a time constant of 600 fs. In the case of the free-base porphyrin, the system relaxes from the B state to the Q_y state within 40 fs, followed by internal conversion between the Q_y and Q_x states in 90 fs. Subsequently, IVR occurs in the Q_x state in 1.5 ps before the porphyrin returns to its ground state in 12 ns.

The S_2 lifetime of water-soluble ZnTPPS (TPPS: meso-tetrasulfonatophenyl porphyrin) was previously determined to be 1.3–1.4 ps [120,129]. However, the S_2 lifetime for water-soluble ZnTMPyP(4) (TMPyP(4): tetrakis(N-methyl-4-pyridyl)-porphyrin) is much shorter (< 100 fs) [126]. Its transient absorption reveals that following rapid S_2–S_1 internal conversion, IVR occurs within the S_1 state in *ca.* 150 fs. A 3 ps time constant present in the transient absorption decay was attributed to either an excited state conformational change or EVR.

In addition to direct one photon excitation in the Soret band, the S_2 state of metalloporphyrin can also be excited by either coherent or sequential two-photon absorption in the visible. The early work of Stelmakh and Tsvirko showed that the S_2 state can be populated sequentially either directly via "prompt" two-photon absorption at high laser fluence (*i.e.*, $S_0 + h\nu \rightarrow S_1$; $S_1 + h\nu \rightarrow S_2$; $S_2 \rightarrow S_0 + h\nu$ (prompt fluorescence)) [130] or indirectly by excited state triplet-triplet annihilation at lower laser fluence (*i.e.*, $S_0 + h\nu \rightarrow S_1 \rightarrow T_1$; $2T_1 \rightarrow S_2 + S_0$; $S_2 \rightarrow S_0 + h\nu$ (delayed fluorescence)) [131]. Tobita *et al.* have employed two visible photon excitation and optical-optical double resonance techniques to determine the absorption cross section of the $S_n \leftarrow S_1$ transitions for ZnTPP in EPA solvent [132,133]. Values ranging between 4.74×10^{-16} and 1.02×10^{-15} cm^2/molecule were determined in the wavelength region of 490 to 560 nm, while the $S_2 \leftarrow S_0$ absorption cross section was reported to be approximately 2.3×10^{-15} cm^2/molecule [132,133].

1.4.2 INTERMOLECULAR AND INTRAMOLECULAR ENERGY AND ELECTRON TRANSFER

Chosrowjan *et al.* were the first to report intermolecular electron transfer from the S_2 state of ZnTPP to dichloromethane in fluid solution [134]. The existence of the electron transfer process was inferred from Stern–Volmer plots of the intensities of $S_2 - S_0$ fluorescence from ZnTPP in varying concentrations of CH_2Cl_2 in acetonitrile, and from the observation by epr and transient absorption spectroscopy of a ZnTPP$^+$ transient. The existence of this intermolecular electron transfer process has been confirmed by Yu *et al.* who also provided considerably more detail concerning the dynamics of the process [124]. The 0.75 ps fluorescence lifetime reported by Chosrowjan *et al.* for the S_2 state of ZnTPP in CH_2Cl_2 differs from the values of 2 ps and 1.9 ps obtained by Gurzadyan *et al.* [119] and Yu *et al.* [124], respectively. A likely explanation for the distinctly shorter lifetime measured by Chosrowjan *et al.* at relatively high solute concentration is the effect of reabsorption in the Soret band region. A detailed analysis of the effects of reabsorption on ϕ_{S2} (and to some extent on τ_{S2}), together with a description of the procedures needed to correct for inner filter effects may be found in Reference [127]. Note that electron transfer between ZnTPP and CH_2Cl_2 occurs only when a favourable orientation exists between donor and acceptor, and an acceptable quenching radius (average of 15 Å) is achieved (*cf.* Perrin's equation).

Morandeira *et al.* have investigated the dynamics of electron transfer and subsequent charge-recombination between the S_2 state of ZnTPP and molecules that can quench *via* weakly bound complexes (such as 1,2,4-trimethoxybenzene and acetophenone) in solvents of vastly different polarity (*e.g.,* toluene and acetonitrile) [135]. Significant reductions in the steady-state values of ϕ_{S2} were observed upon adding the quenchers; these decreases did not correlate with the corresponding decrease in τ_{S2} measured by fluorescence up-conversion. In addition to electron transfer between donor and acceptor chromophores appropriately oriented at reasonable separation, a quenching mechanism involving an even faster charge-transfer process (< 100 fs) and complex formation may be operative. The same study revealed that these quenchers had a negligible effect on the steady-state ZnTPP S_1 fluorescence yield, suggesting that charge-recombination of the ion pairs leads to the barrierless reformation of the metalloporphyrin in its S_1 state.

Mataga and coworkers have reported the first unequivocal observation of the whole k_{CS} vs. $-\Delta G_{CS}$ curve (including the inverted region) for intramolecular charge separation (CS) in the S_2 excited states of covalently tethered Zn-5,15-bis(3,5-di-t-butylphenyl)porphyrin-imide dyads [136–139]. The following observations and conclusions were made:

1. The inverse and top regions of the energy gap law curve are clearly observable in both polar (THF, acetonitrile, and triacetin) and nonpolar (toluene and cyclohexane) solvents, whereas only the normal region is prominent in polar media. This behaviour was attributed to the presence of a small activation barrier in polar solvents.

2. Electron transfer occurs within 100 fs in most solvents, including viscous triacetin, with a solvent relaxation time (τ_L) of *ca.* 100 ps at 20°C. This suggests that charge-transfer between the S_2 state of the zinc porphyrin (ZnP) and the imide (I) is controlled, in part, by the ultrafast impulsive response component of the solvent's relaxation ($> \tau_L^{-1}$) and the coupling of CT with intramolecular high frequency modes.

3. Charge-recombination (CR) in polar solvents occurs from the vibrationally unrelaxed electron transfer state, ZnP^+/I^-, while in nonpolar solvents CR takes place from the vibrationally relaxed ZnP^+/I^- state only after IVR. In both cases, the vibrationally relaxed S_1 state of ZnP is generated after charge-recombination.

Electron transfer following excitation in the Soret band of ZnTPPS to methyl viologen (MV^{2+}) in a ZnTPPS-MV^{2+} complex has been reported to take place in aqueous solutions [129]. The charge-separated species are formed in less than 0.2 ps, while charge-recombination is completed within 0.8 ps. Other reports of photophysical processes involving the S_2 state of porphyrin aggregates can be found in References [140] and [141].

The photoinduced electron and energy transfer processes from higher excited states of the porphyrin have also been studied using dyads consisting of the same chromophores, ZnTPP and ruthenium(II) tris-bipyridine (Ru(bpy)$_3$), tethered by different spacer groups [142,143]. When the donor and acceptor moieties are linked by an amide spacer, through-bond electron transfer, in competition with internal conversion, occurs from the S_2 state of ZnTPP to the ruthenium complex [142]. This conclusion is based on the significant decrease in the S_2 lifetime of the zinc porphyrin unit in the dyad (0.8 ps in DMF) compared to the monomer (1.6 ps). Of interest, fluorescence quenching by Förster energy transfer is negligible (no sensitized ruthenium emission upon excitation of ZnP into its Soret band), even though the center-to-center distance between the chromophores is only 17 Å. This inefficiency of electronic energy transfer may be due to an unfavourable spatial orientation of the transition dipole moments [144].

In Harriman and coworkers' dyad [143], the ZnTPP and Ru(bpy)$_3$ are covalently linked by a Pt^{II} bis-σ-acetylide. When the dyad is irradiated at 390 nm in acetonitrile, the excited singlet S_2 state of the zinc porphyrin undergoes rapid energy transfer to Ru(bpy)$_3$, generating the MLCT excited singlet state of the ruthenium complex.

The experimental energy transfer rate (4×10^{11} s^{-1}) is consistent with the Förster energy transfer rate (3×10^{11} s^{-1}). Unlike LeGourriérec et al. [142], charge separation between the zinc porphyrin and Ru(bpy)$_3$ is not observed in this dyad, suggesting that through-bond superexchange interaction is not significant here.

Efficient quenching of the excited S$_2$ state of zinc porphyrin (ZnP) attached to a fullerene C$_{70}$ via an amido group has been observed in both toluene and benzonitrile [145]. The primary quenching mechanism proposed by Kesti et al. is electronic energy transfer from the S$_2$ state of ZnP to an upper singlet state, S$_n$(n > 1), of C$_{70}$. An additional decay pathway involving an activationless electron transfer from the initially excited S$_2$ state of ZnP to an excited charge-transfer state or an exciplex is needed to fully account for the observed relaxation dynamics. Unfortunately, the direct detection of the C$_{70}$ anion radical by transient absorption was not possible with the instrumentation used. Less efficient quenching of the ZnP S$_2$ state is observed when the C$_{70}$ is replaced with a C$_{60}$ fullerene, possibly due to a smaller Förster spectral overlap integral.

A study of the excited state dynamics of a directly meso, meso-linked porphyrin dimer (Z2) and trimer (Z3) containing Zn-5,15-di(3,5-di-t-butylphenyl) porphyrin (Z1) chromophores in THF and toluene has been reported by Kim and coworkers [122,146]. The ground-state absorption spectra of Z2 and Z3 exhibit two bands in the Soret region: a band at the monomer Z1 band position (B_m), and a red-shifted exciton-split Soret band (B_e). The fluorescence rise times for the Q state when the B_m band is excited are 1.5 ps, 300 fs, and 500 fs for Z1, Z2, and Z3, respectively. This clearly illustrates the operation of a ladder-type deactivation channel via the B_e state following $B_m \rightarrow Q$ relaxation. From transient anisotropy measurements, the time constants for the $B_m \rightarrow B_e$ transition were measured to be 29 and 63 fs for Z2 and Z3, respectively [147]. This step results in the coherent delocalization of the excitation energy among the chromophores which subsequently undergo vibronic relaxation, dynamic Stokes shift, and finally energy localization by $B_e \rightarrow Q$ energy transfer.

Osuka et al. recently synthesized a series of phenylene-bridged linear and crossed-linear porphyrin arrays, and studied the ultrafast S$_2$ – S$_2$ energy transfer dynamics between the porphyrin chromophores [148–150]. When the peripheral energy-donating porphyrins (e.g., ZnTPP or ZnOEP) are excited into the Soret band, the excitation energy is transferred to the S$_2$ state of the energy-accepting central porphyrin (e.g., diphenylethynyl porphyrin). Thereafter, energy is either transferred to the S$_1$ state of the peripheral porphyrins or the central chromophore undergoes fast interconversion to the S$_1$ state. S$_1$ – S$_1$ energy transfer from the peripheral to the central porphyrins has also been shown to take place.

The fluorescence quenching efficiencies of the S$_1$ state ($\tau = 1.93$ ns) of ZnTPPS in water by halide ions (X$^-$) increase in the order Cl$^-$ < Br$^-$ < I$^-$, while quenching of the short-lived S$_2$ state ($\tau = 1.3$ to 1.4 ps) is only possible using I$^-$ ions [151]. Positive values for the calculated Gibbs free energy change (ΔG°) for electron transfer involving the S$_1$ state indicate that charge-transfer between the X$^-$ and excited ZnTPPS (S$_1$) is not a feasible quenching mechanism. It was suggested that the quenching of the S_1 state by Cl^-, Br^-, and I^- proceeds instead *via* the Watkins mechanism [152], in which the interaction between the loosely associated ZnTPPS*⋯ X^- and higher CT states is important.

On the other hand, quenching of the S_2 state of ZnTTPS proceeds *via* electron transfer between the fluorophore and I^- within the quenching sphere of action, consistent with the $\Delta G°$ of the charge-separation process. An important outcome of this work is the demonstration that either the S_1 state only or both the S_1 and S_2 states can be quenched by judiciously selecting the quenchers used. This has pertinent implications for the design of efficient molecular logic devices.

Molecular systems utilizing electronic energy transfer between higher electronic states to perform Boolean logic operations have been designed and characterized [5,6]. One such system consists of azulene (Az) and ZnTPP chromophores, in which $S_2 - S_2$ energy transfer from Az to ZnTPP has been shown to occur in both CH_2Cl_2 and CTAB micellar solutions [5]. The high energy transfer efficiency observed in CH_2Cl_2 cannot be explained solely from Förster theory, and short-range exchange interaction based on an inhomogeneous distribution of acceptors (ZnTPP) surrounding the donor (Az) is required to completely rationalize the quenching mechanism. In CTAB

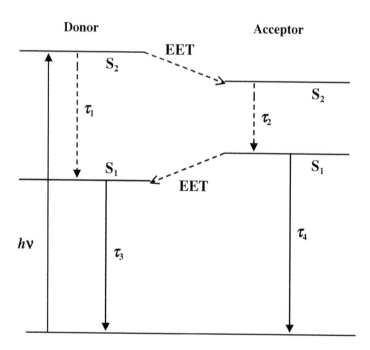

FIGURE 1.5 Schematic energy levels of the donor and acceptor in a porphyrin-azulene dyad. Porphyrin is on the right, azulene on the left. Solid lines are radiative processes and broken lines are nonradiative processes. (From R. Steer et al., *PCCP*, RSC, 2003. With permission.)

micelles, Förster's mechanism is found to agree well with the observed energy transfer efficiency when a surface-uniform distribution between donor and acceptor is assumed. A significant spectral overlap between the S_1 emission spectrum of ZnTPP and the S_1 absorption spectrum of Az leads to back $S_1 - S_1$ energy transfer after rapid internal conversion from the S_2 level of ZnTPP, see Figure 1.5. A molecular logic circuit based on sequential forward $S_2 - S_2$ energy transfer and back $S_1 - S_1$ energy transfer (cyclic energy transfer) has been designed [6].

1.5 CONCLUSION

We have surveyed the literature, concentrating on the last 10 years of research activity, concerning the photophysics and photochemistry of three classes of molecules that exhibit intense "anomalous" fluorescence from their second excited singlet states. Azulene and other nonalternant aromatic compounds, thiones (particularly aromatic thiones), and diamagnetic metalloporphyrins all possess a number of spectroscopic and structural properties that cause the rates of their radiationless relaxation to be relatively slow, enabling electric dipole allowed radiative transitions to the ground state to be observed with relatively large quantum yields. The metalloporphyrins have the greatest structural rigidity of these three classes of compounds, and the failure to observe a distinct energy gap law relationship between their rates of nonradiative decay and their $S_2 - S_1$ electronic energy spacing may be related to recent observations of an underlying dark state at about the same energy as S_2. The aromatic thiones exhibit a very large structural distortion on excitation to S_2 and consequently undergo not only $S_2 - S_1$ internal conversion characteristic of the weak coupling case, but also decay radiationlessly by an aborted H-transfer process that appears to provide the equivalent of a conical intersection between these two surfaces. Conical intersection also provides a rapid $S_1 - S_0$ relaxation pathway in azulene and its derivatives, but $S_2 - S_1$ internal conversion appears to function in the expected weak-coupling fashion provided that no additional states are interposed between S_2 and S_1.

Further research aimed at quantifying more fully the factors that control the lifetimes of these highly excited molecules is expected to be of considerable interest.

ACKNOWLEDGMENTS

GB, JK, and AM gratefully acknowledge support of their research through grant 4 T09A 166 24 from the Polish State Committee for Scientific Research. RPS, SV, and EKLY are grateful to the Natural Sciences and Engineering Research Council of Canada for financial support of this research program.

REFERENCES

1. M Kasha, *Disc Faraday Soc* 9: 14–19, 1950.
2. M Beer, HC Longuet-Higgins, *J Chem Phys* 23: 1390–1391, 1955.
3. D Holten, DF Bocian, JS Lindsey, *Acc Chem Res* 35: 57–69, 2002.
4. F Remacle, RD Levine, *J Phys Chem B* 105: 2153–2162, 2001.
5. EKL Yeow, RP Steer, *Phys Chem Chem Phys* 5: 97–105, 2003.
6. EKL Yeow, RP Steer, *Chem Phys Lett* 377: 391–398, 2003.
7. JLM Hale, *Molecular Spectroscopy,* Prentice Hall, 1999.
8. SJ Strickler, RA Berg, *J Chem Phys* 37: 814–822, 1962.
9. KF Freed, *Top Curr Chem* 31: 105–139, 1972.
10. J Jortner, S Mukamel, in R Daudel, B Pullman, Eds., *The World of Quantum Chemistry,* Dordrecht: Reidel Publishing, 1974, pp. 145–209.
11. DC Moule, EC Lim, *J Phys Chem A* 106: 3072–3076, 2002.
12. T Fujiwara, DC Moule, EC Lim, *J Phys Chem A* 107: 10223–10227, 2003.
13. W Siebrand, *J Chem Phys* 46: 440–447, 1967.
14. R Englman, J Jortner, *Mol Phys* 18: 145–164, 1970.
15. G Viswanath, M Kasha, *J Chem Phys* 24: 574–577, 1956.
16. BD Wagner, D Tittelbach-Helmrich, RP Steer, *J Phys Chem* 96: 7904–7908, 1992.
17. PM Rentzepis, J Jortner, RP Jones, *Chem Phys Lett* 4: 599–602, 1970.
18. D Huppert, J Jortner, PM Rentzepis, *Chem Phys Lett* 13: 225–228, 1972.
19. S Murata, C Iwanaga, T Toda, H Kokubun, *Ber Bunsenges Phys Chem* 76: 1176–1183, 1972.
20. S Murata, C Iwanaga, T Toda, H Kokubun, *Chem Phys Lett* 13: 101–104, 1972.
21. JR Small, JJ Hutchings, EW Small, *Proc SPIE* 1054: 26–35, 1989.
22. MJ Bearpark, F Bernardi, S Clifford, M Olivucci, MA Robb, BR Smith, T Vreven, *J Am Chem Soc* 118: 169–175, 1996.
23. GR Hunt, EF McCoy, IG Ross, *Aust J Chem* 15: 591–604, 1962.
24. GR Hunt, IG Ross, *J Mol Spec* 24: 574–577, 1962.
25. EW-G Diau, SD Feyter, AH Zewail, *J Chem Phys* 110: 9785–9788, 1999.
26. E Heilbronner, *Tetrahedron* 19: 289–313, 1963.
27. G Binsch, E Heilbronner, R Jankow, D Schmidt, *Chem Phys Lett* 1: 135–138, 1967.
28. RSH Liu, AE Asato, *J Photochem Photobiol C: Reviews* 4: 179–194, 2003.
29. SV Shevyakov, H Li, R Muthyala, AE Asato, JC Croney, DM Jameson, RSH Liu, *J Phys Chem A* 107: 3295–3299, 2003.
30. N Tétreault, RS Muthyala, RSH Liu, RP Steer, *J Phys Chem A* 103: 2524–2531, 1999.
31. G Eber, S Schneider, F Dorr, *Chem Phys Lett* 52: 59–62, 1977.
32. SY Kim, GY Lee, SY Han, M Lee, *Chem Phys Lett* 318: 63–68, 2000.
33. RSH Liu, RS Muthyala, X-S Wang, AE Asato, P Wang, C Ye, *Org Lett* 2: 269–271, 2000.
34. OK Abou-Zied, HK Sinha, RP Steer, *J Mol Spectosc* 183: 42–56, 1997.
35. OK Abou-Zied, HK Sinha, RP Steer, *J Phys Chem* 100: 4375–4381, 1996.
36. OK Abou-Zied, DRM Demmer, SC Wallace, RP Steer, *Chem Phys Lett* 266: 75–85, 1997.
37. DRM Demmer, JW Hager, GW Leach, SC Wallace, *Chem Phys Lett* 136: 329–334, 1987.
38. M Fuji, T Ebata, N Mikami, M Ito, *Chem Phys* 77: 191–200, 1983.

39. P Foggi, FVR Neuwahl, L Moroni, PR Salvi, *J Phys Chem A* 107: 1689–1696, 2003.
40. BF Plummer, ZY Al-Saigh, *J Phys Chem* 87: 1579–1582, 1983.
41. C Gooijer, I Kozin, NH Velthorst, M Sarobe, LW Jenneskens, EJ Vlietstra, *Spectrochim Acta A* 54: 1443–1449, 1998.
42. H Yamaguchi, M Higashi, K Fujimuri, *Spectrochim Acta A* 46A: 1719–1720, 1990.
43. H Yamaguchi, M Higashi, K Fujimuri, *Spectrochim Acta A* 43A: 1377–1378, 1987.
44. M Higashi, H Yamaguchi, K Fujimuri, *Spectrochim Acta A* 43A: 1067–1068, 1987.
45. DL Phillen, RM Hedges, *Chem Phys Lett* 43: 358–362, 1976.
46. JB Birks, *Photophysics of Aromatic Molecules,* London: Wiley Interscience, 1970, pp. 169–170.
47. J Kolc, EW Thulstrup, J Michl, *J Am Chem Soc* 96: 7188–7202, 1974.
48. B Nickel, *Helv Chim Acta* 61: 198–222, 1978.
49. JW Hofstraat, GP Hoornweg, C Gooijer, NH Velthorst, *Spectrochem Acta A* 41A: 801–807, 1985.
50. A Maciejewski, RP Steer, *Chem Rev* 93: 67–98, 1993.
51. P De Mayo, *Acc Chem Res* 9: 52–59, 1976.
52. NJ Turro, V Ramamurthy, W Cherry, W Farneth, *Chem Rev* 78: 125–145, 1978.
53. RP Steer, *Rev Chem Intermed* 4: 1–41, 1981.
54. V Ramamurthy, RP Steer, *Acc Chem Res* 21: 380–386, 1988.
55. JD Coyle, *Tetrahedron* 41: 5393–5425, 1985.
56. WG McGimpsey, in V Ramamurthy, KS Schanze, Eds., *Organic and Inorganic Photochemistry,* Vol. 2, Marcel Dekker, New York, 1998, pp. 249–306.
57. V Ramamurthy, *Org Photochem* 7: 231–338, 1985.
58. A Maciejewski, DR Demmer, DR James, A Safarzadeh-Amiri, RE Verrall, RP Steer, *J Am Chem Soc* 107: 2831–2837, 1985.
59. CJ Ho, AL Motyka, MR Topp, *Chem Phys Lett* 158: 51–59, 1989.
60. A Maciejewski, A Safarzadeh-Amiri, RE Verrall, RP Steer, *Chem Phys* 87: 295–303, 1984.
61. A Maciejewski, M Szymanski, RP Steer, *J Phys Chem* 92: 6939–6944, 1988.
62. M Szymanski, A Maciejewski, RP Steer, *J Phys Chem* 92: 2485–2489, 1988.
63. A Maciejewski, *J Photochem Photobiol A* 51: 87–131, 1990.
64. A Maciejewski, M Szymanski, RP Steer, *J Photochem Photobiol A* 100: 43–52, 1996.
65. VP Rao, RP Steer, *J Photochem Photobiol A* 47: 277–286, 1989.
66. M Milewski, J Baksalary, P Antkowiak, W Augustyniak, M Binkowski, J Karloczak, D Komar, A Maciejewski, M Szymanski, W Urjasz, *J Fluorescence* 4: 89–97, 2000.
67. J Kubicki, A Maciejewski, M Milewski, T Wrozowa, RP Steer, *Phys Chem Chem Phys* 4: 173–179, 2002.
68. SR Abrams, M Green, RP Steer, M Szymanski, *Chem Phys Lett* 139: 182–186, 1987.
69. MH Hui, P de Mayo, R Suau, WR Ware, *Chem Phys Lett* 31: 257–263, 1975.
70. A Maciejewski, RP Steer, *Chem Phys Lett* 100: 540–545, 1983.
71. B Bigot, *Isr J Chem* 23: 116–123, 1983.
72. J Kubicki, G Burdzinski, A Maciejewski, in preparation.
73. J Kozlowski, M Szymanski, A Maciejewski, RP Steer, *J Chem Soc Faraday Trans* 88: 557–562, 1992.
74. A Wittmeyer, AJ Kaziska, MI Shchuka, AL Motyka, MR Topp, *Chem Phys Lett* 151: 384–390, 1988.
75. AL Motyka, MR Topp, *Chem Phys* 121: 405–412, 1988.

76. AJ Kaziska, MI Shchuka, SA Wittmeyer, MR Topp, *J Photochem Photobiol A* 57: 383–403, 1991.
77. HK Sinha, RP Steer, *J Mol Spectrosc* 181: 194–206, 1997.
78. HK Sinha, OK Abou-Zied, M Ludwiczak, A Maciejewski, RP Steer, *Chem Phys Lett* 230: 547–554, 1994.
79. AA Ruth, WG Doherty, RP Brint, *Chem Phys Lett* 352: 191–201, 2002.
80. M Szymanski, A Maciejewski, J Kozlowski, J Koput, *J Phys Chem A* 102: 677–683, 1998.
81. JA Kerr, in DR Lide, Ed., *Handbook of Chemistry and Physics,* 79th ed., Boca Raton, FL: CRC Press, 1998, pp. 9–64.
82. A Maciejewski, M Milewski, M Szymanski, *J Chem Phys* 111: 8462–8468, 1999.
83. B Nickel, H Eisenberger, M Wick, RP Steer, *J Chem Soc Faraday Trans* 92: 1101–1104, 1996.
84. G Burdzinski, A Maciejewski, G Buntinx, O Poizat, P Toele, H Zhang, M Glasbeek, *Chem Phys Lett* 393:102–106, 2004.
85. D Komar, A Maciejewski, J Kubicki, J Karolczak, in preparation.
86. M Lorenc, A Maciejewski, M Ziolek, R Naskrecki, J Karolczak, J Kubicki, B Ciesielska, *Chem Phys Lett* 346: 224–232, 2001.
87. F Elisei, JC Lima, F Ortica, G Aloisi, M Costa, E Leitao, I Abreu, A Dias, V Bonifacio, J Medeiros, AL Macanita, RS Becker, *J Phys Chem A* 104: 6095–6102, 2000.
88. U Brühlmann, JR Huber, *Chem Phys Lett* 54: 606–610, 1978.
89. U Brühlaman, JR Huber, *J Photochem* 10: 205–213, 1979.
90. J Kozlowski, A Maciejewski, M Milewski, W Urjasz, *J Phys Org Chem* 12: 47–52, 1999.
91. D Tittelbach-Helmrich, RP Steer, *Chem Phys Lett* 262: 369–373, 1996.
92. G Burdzinski, A Maciejewski, G Buntinx, O Poizat, C Lefumeux, *Chem Phys Lett* 368: 745–753, 2003.
93. G Burdzinski, A Maciejewski, G Buntinx, O Poizat, C Lefumeux, *Chem Phys Lett* 384: 332–338, 2004.
94. K Dobek, A Maciejewski, J Karolczak, W Augustyniak, *J Phys Chem A* 106: 2789–2794, 2002.
95. KY Law, P de Mayo, *J Am Chem Soc* 101: 3251–3260, 1979.
96. KY Law, P De Mayo, SK Wong, *J Am Chem Soc* 99: 5813–5815, 1977.
97. DSL Blackwell, K Lee, P de Mayo, GLR Petrasunias, G Reverdy, *Nouv J Chim* 3: 123–131, 1979.
98. ZB Alfassi, Ed., *S-Centered Radicals,* Chichester: Wiley, 1999.
99. WM Nau, W Adam, JC Scaiano, *Chem Phys Lett* 253: 92–96, 1996.
100. WM Nau, *Ber Bunsenges Phys Chem* 102: 476–485, 1998.
101. M Szymanski, M Balicki, M Binkowski, J Kubicki, A Maciejewski, E Pawlowska, T Wrozowa, *Acta Phys Po, A* 89: 527–546, 1996.
102. RWJ Anderson, RM Hochstrasser, HJ Pownall, *Chem Phys Lett* 43: 224–227, 1976.
103. P-A Muller, E Vauthey, *J Phys Chem A* 105: 5994–6000, 2001.
104. M Gouterman, in D Dolphin, Ed., *The Porphyrins,* Vol. 3, New York: Academic Press, 1978.
105. K Kalyanasundaram, *Photochemistry of Polypyridine and Porphyrine Complexes,* London: Academic Press, 1992.
106. S Takagi, H Inoue, in V Ramamurthy, KS Schanze, Eds., *Molecular and Supramolecular Photochemistry,* Vol. 5, New York: Marcel Dekker, 2000, p, 215.

107. K Tokumaru, *J Porphyrins Phthalocyanines* 5: 77–86, 2001.
108. F Remacle, S Speiser, RD Levine, *J Phys Chem A* 105: 5589–5591, 2001.
109. Y Kurabayashi, K Kikuchi, H Kokubun, Y Kaizu, H Kobayashi, *J Phys Chem* 88: 1308–1310, 1984.
110. O Ohno, Y Kaizu, H Kobayashi, *J Chem Phys* 82: 1779–1787, 1985.
111. H Kobayashi, Y Kaizu, in M Gouterman, PM Rentzepis, KD Straub, Ed., *Porphyrins Excited States and Dynamics,* American Chemical Society, Washington, 1986, pp. 105–117.
112. U Even, J Magen, J Jortner, J Friedman, H Levanon, *J Chem Phys* 77: 4374–4383, 1982.
113. Q Zhong, Z Wang, Y Liu, Q Zhu, F Kong, *J Chem Phys* 105: 5377–5379, 1996.
114. MP Tsvirko, GF Stelmakh, VE Pyatosin, KN Solovyov, TF Kachura, *Chem Phys Lett* 73: 80–83, 1980.
115. MP Tsvirko, KN Solovev, GF Stelmakh, VE Pyatosin, TF Kachura, *Opt Spektrosk* 50: 555–560, 1981.
116. J Aaviksoo, A Freiberg, S Savikhin, GF Stelmakh, MP Tsvirko, *Chem Phys Lett* 111: 275–278, 1984.
117. Y Kaizu, M Asano, H Kobayashi, *J Phys Chem* 90: 3906–3910, 1986.
118. HS Eom, SC Jeoung, D Kim, J-H Ha, Y-R Kim, *J Phys Chem A* 101: 3661–3669, 1997.
119. GG Gurzadyan, T-H Tran-Thi, T Gustavsson, *J Chem Phys* 108: 385–388, 1998.
120. GG Gurzadyan, T-H Tran-Thi, T Gustavsson, *Proc SPIE: Int Soc Opt Eng* 4060: 96–104, 2000.
121. N Mataga, Y Shibata, H Chosrowjan, N Yoshida, A Osuka, *J Phys Chem B* 104: 4001–4004, 2000.
122. NW Song, HS Cho, M-C Yoon, SC Jeoung, N Yoshida, A Osuka, D Kim, *Bull Chem Soc Jpn* 75: 1023–1029, 2002.
123. A Vlcek, *Chemtracts: Inorg Chem* 13: 776–779, 2000.
124. H-Z Yu, JS Baskin, AH Zewail, *J Phys Chem A* 106: 9845–9854, 2002.
125. EJ Baerends, G Ricciardi, A Rosa, SJA van Gisbergen, *Coord Chem Rev* 230: 5–27, 2002.
126. M Enescu, K Steenkeste, F Tfibel, M-P Fontaine-Aupart, *Phys Chem Chem Phys* 4: 6092–6099, 2002.
127. J Karolczak, D Kowalska, A Lukaszewicz, A Maciejewski, RP Steer, submitted.
128. S Akimoto, T Yamazaki, I Yamazaki, A Osuka, *Chem Phys Lett* 309: 177–182, 1999.
129. M Andersson, J Davidsson, L Hammarström, J Korppi-Tommola, T Peltola, *J Phys Chem B* 103: 3258–3262, 1999.
130. GF Stelmakh, MP Tsvirko, *Opt Spektrosk* 48: 185–188, 1980.
131. GF Stelmakh, MP Tsvirko, *Opt Spektrosk* 49: 511–516, 1980.
132. S Tobita, I Tanaka, *Chem Phys Lett* 96: 517–521, 1983.
133. S Tobita, Y Kaizu, H Kobayashi, I Tanaka, *J Chem Phys* 81: 2962–2969, 1984.
134. H Chosrowjan, S Tanigichi, T Okada, S Takagi, T Arai, K Tokumaru, *Chem Phys Lett* 242: 644–649, 1995.
135. A Morandeira, L Engeli, E Vauthey, *J Phys Chem A* 106: 4833–4837, 2002.
136. N Mataga, H Chosrowjan, Y Shibata, N Yoshida, A Osuka, T Kikuzawa, T Okada, *J Am Chem Soc* 123: 12422–12423, 2001.
137. N Mataga, H Chosrowjan, S Taniguchi, Y Shibata, N Yoshida, A Osuka, T Kikuzawa, T Okada, *J Phys Chem A* 106: 12191–12201, 2002.
138. N Mataga, S Taniguchi, H Chosrowjan, A Osuka, N Yoshida, *Photochem Photobiol Sci* 2: 493–500, 2003.

139. N Mataga, S Taniguchi, H Chosrowjan, A Osuka, N Yoshida, *Chem Phys* 295: 215–228, 2003.
140. GA Schick, *Thin Solid Films* 179: 521–527, 1989.
141. H Kano, T Kobayashi, *J Chem Phys* 116: 184–195, 2002.
142. D LeGourriérec, M Andersson, J Davidsson, E Mukhtar, L Sun, L Hammarström, *J Phys Chem A* 103: 557–559, 1999.
143. A Harriman, M Hissler, O Trompette, R Ziessel, *J Am Chem Soc* 121: 2516–2525, 1999.
144. A Vlcek, *Chemtracts: Inorg Chem* 12: 863–869, 1999.
145. T Kesti, N Tkachenko, H Yamada, H Imahori, S Fukuzumi, H Lemmetyinen, *Photochem Photobiol Sci* 2: 251–258, 2003.
146. HS Cho, NW Song, YH Kim, SC Jeoung, S Hahn, D Kim, SK Kim, N Yoshida, A Osuka, *J Phys Chem A* 104: 3287–3298, 2000.
147. C-K Min, T Joo, M-C Yoon, CM Kim, YN Hwang, D Kim, N Aratani, N Yoshida, A Osuka, *J Chem Phys* 114: 6750–6758, 2001.
148. S Akimoto, T Yamazaki, I Yamazaki, A Nakano, A Osuka, *Pure Appl Chem* 71: 2107–2115, 1999.
149. A Nakano, A Osuka, T Yamazaki, Y Nishimura, S Akimoto, I Yamazaki, A Itaya, M Murakami, H Miyasaka, *Chem Eur J* 7: 3134–3151, 2001.
150. A Nakano, Y Yasuda, T Yamazaki, S Akimoto, I Yamazaki, H Miyasaka, A Itaya, M Murakami, A Osuka, *J Phys Chem A* 105: 4822–4833, 2001.
151. TP Lebold, EKL Yeow, RP Steer, *Photochem Photobiol Sci* 3: 160–166, 2004.
152. AR Watkins, *J Phys Chem* 78: 1885–1890, 1974.

2 Proton Transfer Reactions in the Excited States

Haruo Shizuka and Seiji Tobita

CONTENTS

2.1 INTRODUCTION

Proton transfer reactions (proton association and dissociation) in the excited state of aromatic compounds are elementary processes in both chemistry and biochemistry. The acid-base properties in the excited state of aromatic compounds are closely related to electronic structure, which is considerably different from that in the ground state. A large number of studies on the acidity constants pK_a* in the excited state of aromatic compounds have shown that the pK_a* values are markedly different from the acidity constants pK_a in the ground state [1–31].

It is well-known that the pK_a* values can be estimated by means of the Förster cycle [16–20], the fluorescence titration curve [18,21–26], and the triplet-triplet absorbance titration curve [27]. These methods involve the assumptions that proton transfer in the excited state is very fast and that acid-base equilibrium may be established during the lifetime in the excited state.

Weller [21–23] has pointed out the competition between the rates of proton transfer and the deactivation in the excited state. In fact, it has been shown that proton-induced fluorescence quenching competitive with the proton-transfer reaction is present in the excited state of naphthylamines, that is, simple acid-base equilibrium cannot be accomplished in the excited state of aromatic amines, and that a dynamic analysis containing the quenching process is, therefore, needed in order to obtain the correct pK_a^* values [32,33]. The dynamic analyses by means of nanosecond time-resolved spectroscopy with fluorimetry have been applied to 1-aminopyrene [34,35], 1-aminoanthracene [36], phenanthrylamine [37,38], and naphthols [39]. This method to determine the pK_a^* values of naphthylamines has been used by Hafner et al. [40], and similar experiments for excited naphthols have been carried out by Harris and Selinger [41]. On the other hand, establishment of prototropic equilibrium has been reported in the case of 2-hydroxynaphthalene-6,8-disulfonate [42].

For the proton-induced quenching mechanism, a complex in which a proton is shared between excited 2-naphthylamine and one water molecule [43] or a hydrated naphthylammonium cation in the ground state [44] is assumed as an intermediate for the quenching. It has been shown that the proton-induced quenching at moderate acid concentrations is caused by electrophilic protonation at one of the carbon atoms of the aromatic ring in the excited singlet state (S_1), leading to proton exchange (or isotope exchange) [45,46].

For the studies of the excited-state proton transfer reactions of aromatic compounds, kinetic analyses by means of fluorimetry, single-photon counting, and laser photolysis methods are very important to obtain the exact data. Their acid-base properties in the excited states can be understood on the bases of thermodynamic analyses and electronic structures. Large changes in the acidity constant of organic compounds upon electronic excitation may be applicable to various fields, especially to biochemistry.

In this chapter, excited-state proton-transfer reactions of aromatic compounds are described from the following viewpoints:

- Determination of pK_a^* values in the excited state involving proton-induced quenching
- Proton-induced quenching mechanism
- Effect of electronic structures on proton transfer reactions
- Environmental effects on excited-state proton transfer reactions
- Proton and hydrogen atom transfer reactions in the excited states
- Ultrafast proton transfer including geminate proton recombination in the excited states
- Application of excited-state proton transfer reactions

2.2 DETERMINATION OF pK_a* VALUES IN THE EXCITED STATE

It is well-known that the pK_a* values can be estimated by means of the Förster cycle [10,17,19–21,47], the fluorescence titration curve [21,47], and the triplet-triplet absorption curve [27]. These methods involve the assumptions that proton transfer in the excited state is very fast and acid-base equilibrium may be established during the lifetime in the excited state.

However, it has been shown that proton-induced fluorescence quenching (k_q) competitive with the proton-transfer reaction is present in the excited state of aromatic compounds such as naphthylamines and naphthols (i.e., simple acid-base equilibrium cannot be accomplished in the excited state of aromatic amines) and a dynamic analysis including the quenching process is, therefore, needed in order to obtain the correct pK_a* values [32,33].

2.2.1 THE FÖRSTER CYCLE

In the Förster cycle method, the thermodynamic cycle as shown in Scheme 2.1 is utilized to determine the acidity constant in the excited state from the corresponding ground-state value (pK_a) and electronic transition energies ($h\nu_{AH}$ and $h\nu_{A^-}$) of the acid (protonated) and base (deprotonated) forms [17,47]. This method is based on the determination of the energy gap between the ground and excited states of the acid and base forms of a molecule (AH). ΔH and ΔH* are, respectively, the enthalpy changes of reactions in the ground and excited states. $\Delta E_{AH}(=h\nu_{AH})$ and $\Delta E_{A^-}(=h\nu_{A^-})$ are the energy differences between the two states whose values can be determined by absorption and emission measurements. We have $\Delta H - \Delta H^* = \Delta E_{AH} - \Delta E_{A^-}(=h\nu_{AH} - h\nu_{A^-})$. If the entropy changes in both reactions are the same ($\Delta S = \Delta S^*$), $\Delta H - \Delta H^* = \Delta G - \Delta G^*$, where ΔG and ΔG^* are the free energy changes in the corresponding reactions.

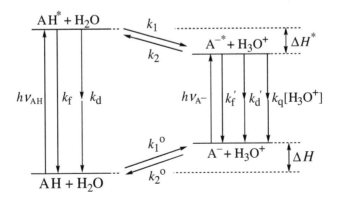

SCHEME 2.1

When the k_2 value is very much greater than that of k_q (i.e., $k_2 \gg k_q$) in the protonation reactions $(A^{-*} + H_3O^+ \rightarrow AH^* + H_2O$; $A^{-*} + H_3O^+ \rightarrow AH + H_2O)$, the prototropic equilibrium in the excited state is established; thus, $pKa^* = -\log K^* = -\log(k_1/k_2)$. In this case, we can obtain the Förster cycle equation:

$$pK_a - pK_a^* = (h\nu_{AH} - h\nu_{A^-}) / (2.303RT) \tag{2.1}$$

where R and T denote the gas constant and absolute temperature, respectively. However, in most cases, the back-protonation process (k_2) is competitive with proton-induced quenching process (k_q) due to proton attack to one of the carbon atoms of aromatic ring as described later. In these cases, it is impossible to obtain correct pKa^* value by use of the Förster cycle method.

2.2.2 FLUORESCENCE TITRATION CURVE

The changes in fluorescence intensities of the acid and base forms give useful information about the prototropic behavior in the excited singlet state of the compound.

If the prototropic equilibrium is established within the lifetime of the S_1 state, i.e., the fluorescence lifetimes of AH and A⁻ are equal, the pKa^* value can be estimated from the midpoint of fluorescence titration curve as shown in Figure 2.1, where the fluorescence quantum yields for AH and A⁻ are plotted as a function of pH values of the solution [21,47]. However, such a case is rather rare for the acid-base reactions in the excited singlet state of aromatic compounds.

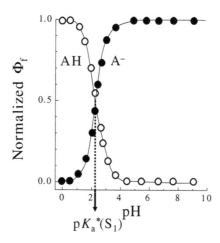

FIGURE 2.1 Typical fluorescence titration curves without significant proton-induced quenching.

2.2.3 DYNAMIC ANALYSIS

The Förster cycle and the fluorescence titration methods assume that the dynamic equilibrium between prototropic species is established in the excited state. When fast proton-induced fluorescence quenching (k_q) competing with proton transfer process (k_2) is involved (see Scheme 2.1), dynamic analysis as described below is required to determine the accurate pK_a^* values.

The pK_a^* values for 1- and 2-naphthylamines (1-NA and 2-NA) have been determined by use of the dynamic analysis [33]. Figure 2.2 shows the fluorescence titration curves, i.e., fluorescence quantum yields for neutral amine A (Φ_A) and ammonium ion AH$^+$ (Φ_{AH^+}) as a function of log[H_3O^+], for 1-NA and 2-NA. It clearly shows that the midpoints of the fluorescence titration curves appear at completely different positions for the neutral amine (A) and the ammonium ion (AH$^+$). This

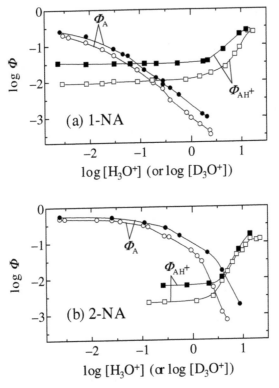

FIGURE 2.2 Logarithmic plots of the fluorescence quantum yields of neutral amines Φ_A (circles) and cations Φ_{AH^+} (squares) of (a) 1-NA and (b) 2-NA as a function of [H_3O^+]. (Open symbols: H_2SO_4/H_2O; closed symbols: D_2SO_4/D_2O). (From Tsutsumi, K. and Shizuka, H., *Z. Phys. Chem.* (*Wiesbaden*), 1978, 111, 129. With permission.)

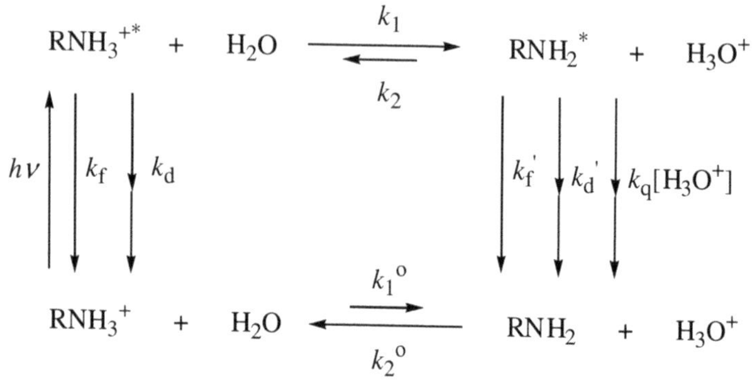

$$RNH_3^{+*} + H_2O \xrightleftharpoons[k_2]{k_1} RNH_2^* + H_3O^+$$

$$hv \quad k_f \downarrow k_d \qquad\qquad\qquad k_f' \downarrow k_d' \downarrow k_q[H_3O^+]$$

$$RNH_3^+ + H_2O \xrightleftharpoons[k_2^0]{k_1^0} RNH_2 + H_3O^+$$

SCHEME 2.2

discrepancy, which is seen more remarkably for 1-NA, can be attributed to the existence of rapid proton-induced fluorescence quenching (k_q) in the excited singlet state (A*), and the dynamic equilibrium is not established within the lifetimes of the excited singlet states of the neutral amine and the ammonium ion. Therefore, the fluorescence titration and Förster cycle methods cannot be applicable to these cases, and the dynamic analysis is required to obtain the correct pK_a^* value.

A kinetic scheme for prototropism of 1-NA or 2-NA in aqueous solution is given in Scheme 2.2, where k_1 and k_2 denote proton-dissociation and recombination rate constants in the excited state, k_f (k_f') and k_d (k_d') are radiative and nonradiative rate constants, respectively. By applying the steady-state approximation for the fluorescence quantum yields in the absence (Φ_A^0) and presence (Φ_A) of acid, the following equation can be derived:

$$\frac{\Phi_A^0}{\Phi_A} = 1 + \frac{1}{k_1 \tau_{AH}^0} + \frac{\{k_2 + k_q'(1 + k_1 \tau_{AH}^0)\}\tau_A^0}{k_1 \tau_{AH}^0}[H_3O^+] \qquad (2.2)$$

where τ_{AH}^0 and τ_A^0, respectively, represent the fluorescence lifetimes in the presence of high concentrations of acid (for 1-NA $\tau_{AH}^0 = 47.2$ ns when $[H_3O^+] = 13.6$ M, and for 2-NA $\tau_{AH}^0 = 28$ ns when $[H_3O^+] = 16.4$ M) and in the absence of acid ($\tau_A^0 = 27$ ns for 1-NA and $\tau_A^0 = 19$ ns for 2-NA). When τ_{AH}^0 is large and $[H_3O^+] < 0.2$ M, k_1 becomes larger than $k_2[H_3O^+]$ and the relations $k_1 \tau_{AH}^0 \gg 1$ and $k_2(k_1 \tau_{AH}^0)^{-1} \ll k_q$ hold. Hence Equation (2.2) can be simplified as follows:

$$\frac{\Phi_A^0}{\Phi_A} = 1 + k_q' \tau_A^0 [H_3O^+] \qquad (2.3)$$

From the Stern–Volmer plot of Φ_A^0/Φ_A against $[H_3O^+]$, linear relations are obtained for 1-NA and 2-NA:

$$\left(\frac{\Phi_A^0}{\Phi_A}\right)_{\text{1-RNH}_2} = 1.0 + 2.4 \times 10^2 [\text{H}_3\text{O}^+] \tag{2.4}$$

$$\left(\frac{\Phi_A^0}{\Phi_A}\right)_{\text{2-RNH}_2} = 1.0 + 6.3 [\text{H}_3\text{O}^+] \tag{2.5}$$

From Equations (2.4) and (2.5), the proton-induced fluorescence quenching rate constants (k_q) of 1-NA and 2-NA are obtained to be 8.9×10^9 M^{-1} s^{-1} and 3.3×10^8 M^{-1} s^{-1}, respectively.

The fluorescence response functions of RNH$_2$* and RNH$_3^+$* (F_A(t) and F_{AH}(t)) are given by

$$F_A(t) = \frac{k'_f k_1}{\lambda_2 - \lambda_1} (e^{-\lambda_1 t} - e^{-\lambda_2 t}) \tag{2.6}$$

$$F_{AH}(t) = \frac{k_f (\lambda_2 - X)}{\lambda_2 - \lambda_1} (e^{-\lambda_1 t} + A e^{-\lambda_2 t}) \tag{2.7}$$

where $A = (X - \lambda_1) / (\lambda_2 - X)$, $X = (\tau_{AH}^0)^{-1} + k_1$, and $Y = (\tau_A^0)^{-1} + (k_q + k_2)[\text{H}_3\text{O}^+]$. The deactivation parameters (λ_1 and λ_2) are given by

$$\lambda_{1,2} = \frac{1}{2} [X + Y \mp \{(Y - X)^2 + 4k_1 k_2 [\text{H}_3\text{O}^+]\}^{1/2}] \tag{2.8}$$

The observed fluorescence response functions (I_A(t) and I_{AH}(t)) of RNH$_2$* and RNH$_3^+$* are related to the undistorted fluorescence response functions (F_A(t) and F_{AH}(t)) by the following convolution integrals

$$I_A(t) = \int_0^t F_A(t') I_L(t - t') dt' \tag{2.9}$$

$$I_{AH}(t) = \int_0^t F_{AH}(t') I_L(t - t') dt' \tag{2.10}$$

where I_L is the corresponding excitation pulse function. From the deconvolution analyses, the λ_1 and λ_2 can be determined, and finally k_1 and k_2 values are obtained from the equations on X and Y.

In Table 2.1, the proton dissociation (k_1) and protonation (k_2) rate constants determined by dynamic analyses and the proton-induced quenching rate constants (k_q) are shown for H$_2$O/H$_2$SO$_4$ and D$_2$O/D$_2$SO$_4$ systems. It is noted that the magnitude

TABLE 2.1
Proton Dissociation (k_1), Protonation (k_2), and Quenching (k_q) Rate Constants and the Acidity Constants (pK_a* and pK_a in the Excited Singlet and Ground States, Respectively) of Naphthylamines at 27°C

Compound	Quencher	$k_1/10^9 s^{-1}$	$k_2/10^8 M^{-1} s^{-1}$	$k_q/10^9 M^{-1} s^{-1}$	pK_a* a	b	c	pK_a
1-NA	H$^+$	1.3	1.2	8.9	−1.0	2.7	−5.9	3.9
	D$^+$	1.1	0.75	6.0	−1.2	2.6	−5.9	
2-NA	H$^+$	1.0	1.5	0.33	−0.8	0.8	−4.0$_5$	4.1
	D$^+$	0.85	0.094	0.19$_6$	−0.96	0.3	−4.0	

Note: pK_a*, a = Determined by the dynamic analyses with fluorimetry; pK_a*, b = Estimated by the midpoint of fluorimetric titration of RNH$_a$*; pK_a*, c = Estimated by the Förster cycle.
Source: From Tsutsumi, K. and Shizuka, H., *Z. Phys. Chem.* (*Wiesbaden*), 1978, 111, 129. With permission.

of the proton-induced fluorescence quenching rate constants k_q for 1-NA is larger than that of the protonation rate constant (k_2). By using k_1 and k_2, accurate pK_a* values can be obtained for 1-NA (pK_a* = −1.0) and 2-NA (pK_a* = −0.8). These values are very different from those obtained based on the Förster cycle and fluorescence titration methods.

The dynamic analysis method with nanosecond time-resolved fluorimetry has been applied to determine the pK_a* values for 1-aminopyrene [34,35], 1-aminophenanthrene [36], phenanthrylamines [37], and naphthols [39]. The Stuttgart group has supported this method to determine the pK_a* values of naphthylamines [40]. The detailed kinetics of proton transfer reaction of 1-naphthol in H$_2$O and D$_2$O have been investigated by using picosecond time-resolved fluorescence spectroscopy [48].

2.2.4 APPARENT ACIDITY CONSTANT IN THE EXCITED STATE

The excited-state proton-transfer reactions of 4-(9-anthryl)-N,N-dimethylaniline (A) in EtOH-H$_2$O (4:1 by weight) mixtures at 300 K have been studied by fluorimetry and the single-photon counting method [49]. The excited acid and base forms of A decay independently with single exponential functions. The pK_a* value (2.4 ± 0.2) of A obtained from the usual fluorescence titration curves is, therefore, an apparent value, reflecting the pK_a value (2.5 ± 0.2) in the ground state. This is mainly caused by the extremely slow proton-dissociation rate (k_1) of $^1A^+H^*$ compared to its decay rate (2×10^8 s^{-1}). No proton-induced quenching of A is observed under [H$_2$SO$_4$] < 0.01 M, though the intramolecular CT state $^1A_{CT}$* (τ_{CT}^0 = 9.7 ± 0.2 ns) is produced very rapidly via $^1A^*$ under experimental conditions. The experimental results can be accounted for by Scheme 2.3 [49], where τ_{AH}^0, τ_A^0, and τ_{CT}^0 represent the lifetimes of $^1A^+H^*$, $^1A^*$, and $^1A_{CT}^*$, respectively; k_1 and k_{-1} the rate constants for

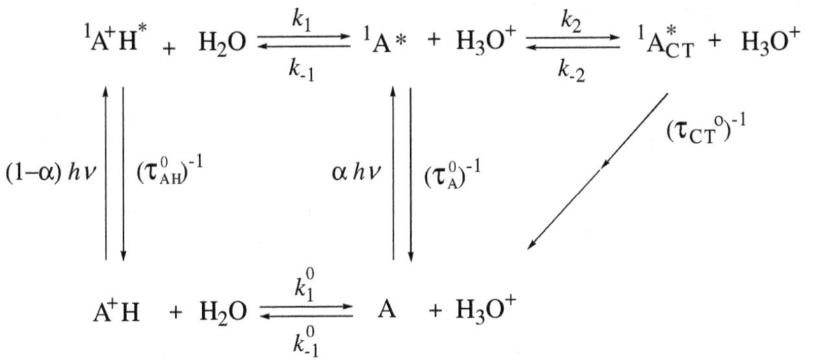

$$^1A^+H^* + H_2O \underset{k_{-1}}{\overset{k_1}{\rightleftharpoons}} {}^1A^* + H_3O^+ \underset{k_{-2}}{\overset{k_2}{\rightleftharpoons}} {}^1A^*_{CT} + H_3O^+$$

$(1-\alpha)\,h\nu \quad (\tau^0_{AH})^{-1} \qquad \alpha\,h\nu \quad (\tau^0_A)^{-1} \qquad\qquad (\tau^0_{CT})^{-1}$

$$A^+H + H_2O \underset{k^0_{-1}}{\overset{k^0_1}{\rightleftharpoons}} A + H_3O^+$$

SCHEME 2.3

the proton dissociation and association, respectively; k_2 and k_{-2} the rate constants for the CT formation and the back CT reaction, respectively; and α the ratio of the absorbance of A^+H to those of A plus A^+H at 360 nm (an isosbestic point). The k_1 value is extremely small compared to the decay rate of $^1A^+H^*$ $(\tau_{AH}^0)^{-1}$. The k_2 value is very large ($>3 \times 10^{10}\,s^{-1}$ [50]) and greater than that of $k_{-1}[H_3O^+]$. The equilibrium between $^1A^*$ and $^1A_{CT}^*$ is extremely shifted to $^1A_{CT}^*$ (i.e., $k_2 \gg k_{-2}$). Thus, the ratio of [A] to $[A^+H]$ in the ground state may be kept even in the excited state, leading to the apparent pK_a^* value. The proton-induced quenching for $^1A_{CT}^*$ is very slow compared to the decay rate of $^1A_{CT}^*$ [i.e., $(\tau_{CT}^0)^{-1}$] at low concentrations of protons less than ~10^{-2} M. Therefore, the excitation and decay processes in A and A^+H take place independently as shown by

$$A \xrightarrow{\alpha h\nu} {}^1A^* \longrightarrow {}^1A_{CT}^* \longrightarrow A$$

and

$$A^+H \xrightarrow{(1-\alpha)h\nu} {}^1A^+H^* \longrightarrow A^+H$$

2.3 PROTON-INDUCED QUENCHING

2.3.1 Proton-Induced Quenching Mechanism

Proton-induced quenching often plays an important role in acid-base equilibrium in the excited state of aromatic molecules as described above. For the proton-induced quenching mechanism of naphthylamines, a complex in which a proton is shared between excited 2-naphthylamine and one water molecule [43] or a hydrated naphthylammonium cation in the ground state [44] was assumed as an intermediate for the quenching. To clarify the quenching mechanism, proton-induced fluorescence quenching of various aromatic compounds has been studied in polar solvents at moderate acid concentrations [45,46]. In Table 2.2, the values of bimolecular quenching rate constant k_q due to protons are listed for typical aromatic compounds together with their photophysical properties. From an inspection of Table 2.2, one can understand that excited aromatic molecules having an intramolecular charge-transfer structure in the fluorescent state (e.g., 1-methoxynaphthalene, 1-naphthol [39], naphthylamines [32,33], 1-aminoanthracene [36], 1-aminopyrene, and 9,9′-bianthryl [51]) are quenched appreciably by protons. The magnitude of k_q is not

---

TABLE 2.2
Fluorescence Lifetimes (τ_0) and Quantum Yields (Φ_f^0) without Acids, Quenching Rate Constants (k_q), and Ionization Potentials (I_p^*) in the Excited State of Aromatic Compounds at 300 K

Sample	Solvent	Quencher	τ_0/ns	Φ_f^0	$^1k_q/10^9$ M^{-1} s^{-1}	I_p^*/eV
1-methoxynaphthalene	20% CH$_3$CN in H$_2$O	H$^+$	8.9	0.43	1.08	3.91
	20% CH$_3$CN in D$_2$O	D$^+$	9.4	0.41	0.51	
	80% CH$_3$CN in H$_2$O	H$^+$	12.6	0.36	0.16	
	80% CH$_3$CN in D$_2$O	D$^+$	12.8	0.42	9.0×10^{-2}	
2-methoxynaphthalene	20% CH$_3$CN in H$_2$O	H$^+$	10.0	0.33	$<3 \times 10^{-3}$	4.24
1-cyanonaphthalene	20% CH$_3$CN in H$_2$O	H$^+$	4.4	0.37	6.4×10^{-2}	5.43
2-cyanonaphthalene	20% CH$_3$CN in H$_2$O	H$^+$	12.0	0.53	2.4×10^{-2}	
1-naphthylamine	5% CH$_3$CN in H$_2$O	H$^+$	27.0	0.49	8.90	4.06
	5% CH$_3$CN in D$_2$O	D$^+$	29.0	0.56	6.00	
2-naphthylamine	5% CH$_3$CN in H$_2$O	H$^+$	19.0	0.45	0.33	3.93
	5% CH$_3$CN in D$_2$O	D$^+$	20.0	0.47	0.19	
1-naphthol	H$_2$O	H$^+$	7.8	1.3×10^{-3}	0.75	4.13
	D$_2$O	D$^+$	9.0	2.1×10^{-3}	0.46	
2-naphthol	H$_2$O	H$^+$	7.9	0.21	<0.03	4.18
	D$_2$O	D$^+$	8.6	0.23	<0.02	
naphthalene	20% EtOH in H$_2$O	H$^+$	43.0		$<1 \times 10^{-3}$	4.20
1-aminoanthracene	20% CH$_3$CN in H$_2$O	H$^+$	4.1	5.1×10^{-3}	0.12	4.3
	20% CH$_3$CN in D$_2$O	D$^+$	20.0	7.8×10^{-2}	0.02	
anthracene	MeOH	H$_2$SO$_4$	3.8		$<1 \times 10^{-2}$	4.24
9,9′-bianthryl	MeOH	H$_2$SO$_4$	37.0	0.21	1.54	
1-aminopyrene	50% CH$_3$CN in H$_2$O	H$^+$	5.4	0.54	1.3×10^{-2}	3.91
	50% CH$_3$CN in D$_2$O	D$^+$	8.0	0.61	0.68×10^{-2}	
pyrene	30% MeOH in H$_2$O	H$^+$	3.5×10^2		$<1 \times 10^{-4}$	4.22

Source: From Shizuka, H. and Tobita, S., *J. Am. Chem. Soc.*, 1982, 104, 6919. With permission.

correlated with the ionization potential for the excited state I_p^*, which is estimated from the difference between the ionization potential for the ground state (I_p) and the $S_1 \leftarrow S_0$ transition energy. This suggests that the proton-induced quenching is not due to electron transfer between excited aromatic molecules and the hydronium ion, which is in contrast with the fluorescence quenching of excited aromatic molecules by inorganic anions [52,53].

With photochemical isotope-exchange reactions of 1- and 2-methoxynaphthalenes (1-RH and 2-RH) in H_2O (or D_2O)-CH_3CN mixtures at moderate acid concentrations, the proton-induced fluorescence quenching mechanism has been attributed to electrophilic protonation at the proper carbon atom of the naphthalene ring in the lowest excited singlet state [45,46]. Similar quenching mechanism for 1-naphthol caused by a nonadiabatic proton attack on a ring site has been proposed by Harris and Selinger [41]. The H-D isotope exchange reactions of methoxynaphthalenes have been carried out with 5×10^{-3} M D_2O-CH_3CN (1:4) solution of RH with $[D_2SO_4] = 0.05$ M at 300 K using 254 nm light of a low-pressure mercury lamp [46]. The product analyses using ^1H NMR and mass spectroscopy clearly demonstrate the exchange position in the excited state of 1-RH to be mainly position 5, and slightly position 8, of the naphthalene ring (see Figure 2.3). The rate constant for electrophilic protonation to the carbon atom of the aromatic ring in the excited state is almost equal to the rate constant for proton-induced quenching (k_q). Triplet sensitization measurements of 1-RH by benzophenone show that the H-D isotope exchange reaction is predominant for the excited singlet state.

The proton-induced fluorescence quenching is much more significant for 1-RH compared to 2-RH. In 1-RH in polar solvents at room temperature, inversion of the electronic energy levels $(^1L_b$ and $^1L_a)$, which are the Franck–Condon S_1 and S_2 states, respectively, takes place during the lifetime of the excited singlet state [54–57]. Since the ^1L_a state has the transition moment along the molecular short axis of the naphthalene ring, charge migration from the methoxy group to the naphthalene moiety is induced upon electronic excitation of 1-RH. According to MO calculations based on a semiempirical SCF MO CI method, the charge densities of positions 5 and 8 in 1-RH are increased significantly in the ^1L_a state (Table 2.3), which is

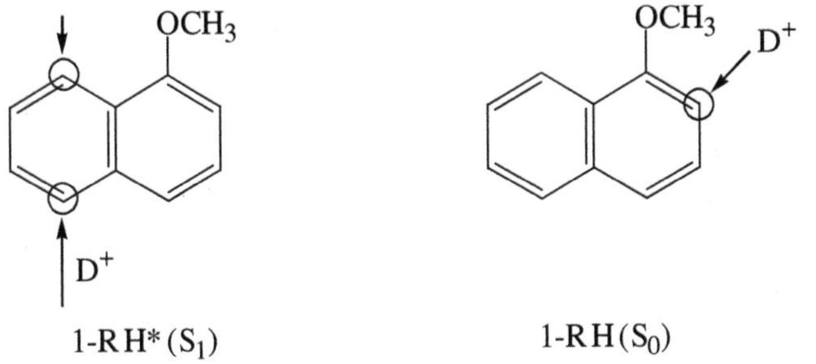

$$1\text{-RH}^*\,(S_1) \qquad\qquad 1\text{-RH}\,(S_0)$$

FIGURE 2.3 Reactive positions of 1-methoxynaphthalene (1-RH) in the S_1 and S_0 states for protons.

TABLE 2.3

Reactive Indices for Electrophilic Reactions in the Ground and Excited States of RH[a]

Reactive Index	Sample	State	Atom (r)								Reactive Position (obsd)
			1	2	3	4	5	6	7	8	
q_r^b	1-RH	^1A	0.9828	1.0470	0.9944	1.0202	1.0017	0.9978	1.0037	0.9976	2
		^1L_b	0.9577	1.0306	1.0013	0.9752	0.9968	1.0212	1.0041	0.9979	
		^1L_a	0.9789	1.0055	0.9999	0.9838	1.0367	1.0192	1.0041	1.0424	5,8
	2-RH	^1A	1.0467	0.9844	1.0276	0.9936	0.9989	1.0045	0.9986	1.0035	1
		^1L_b	1.0284	0.9838	1.0202	1.0120	1.0102	1.0207	0.9992	1.0030	
f_r^b	1-RH	^1A	0.3466	0.2048	0.1320	0.4022	0.2968	0.1094	0.1408	0.2817	2
		^1L_b	0.0698	0.1053	0.1147	0.0665	0.0777	0.1165	0.1128	0.0789	
		^1L_a	0.1534	0.0631	0.0757	0.1461	0.1706	0.0823	0.0753	0.1734	5,8
	2-RH	^1A	0.4190	0.2010	0.0588	0.3100	0.3139	0.1908	0.0922	0.3504	1
S_r^b	1-RH	^1A	0.0423	0.0591	0.0482	0.0490	0.0362	0.0463	0.0506	0.0344	2
		^1L_b	0.0944	0.2798	0.2876	0.0960	0.0976	0.2857	0.2844	0.0976	
		^1L_a	0.1655	0.1224	0.1282	0.1622	0.1786	0.1337	0.1293	0.1803	5,8
	2-RH	^1A	0.0502	0.0485	0.0533	0.0386	0.0403	0.0427	0.0491	0.0439	1
$L_r^{\ddagger c}$	1-+RHD			1.2626	2.3439	1.0385	1.5768	2.2725	1.8719	1.9496	2
$(L_r^{\ddagger})^*$				−0.0202	0.4849	0.3002	0.2505	0.2318	0.1110	0.0787	5,8
$L_r^{\ddagger c}$	2-+RHD		1.0803		1.7718	1.9833	1.9218	1.7344	2.2469	1.4929	1
$(L_r^{\ddagger})^*$			0.3575		0.3390	0.9199	0.2531	0.6016	0.4863	0.7930	

[a] For details, see the text. [b] Calculated by a semiempirical SCF MO CI method. [c] Calculated by an extended-Hückel MO method. The total energies of 1- and 2-RH were −1098.5225 and −1098.3359 eV, respectively.

Source: From Shizuka, H. and Tobita, S., *J. Am. Chem. Soc.*, 1982, 104, 6919. With permission.

consistent with the experimental fact that the H-D isotope exchange reaction occurs preferentially at positions 5 and 8. These results demonstrate that the proton-induced fluorescence quenching of 1-RH is caused by electrophilic protonation to the carbon atom of the aromatic ring in the excited state.

In addition to the photochemical reactions, thermal H-D isotope exchange reactions have also been investigated for 1-RH in H_2O (or D_2O)-CH_3CN mixtures. The H-D exchange rate in the thermal reaction is negligibly small in comparison with that of the photochemical reaction, and the main reactive site in the H-D isotope exchange reaction in the ground state is position 2 in the naphthalene ring. As shown in Table 2.3, the specific reactivity of protons against 1-RH in the S_0 (^1A) and S_1 (^1L_a) states is well supported by various reactive indices (q_r: charge density [58]; f_r: frontier electron density [59]; S_r: super delocalizability [60]; and L_r: the localization energy [61]) for electrophilic reactions.

2.3.2 INTRAMOLECULAR PROTON-INDUCED QUENCHING

As an example of intramolecular proton-induced quenching, there is the excited-state behavior of tryptophan (Trp) and tryptamine (see Figure 2.4). The mechanistic study on the fluorescence decay of tryptophan in polar media has been an interesting subject in photochemistry and biophotochemistry [62,63]. A number of mechanisms of the decay process of the excited Trp have been proposed [62]. Two types of quenching mechanisms (internal and external) have been proposed. The internal quenching of Trp has been attributed to:

- Simultaneous emission from uncoupled ^1L_a and ^1L_b states (this explanation was later discarded by the authors) [64]
- Intramolecular charge-transfer quenching caused by the interaction between the excited indole moiety and the alanyl side chain [65-67]
- Intramolecular charge-transfer quenching arising from different ground-state C_α-C_β rotamers (the conformer model [68]) or the conformer model containing both C_α-C_β and C_β-C_γ rotamers in the ground state [69]
- Proton-transfer quenching by the ammonium group [70-75]

In the early stage, the C-2 position of the indole ring was assumed to be the reactive position [73]. Saito et al. [74,75] have shown by a photochemical H-D isotope exchange reaction that the major reactive position of the indole ring is not the C-2, but the C-4 position. The external quenching mechanism was assumed to be caused by:

- The formation of an exciplex between indole and a polar solvent molecule [76,77]
- The charge transfer to solvent (CTTS) [78]
- Photoelectron ejection [79] from the excited indole moiety

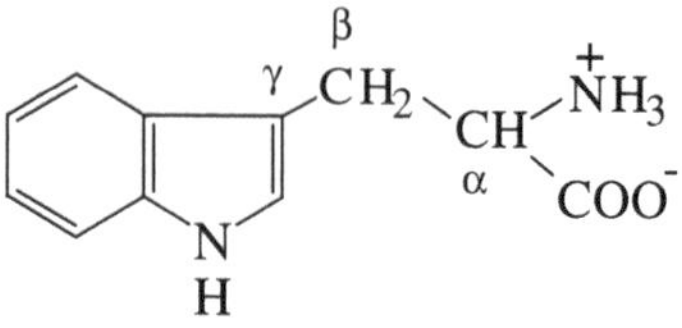

Tryptophan (Trp)

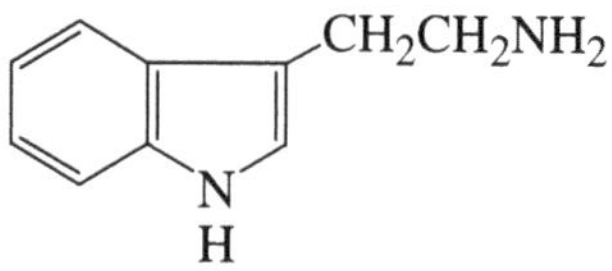

Tryptamine

FIGURE 2.4   Molecular structures of tryptophan (Trp) and tryptamine.

The fluorescence decay of Trp is complicated. The multiexponential decay functions (double or triple) have been observed in measuring the fluorescence decay of Trp in polar media [68,69,73,80], suggesting that the rate for rotamer interconversion (rotational conversion) is comparable to that for the fluorescence decay.

It has been shown that the relatively fast fluorescence decay of tryptamine is caused completely by internal quenching via the electrophilic protonation by the ammonium ion at the C-4 position of the excited indole ring [81]. It is known that the proton-induced fluorescence quenching of aromatic compounds plays a very important role for excited-state acid-base reactions, as described above. The proton-induced fluorescence quenching is caused by electrophilic protonation at one of the carbon atoms of the aromatic ring in the excited state, leading to hydrogen exchange (or deuterium exchange). In the case of Trp [82], fluorescence quenching is not due to external quenching, but to internal quenching; the internal quenching also originates from the electrophilic protonation of the $^+NH_3$ (or $^+ND_3$) group of Trp at the C-4 position of the indole ring plus the charge-transfer interaction between the excited indole ring and the ammonium group as shown in Scheme 2.4.

2.3.3 GEMINATE PROTON-INDUCED QUENCHING

In the low concentration of acid, Pines and Fleming [83,84] have demonstrated that the geminate proton itself may quench the excited singlet (S_1) state of 1-naphthol. The geminate proton-induced quenching has been explained theoretically by Agmon [85]. The ultrafast phenomena in the time-dependent geminate recombination are described in Section 2.7.

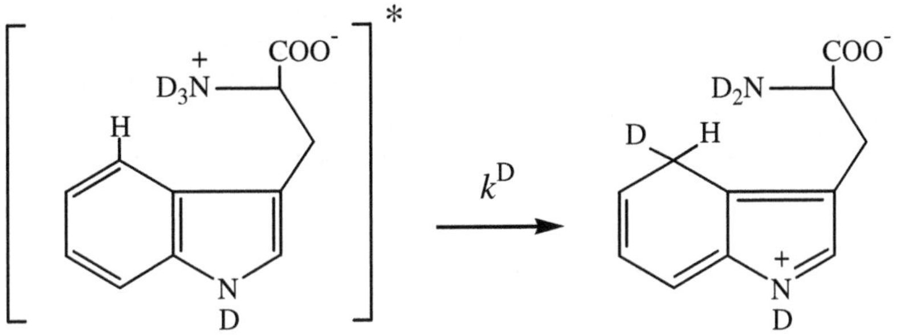

SCHEME 2.4

2.4 EFFECT OF ELECTRONIC STRUCTURES ON PROTON TRANSFER REACTIONS

It is well known that the acidity of aromatic acids such as phenol and aniline is strongly affected by introduction of substituents to the aromatic ring [86]. The acid-base properties of aromatic compounds are also modified remarkably upon electronic excitation. These indicate that the proton-emitting or accepting ability of aromatic compounds is greatly influenced by their electronic structures. Earlier studies by Wehry and Rogers [87] have revealed that excited-state acidities of phenol and monosubstituted phenols are correlated well with ground-state substituent constants evaluated by the Hammett and Taft equations, and that conjugative effects are much more important, relative to inductive effects, in the excited states than in the ground state. The influence of substituent orientation in acidity of three isomeric cyanophenols has been investigated by Schulman et al. [88] on the basis of fluorimetric titrimetry. Both in the S_1 and T_1 states the ortho isomer is the strongest acid and the para isomer the weakest. The influence of substituent orientation has been interpreted in terms of the degree of coupling of the $^1L_b \leftarrow {}^1A$ transition moment with the moment corresponding to the direction of electronic interaction of the substituents with the benzene ring in each isomer. Proton-transfer reactions in the excited state of protonated phenanthrylamines have been studied by means of nanosecond time-resolved spectroscopy and fluorimetry [37]. Linear relations between the acidity constants and formal charges on the nitrogen atom of phenanthrylamines are found for both ground and excited singlet states, showing the importance of charge migration character in the acid-base properties.

Recent developments in ultrafast spectroscopy have enabled us to investigate directly the ultrafast proton-transfer reactions in the excited state of aromatic compounds. The effect of electronic structure on proton transfer rate is of great interest not only from fundamental aspects in reaction dynamics, but also from the viewpoint of developing new photoacids. Among a number of photoacids investigated so far, 1- and 2-naphthols (1-NL and 2-NL) are representative compounds for investigating proton transfer reactions in solution. The proton dissociation rate constant (k_{dis}) of 1-NL and 2-NL in water are shown in Table 2.4 along with the pK_a and pK_a^* values. As can be seen from the k_{dis} values and the pK_a^* values in the S_1 state, the excited-state acidity of 1-NL is much stronger than that of 2-NL in spite of the same atomic composition. The stronger photoacidity of 1-NL can be attributed to an intramolecular

TABLE 2.4
pK_a, pK_a^*, and k_{dis} Values of Aromatic Hydroxy Compounds

Compound		pK_a	pK_a^*	k_{dis} (10^{10} s^{-1})
	1-NL	9.23[a]	2.0[a]	2.5[b]
	5-CN-1-NL	—	−2.8[c]	13.0[c]
	2-NA	9.45[d]	2.8[a]	1.0×10^{-2e}
	5-CN-2-NA	8.75[d]	−1.2[d]	7.0[f]
	6-CN-2-NA	8.40[d]	0.2[d]	1.1[f]
	7-CN-2-NA	8.75[d]	−1.3[d]	5.5×10^{-1f}
	8-CN-2-NA	8.35[d]	−0.4[d]	2.7×10^{-1f}

[a] Ireland, J.F., Wyatt, P.A.H., *Adv. Phys. Org. Chem.*, 1976, 12, 131. [b] Agmon, N., Huppert, D., Masad, A., Pines, E., *J. Phys. Chem.*, 1991, 95, 10407. [c] Pines, E., Pines, D., Barak, T., Magnes, B.-Z., Tolbert, L.M., Haubrich, J.E., *Ber. Bunsenges. Phys. Chem.*, 1998, 102, 511. [d] Tolbert, L.M., Haubrich, J.E., *J. Am. Chem. Soc.*, 1990, 112, 8163. [e] Formosinho, S.J., Alnaut, L.G., *J. Photochem. Photobiol. A: Chem.*, 1993, 75, 21. [f] Huppert, D., Tolbert, L.M., and Linares-Samaniego, S., *J. Phys. Chem. A*, 1997, 101, 4602.

charge transfer character of 1-NL in the fluorescent ^1L_a state having the transition moment parallel to the molecular short axis in contrast to the ^1L_b state of 2-NL.

From the results of H-D isotope exchange reactions described in Section 2.3, it is expected that the introduction of electron-withdrawing groups at the C-5 and/or C-8 positions of naphthols lowers the energy of the conjugate base, and enhances the acidity in the excited state. Actually, Tolbert and coworkers [14,89] have reported that naphthols with electron-withdrawing groups at the C-5 and/or C-8 positions exhibit greatly enhanced photoacidity, and excited-state proton transfer (ESPT) is observed even in nonaqueous solvents such as alcohols and DMSO [14,90,91]. The data reported by Tolbert and coworkers, compiled in Table 2.4, clearly indicate that the introduction of an electron-withdrawing CN group remarkably accelerates the proton dissociation rate in the excited state. It is also recognized that the position of the substituent is an important factor to determine the magnitude of the substituent effect. Position 5 is found to be one of the most effective sites to increase the photoacidity, as expected from the higher electron density in the excited state compared to the other positions (see Table 2.3).

Substituent effects on proton transfer to water of protonated aniline derivatives have been investigated by picosecond time-resolved fluorescence measurements [92–94]. Protonated aniline in the S_1 state releases proton to water with a rate constant of $1.3 \times 10^{10} s^{-1}$ in aqueous solution. The proton transfer rate is significantly increased by substitution of cyano group at the *meta*-position ($k_{dis} = 3.7 \times 10^{11} s^{-1}$). In contrast, the methoxy substitution at the *meta*-position decreases the rate remarkably ($k_{dis} < 2.5 \times 10^7 s^{-1}$). Either substituent at the *para*-position shows only slight influences on the rate. The results of a kinetic study [94] are summarized in Table 2.5 together with the pK_a and pK_a^* values. The less prominent effect of the *para* substituent is qualitatively explained by the direction of $^1L_b \leftarrow {^1A}$ transition moment, which is perpendicular to the direction of the two substituents. A quantitative explanation for the remarkable substituent effects on proton transfer rate can be made in terms of the free energy change (ΔG^*) for the proton transfer reactions in the excited state. In Figure 2.5, the values of $\log(k_{dis})$ are plotted as a function of ΔG^* [94].

The exergonicity is found to depend not only on the electronic character of the substituent, but also on the substituted position, and a clear correlation between $\log(k_{dis})$ and ΔG^* can be seen: the proton dissociation rate increases substantially with an increase of exergonicity.

The traditional interpretation for the enhancement of acidity in the excited state of aromatic compounds is based on charge transfer character in the excited state of protonated form as described above. For the deprotonation rate constant of a series of phenols and naphthols, a good correlation between $\log(k_{dis})$ and the ionic contribution to the O-H bond (E_{ion}) calculated by the following equation has been reported by Lima and coworkers [95].

$$E_{ion} = \frac{1}{4\pi\varepsilon_0} \frac{(\delta_O e)(\delta_H e)}{r_{OH}} \tag{2.11}$$

TABLE 2.5

pK_a, pK_a^*, and k_{dis} Values of Protonated Aniline Derivatives

Compound		pK_a	pK_a^*	k_{dis} $(10^{10}$ s$^{-1})$
	AN	4.6	−6.5	1.3
	m-ANCN	2.7	−10.5	37.0^c
	p-ANCN	1.7	−3.7	0.5
	m-ANMO	4.2	−1.5	6.6×10^{-3}
	p-ANMO	5.3	−5.3	1.9

Source: Shiobara, S., Tajima, S., Tobita, S., *Chem. Phys. Lett.*, 2003, 280, 673. With permission.

where δ_O and δ_H are calculated Mulliken charge densities of O and H atoms of the O-H group in the S_1 state, r_{OH} is the O-H bond length in the S_1 state, and e and ε_0 are the electronic charge and the static dielectric constant, respectively.

Recently, Hynes and coworkers [96–99] have made theoretical investigations on excited-state acidities of phenol and cyanophenols. They have proposed that the n-π* CT on the anion side of the reaction, not the acid side, is the fundamental origin of the enhanced excited state acidity. Similar considerations will be also applicable for the proton dissociation of protonated aromatic amines in the excited state, because

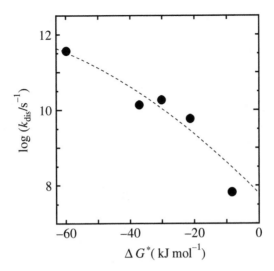

FIGURE 2.5 Free energy (ΔG^*) dependence of $\log(k_{dis})$ for aniline derivatives in the S_1 state. (From Shiobara, S., Tajima, S., and Tobita, S., *Chem. Phys. Lett.*, 2003, 280, 673. With permission.)

in the protonated form charge migration from the ammonium group to the aromatic moiety is not plausible in the excited singlet state.

The molecular mechanism of photoacidity of phenol has been investigated theoretically by Sobolewski and coworkers [100–102]. The *ab initio* calculations on phenol-ammonia cluster predicted that the increased acidity of excited phenol is not due to a property of the optically excited $^1\pi\pi^*$ state, but rather arises from the nonadiabatic interaction of the $^1\pi\pi^*$ state with an optically dark state of $^1\pi\sigma^*$ character. The $^1\pi\pi^*$ potential energy function is crossed by the $^1\pi\sigma^*$ function, and the $^1\pi\sigma^*$ energy is strongly stabilized when the proton moves from the chromophore to the solvent (ammonia) as illustrated in Figure 2.6 [100]. Hence, they consider that $^1\pi\sigma^*$ state plays a key role in the ESPT reaction of phenol-ammonia cluster.

Formosinho and coworkers [103] have applied the intersecting-state model (ISM) to the theoretical calculations of absolute rate constants for proton-transfer reactions of naphthol and substituted naphthols in the lowest excited singlet state and for the ground state. Quantum-mechanical tunneling, zero-point energy corrections and quantitative electronic parameters are incorporated in ISM, which can estimate a large range of rates (ca. 12 orders of magnitude) for a series of naphthol derivatives in an absolute manner and in good quantitative agreement with experiment. It accounts well not only for the changes in the excited-state reactivity, but also for ground states.

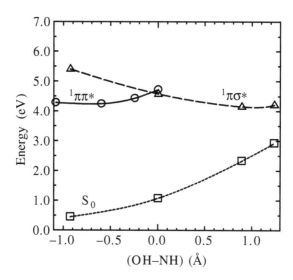

FIGURE 2.6 Calculated potential energy profiles associated with proton transfer reaction in phenol-ammonia cluster. (From Domcke, W. and Sobolewski, A.L., *Science*, 2003, 202, 1693. With permission.)

2.5 ENVIRONMENTAL EFFECTS ON EXCITED-STATE PROTON TRANSFER REACTIONS

ESPT reactions are affected by environmental conditions. The ESPT behaviors of aromatic molecules in various organized media such as liposomes, proteins, cyclodextrins, polymers, sol-gel glasses, monolayers, LB films, and solids have been reviewed by Mishra [3].

Recently, fast ESPT reactions at the surface of anionic sodium dodecyl sulfate (SDS) micelles have been investigated using the photoacid 4-methyl-7-hydroxyflavylium (HMF) chloride as a probe [104]. The acid-base kinetics of excited HMF are straightforward in water, with biexponential fluorescence decays reflecting ultrafast deprotonation of the excited acid (AH^{+}*) (k_d = 1.5 × 10^{11}s^{-1}) [105] and diffusion-controlled protonation of the excited base A* (k_p = 2.3 × 10^{11} M^{-1}s^{-1} at 20°C). In aqueous micellar SDS solutions, the kinetics are much more complex; triple exponential fluorescence decays are observed at all pH values and temperatures examined. The longest decay time (τ_1 = 760 ps at 22°C), observed only for AH^{+}* and uncoupled from the acid-base equilibrium, is assigned to excitation of HMF in orientations incapable of prompt transfer of the proton to water, i.e., that must rotate to expose the acidic OH group to water (k_{rot} = 1.2 × 10^{9}s^{-1} at 22°C). The other two decay times, τ_3 and τ_2, are due to emission from the species involved in the acid-base reaction at the micelle surface. Two processes are operative in the back protonation of A*: (1) pH-independent unimolecular reprotonation in the initially formed geminate compartmentalized pair

$(A^* \cdots H_3O^+)$ $(k_r = 8.8 \times 10^9 s^{-1})$ and (2) pH-dependent bimolecular protonation of A^* via entry of an aqueous phase proton into the micelle $(k_p = 1.6 \times 10^{11}\ M^{-1}s^{-1})$. Dissociation of the geminate pair $(k_{diss} = 1.6 \times 10^9 s^{-1})$ forms A^* at the micellar surface. The results show the complicated features of ESPT reactions in anionic micelles.

Dual emission from ion pairs produced by ESPT has been observed in 1-naphthol (ROH)-triethyl amine (NEt_3) hydrogen-bonded systems in nonpolar rigid matrices at 77 K [105]. Dual fluorescences are ascribable to the contact ion pair (CIP), with an in-phase orientation between excited naphtholate (RO^{-*}) and alkylammonium ions $HNEt_3^+$, and the separated ion pair (SIP) with an out-of-plane between them. Using nanosecond time-resolved emission spectroscopy, temperature effects on the fluorescence-decay kinetics of the ROH-NEt_3 system have been studied in polyethylene film and methylcyclohexane-isopentane (MP, 3:1 in volume) to obtain the excited-state behavior of CIP and SIP. The fluorescence intensities and peaks do not change in the temperature range 77 to 100 K. Above 100 K, a gradual red shift of the total fluorescence peak with loss of intensity is observed, the effect being more pronounced in MP, especially after the glass-softening temperature (ca. 115 K). Two different quenching processes are found to be operative. Rapid quenching competing with ESPT in both hydrogen-bonded systems occurs at high temperatures (>100 K). This results in a considerable decrease of the total fluorescence quantum yield with no significant change in fluorescence lifetimes in the temperature range 100 to 130 K. The rapid quenching is considered to be a fast internal conversion due to the out-of-plane bending motion of OH$\cdots$N bond. Above 130 K, a new quenching process resulting in the decrease of fluorescence lifetimes of both SIP and CIP also appears. This dynamic quenching of excited ion pairs is probably caused by a charge-transfer interaction.

The NaCl effect upon the proton-transfer reactions in the excited state of 2-naphthol has been studied by Harris and Selinger [41]. They have reported that the enhancement of the fluorescence of 2-naphthol is due either to a disruption of the water structure by the high concentration of Na^+ and Cl^- ions or to an increase in the activity coefficient of the excited 2-naphthol. In previous works [53,106,107], it is shown that NaCl is a very weak quencher; there is a weak quenching ability for Cl^-, but no ability for Na^+. The NaCl effect on the proton dissociation reactions in water at 300 K has been studied by means of nanosecond and picosecond spectroscopy with fluorimetry [108]. The proton dissociation rate constant k_1 decreases with an increase of [NaCl] according to the equation

$$k_1 = k_0[(H_2O)_4](1 - \alpha[NaCl]) \qquad (2.12)$$

where k_0 denotes the second-order rate constant for the proton transfer from excited naphthols to water clusters $(H_2O)_4$ and α is a constant value ($\alpha = 0.21$ for 1-NL and $\alpha = 0.20$ for 2-NL). The values of k_0 are estimated to be $1.34 \times 10^9\ M^{-1}s^{-1}$ (1-NL) and $4.86 \times 10^6\ M^{-1}s^{-1}$ (2-NL) at 300 K. This means that the decrease in the k_1 value is caused by NaCl-induced destruction of water clusters resulting in production of the hydrated ions $(Na^+)_{hyd}$ and $(Cl^-)_{hyd}$. The average hydration number of Na^+ or Cl^- in

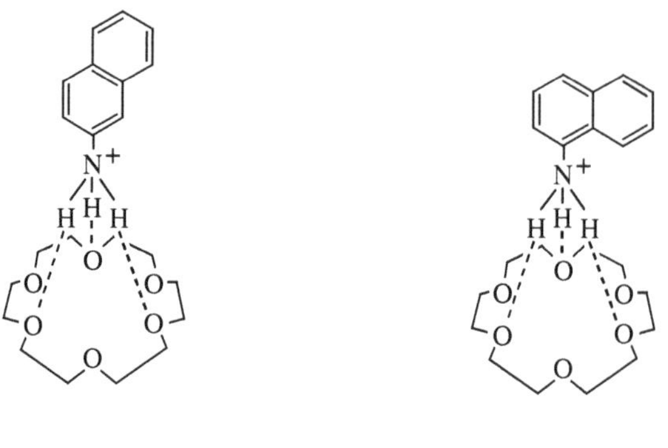

2-R N$^+$H$_3$-18-Crown-6   1-R N$^+$H$_3$-18-Crown-6

FIGURE 2.7 Complex formation of naphthylammonium ions with 18-crown-6 in aqueous solution.

water is obtained to be 4.4 or 6.6, respectively. Therefore, the water cluster acts as a proton acceptor in the excited state.

It has been found that the complex formation of naphthylammonium ions [109] or phenanthrylammonium ions [110] with 18-crown-6 decreases significantly the proton-transfer rate in the excited state resulting in an increase of their lifetimes, and that the back protonation rate in the excited state is negligibly small (Figure 2.7). The values of the ground-state association constants K_g of the complexes can be determined by the fluorimetric titration method. In these complexes, a one-way proton-transfer reaction occurs in the excited state; there is no excited-state prototropic equilibrium. The negligibly small back-protonation is due to a large steric effect on protonation to the amino-group of the excited neutral complex. In contrast, proton-induced quenching occurs effectively in the excited neutral amine-crown complexes.

The environmental effects on the ESPT reactions are very large and complex, as described above.

2.6 PROTON AND HYDROGEN ATOM TRANSFER REACTIONS

A laser study of the prototropic equilibrium of triplet benzophenone (BP) has been reported [111]. Acid-base properties in the triplet state of aromatic ketones in H_2O-CH_3CN (4:1) mixtures have been studied by means of nanosecond laser flash photolysis. The acidity constants $pK_a(T)$ in the triplet state are determined by means of the $T_n \leftarrow T_1$ absorbance titration curve, the Ware plot, and the Rayner–Wyatt plot, whose values agree well among them, showing that the acid-base equilibrium in the

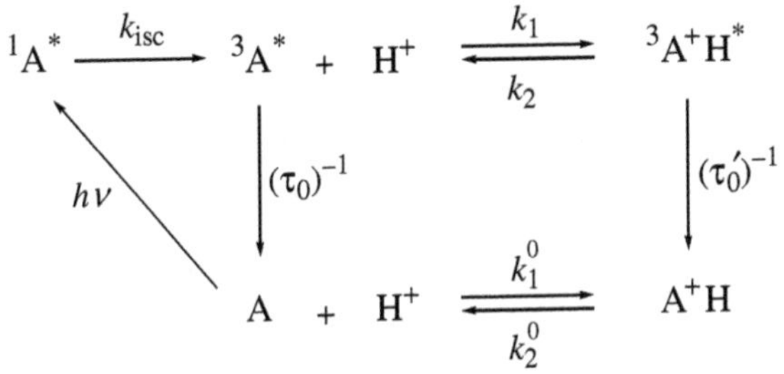

$$^1A^* \xrightarrow{\ k_{isc}\ } {}^3A^* + H^+ \underset{k_2}{\overset{k_1}{\rightleftharpoons}} {}^3A^+H^*$$

(with hv and $(\tau_0)^{-1}$, $(\tau_0')^{-1}$ branches)

$$A + H^+ \underset{k_2^0}{\overset{k_1^0}{\rightleftharpoons}} A^+H$$

SCHEME 2.5

T_1 state of aromatic ketones is established during the triplet lifetime [112]. The acid-base reaction in the triplet state of aromatic ketones ($^3A^*$) can be accounted for by Scheme 2.5, where τ_0 and τ_0' denote the lifetimes of $^3A^*$ (in the absence of protons) and $^3A^+H^*$, respectively; k_1 and k_2 denote the protonation and deprotonation rate constants in the triplet state, respectively; and A^+H denotes protonated ketones. When the acid-base equilibrium is established, that is, the proton-induced quenching is small, and the intersystem crossing rate (k_{isc}) from $^1A^*$ to $^3A^*$ is very large according to the El-Sayed rule [113-115], the Ware equation [116] reads

$$\{\tau^{-1} - \tau_0^{-1}\}^{-1} = \{(\tau_0')^{-1} - \tau_0^{-1}\}^{-1}\left\{1 + \frac{K_a}{[H^+]}\right\} \tag{2.13}$$

and the Rayner–Wyatt equation [111] reads

$$\frac{1}{[H^+]}\{\tau^{-1} - \tau_0^{-1}\} = \frac{(\tau_0')^{-1}}{K_a} - \frac{\tau^{-1}}{K_a} \tag{2.14}$$

where τ denotes the observed triplet lifetime and K_a ($= k_1/k_2$) the equilibrium constant in the triplet state. Both equations are derived on the assumptions that $k_1[H^+] \gg \tau_0^{-1}$ and $k_2 \gg (\tau_0')^{-1}$. The results are shown in Table 2.6 [112].

The acid-base equilibria in the triplet state of aromatic ketones (PhCOX) are demonstrated, since their lifetimes are long enough to allow the acid-base equilibria. The values of the acidity constants $pK_a(T)$ in the triplet state markedly increase compared to those ($-5.7 \sim -6.3$) [117,118] in the ground state. The plot of $pK_a(T)$ values vs. the Taft σ^* values [119] of the substituents X gives a linear relation, indicating that the charge migration from X to the carbonyl group increases the $pK_a(T)$ value [108]:

$$pK_a(T) = -1.33\,\sigma^* + 0.60 \tag{2.15}$$

$\sigma^*(X)$: -0.100 (CH_2CH_3); 0.00 (CH_3); $+0.215$ (CH_2Ph); $+0.600$ (Ph)

TABLE 2.6
The Taft σ* Values, $pK_a(T)$ Values, the Triplet Lifetimes (τ_0) of Aromatic Ketones, and the Triplet Lifetimes of Protonated Aromatic Ketones (τ'_0)

Compound	X^a	σ^{*b}	$pK_a(T)$ Titration	$pK_a(T)$ Ware	$pK_a(T)$ Rayner–Wyatt	$\tau_0/\mu s^c$	τ'_0/ns^d
BP	Ph	+0.600	0.20±0.02	0.13±0.02	0.10±0.10	4.2_0	17
AP	CH_3	0.000	0.63±0.07	0.60±0.10	0.60±0.15	1.0_2	30
PP	CH_2CH_3	−0.100	0.75±0.05	0.70±0.10	0.78±0.20	1.1_4	38
BK	CH_2Ph	+0.215	0.35±0.05	0.30±0.05	—^{e}	0.45_3	25

^{a} The substituent of aromatic ketones (PhCOX). ^{b} Taken from Taft, R.W., in *Steric Effects in Organic Chemistry*, M.S. Newman, Ed., Wiley, New York, 1956, 619. ^{c} Experimental errors within ±5%. ^{d} Experimental errors within ±20%. ^{e} The $pK_a(T)$ value of BK could not be determined by the Rayner–Wyatt plot.
Source: Shizuka, H., Kimura, E., *Can. J. Chem.*, 1984, 62, 2041. With permission.

However, the $pK_a(T)$ value (0.2) of benzophenone (X=Ph) slightly deviated from the line in a positive direction. This may be due to a positive mesomeric effect from the phenyl group.

The proton transfer reactions in the triplet state of 2-naphthylammonium ions (2-RNH_3^+) sensitized by triplet benzophenone (or acetophenone) are examined by laser flash photolysis at 371 nm. In this case, the photochemical reaction of $^3(2\text{-}RNH_3^+)^*$ is not the proton transfer, but the hydrogen atom transfer (HT) to give the cation radical $2\text{-}RNH_2^{+\cdot}$ and the benzophenone ketyl radical $>\dot{C}OH$ [120]. The mechanism of the hydrogen atom transfer reaction can be accounted for by the intracomplex HT reaction of the triplet exciplex $^3(RNH_3^+\cdots>CO)^*$ with a rate constant of ca. $5 \times 10^6 s^{-1}$ in methanol-water (9:1 v/v) at 290 K. At higher acid concentrations, the triplet exciplex is decomposed by protonation to give free $^3RNH_3^{+*}$ and BP, resulting in suppression of the rate for the HT reaction. It is found that the hydrogen atom transfer reaction from the triplet 1-naphthol $^3ROH^*$ (produced by the triplet sensitization of $^3BP^*$) to the ground state of BP occurs effectively via the triplet exciplex $^3(ROH\cdots>CO)^*$ to give the 1-naphthoxy radical $RO\cdot$ plus $>\dot{C}OH$ [121]. The triplet-triplet energy transfer k_{ET} (4.1×10^9 $M^{-1}s^{-1}$) from $^3BP^*$ to ROH is competitive with both the usual hydrogen atom abstraction k_{HA} (7.4×10^8 $M^{-1}s^{-1}$) of $^3BP^*$ from ROH and the quenching (2.2×10^9 $M^{-1}s^{-1}$) of $^3BP^*$ induced by ROH in methanol at 290 K [118]. These primary processes of $^3BP^*$ are completed within 300 ns, and the equilibrium $^3ROH^* + >CO \rightleftarrows {}^3(ROH\cdots>CO)^*$ is established very quickly, with the equilibrium constant K^* (6.7 M^{-1}), $\Delta H^* = -2.4$ kcal mol^{-1}, and $\Delta S^* = -2.7$ eu. Then, the hydrogen atom transfer reaction takes place via the triplet exciplex with the rate constant k_{HT} ($1.3 \times 10^6 s^{-1}$ at 290 K). It is shown that in the presence of protons (H_2SO_4) in acetonitrile-water (4:1) at 290 K, the HT reaction via the triplet exciplex is considerably enhanced in the presence of protons (H_2SO_4) [122], in contrast to the case of the RNH_3^+-BP system [123]. The protonation to $^3(ROH\cdots>CO)^*$ forms the protonated triplet exciplex $^3(ROH\cdots>COH^+)^*$, and subsequently the intraexciplex electron-transfer reaction of $^3(ROH\cdots>COH^+)^*$ occurs to give $ROH^{+\cdot}$ and $>C\dot{O}H$. The cation radical (ROH^+) rapidly dissociates into $RO^\cdot$ and H^+.

These phenomena have been reviewed in detail [8].

2.7 ULTRAFAST PROTON TRANSFER IN THE EXCITED STATES

Since acid-base reactions are among the most common and elementary chemical reactions, the molecular mechanism of proton-transfer to solvent is of great importance for the understanding of reaction dynamics in solution.

According to general views on bimolecular proton-transfer in solution, the rate of proton transfer to solvent is closely coupled with solvent relaxation, and the measured dissociation constants (k_{dis}) are assumed to follow an Arrhenius relation:

$$k_{dis} = k_{dis}^{\;0}\exp[-\Delta G^{\neq}/RT] \qquad (2.16)$$

where $k_{dis}^{\;0}$ is the frequency factor which mainly depends on solvent relaxation frequency along the proton coordinate and $\Delta G^{\neq}$ is the activation free energy. Robinson and coworkers [124–129] have proposed a similar rate expression for weak acid dissociation:

$$k_{dis} = \Omega \tau_D^{\;-1}\exp[-\Delta G^{\neq}/RT] \qquad (2.17)$$

where Ω is a mobility/steric factor introduced by Eigen and Kustin [130] and τ_D is the Debye rotational relaxation time of pure water solvent. Ω lies between 0.25 and 1.0, and τ_D is nearly equal to the dielectric relaxation time (τ_d) [126]. Robinson et al. performed ultrafast dynamics experiments on neutral photon-initiated weak acid such as the electronically excited 1- and 2-naphthols in which the free energy changes (ΔG) are positive ($\Delta G > 0$) [124–129]. Using a variety of mixed aqueous solvents, e.g., water/methanol or water/acetonitrile, they have found that the rate of proton dissociation in these solvents depended nonlinearly on the water concentration. From the analysis of the nonlinear concentration dependence on the basis of a Markov random walk theory [129], a four-water cluster, $(H_2O)_4$, is identified as the effective proton acceptor.

The electronic excitation of a moderately weak photon-initiated acid by an ultrafast light pulse is followed by vibrational relaxation combined with nonequilibrium solvation which takes place on extremely fast time scales (typically subpicosecond in nonviscous solvents). The proton-transfer step would occur after these ultrafast events are completed [124].

According to Equations (2.16) and (2.17), the proton-transfer rate is expected to become comparable to that of solvent relaxation rate when the free energy change (ΔG) for the reaction is nearly zero or negative. Recently, the direct measurements of the ultrafast proton transfer of strong photoacids to solvent have been carried out by several groups [90–94,131–138]. Pines et al. [133] have reported the proton dissociation time of 5-cyano-1-naphthol in H_2O to be 8 ps by using a picosecond laser system (1 ps pulses) combined with the single-photon counting technique. The pK_a^* value of 5-cyano-1-naphthol in H_2O was estimated to be –2.8 by Förster cycle calculations, which corresponds to ΔG of –16 kJ mol^{-1}. Shiobara et al. [94] reported the k_{dis} value of protonated m-cyanoaniline in D_2O to be 7.4×10^{10}s^{-1} (τ (D) = 14 ps) and estimated the τ (H) in H_2O to be 3 ps by assuming that the isotope effect on the proton dissociation rate of protonated m-cyanoaniline is the same as that ($k_{dis}{}^{H}$/ $k_{dis}{}^{D}$ = 5.0) of the excited anilinium ion. The rate was only slightly larger than that of 5-cyano-1-naphthol in H_2O, in spite of the extremely large exothermicity (ΔG = –60 kJ mol^{-1}) of the reaction. Recently, Lima et al. [138] have reported the deprotonation time of 7-hydroxy-4-methyl-flavylium cation in H_2O as 7 ps. All these observed rates are apparently close to the solvent relaxation rate of water.

The dielectric response of water can be described by a Debye model including two relaxation times, a slow (τ_d) and a fast (τ_2), which are related to the collective reorientation of the hydrogen bonded liquid and the fast reorientation of a single water molecule, respectively [139,140]. According to this treatment, the solvent is modeled as a structureless fluid with a frequency-dependent dielectric constant $\varepsilon(\omega)$. In the case of water, $\varepsilon(\omega)$ is generally expressed in the Debye form:

$$\varepsilon(\omega) = \varepsilon_\infty + \frac{\varepsilon_s - \varepsilon_1}{1 + i\omega\tau_d} + \frac{\varepsilon_1 - \varepsilon_\infty}{1 + i\omega\tau_2} \qquad (2.18)$$

where ε_∞ and ε_0 are the high-frequency and zero-frequency dielectric constants, respectively.

According to recent measurements using THz-time domain spectroscopy [139,141], the magnitude of τ_d and τ_2 in H_2O are reported to be 8.5 ps and 170 fs at 292.3 K. Similar two relaxation times for water have also been observed by time-dependent fluorescence Stokes shift measurements [142–144]. The proton dissociation times of the compounds described previously are found to be comparable or slightly shorter than τ_d, which suggests that the proton transfer takes place through rapid cooperative motions of water molecules in the vicinity of the photoacids.

The spectroscopic and time-resolved studies of ESPT reactions in gas-phase clusters have provided significant information regarding the proton transfer mechanism between an excited molecule and a cluster of solvent molecules at a molecular level [15]. The resonant two-photon ionization method combined with mass spectrometry was used to measure the $S_1 \leftarrow S_0$ spectra of various clusters. Phenol (PhOH) and naphthol (NpOH) complexed with a variety of solvents were investigated by using molecular beam mass spectrometry and molecular beam photoelectron spectroscopy [15]. The picosecond and nanosecond time-resolved studies showed that the ESPT rate in clusters strongly depends on solvent type, size, and structure. Fluorescence spectra of 1-NpOH·(NH$_3$)$_n$ clusters up to n = 3, excited at the respective origin transitions, are dominated by naphthol-like sharp emission bands from the ^1L_b emitting state. In contrast the emission of the n = 4 cluster exhibits an additional broad and strongly red-shifted band, which can be assigned to ^1L_a naphtholate-type ESPT emission [145]. For 1-NpOH clusters with strong bases such as ammonia and piperidine, it has been found that the minimum numbers of solvent molecules necessary to observe ESPT are n = 4 and 2, respectively, and the ESPT reaction occurs when the proton affinity of the base aggregate reaches about 243 kcal/mol [146].

Picosecond pump and probe experiments of PhOH·(NH$_3$)$_n$ [147,148] showed a distinct onset to ESPT reaction at n = 5 ammonia molecules, and the n = 5 and 6 clusters reacted in 55 and 65 ps, respectively. In a mixed solvent system PhOH·(NH$_3$)$_5$(CH$_3$OH), the reaction rate decreased remarkably (750 ps), i.e., the substitution of a single CH$_3$OH molecule for an NH$_3$ molecule caused a substantial decrease in reaction rate. The reduced reactivity in the presence of a single CH$_3$OH molecule was explained by the unfavorable solvent structure for stabilizing a proton as compared to that of pure ammonia clusters [15].

In contrast to ammonia clusters, the minimum size for the ESPT reaction of water clusters is much larger, e.g., n = 20 to 30 for 1-NpOH·(H$_2$O)$_n$. For clusters of less than 100 water molecules, fast (<60 ps) and slow (~0.5 ns) components of the naphtholate anion fluorescence (ESPT emission) have been observed [149,150]. The reaction in 1-NpOH·(H$_2$O)$_n$ clusters is strongly influenced by the cluster temperature (internal energies), suggesting that internal cluster motion or dynamic solvation plays a crucial role in the ESPT reaction [146]. Knochenmuss et al. have considered that solvent-solute interaction induces $^1L_a/^1L_b$ inversion, and after relaxation into the more acidic ^1L_a state actual proton transfer takes places. Since the proton jump may be so fast that the overall kinetics are limited by the inversion of the electronic states.

As mentioned in Section 2.4, the greatly enhanced acidity of hydroxyarene in the excited state compared to its ground state has been traditionally interpreted in terms of partial charge transfer character in the excited singlet state resulting from a direct excitation from a lone π–electron pair of the hydroxylic oxygen into a π^* orbital of the aromatic ring (n-π^* CT); the resulting coulomb repulsion between the hydroxylic oxygen atom and the acidic proton would thus enhance the acidity. Another view, which is primarily discussed for the case of 1-NpOH, is based on inversion of closely-lying ^1L_b (lowest) and ^1L_a states. In this picture, $^1L_b/^1L_a$ level inversion is associated with the ESPT itself.

Recently, Hynes and coworkers have proposed a new perspective (a three-step mechanism) on intermolecular photochemical proton transfer in solution [96–99]. They performed femtosecond fluorescence and absorption measurements on pyranine (PyOH; 8-hydroxy-1,3,6-trisulfonate pyrene) in aqueous solution, and found that the early events of the photoinduced proton transfer from pyranine to water involve three

$$\text{PyOH} \xrightarrow{hv} \underset{(\text{LE}^{FC})}{\text{PyOH*}} \xrightarrow[300\text{fs}]{(1)} \underset{(\text{LE}^{EQ})}{\text{PyOH*}} \xrightarrow[2.2\text{ps}]{(2)} \underset{(\text{n-}\pi^* \text{ CT})}{\text{PyOH*}} \xrightarrow[87\text{ps}]{(3)} \text{PyO}^{-*} + \text{H}^+$$

SCHEME 2.6

successive steps: two ultrafast steps (300 fs and 2.2 ps), which precede the relatively slow (87 ps) proton transfer step. Scheme 2.6 shows a summary of their femtosecond experiments on pyranine in aqueous solution [96–99]. The first step (less than 300 fs) is attributable to solvation dynamics of the Franck–Condon S_1 state (LE^{FC}) of pyranine produced by femtosecond laser pulse excitation to produce the equilibrium S_1 state (LE^{EQ}). The second step, with a time constant of 2.2 ps, accompanies some change in oscillator strength, which is not found for the first or last steps. This suggests that the second step is associated with a change in electronic state, i.e., from locally excited state (or ^1L_b state) to n-π^* charge transfer state (or ^1L_a state). The latter state involves partial charge migration from the oxygen of the OH into the ring system, but no PT is involved in this step. The third step involves n-σ^* charge transfer, i.e., CT from the nonbonding orbital of the base to the antibonding orbital of the acid, and this step corresponds to the PT step itself [99]. Participation of n-σ^* CT in PT reactions has been originally proposed for the ground-state reactions [151]. The PT picture depicted in Scheme 2.6 demonstrates that the electronic rearrangements associated with the proton transfer act is a crucial aspect of PT reactions in solution.

Ultrafast measurements of fluorescence decay curves of photoacids have also revealed the important contribution of geminate recombination (neutralization) of the two separated ions in the overall dissociation processes, by observing nonexponential decay profiles [152–160]. Pines and Huppert [152] have reported the first direct detection of geminate recombination kinetics in ESPT. The fluorescence decay profile of hydroxyarene (ROH) obtained by picosecond laser pulse excitation showed nonexponential behavior with a long-time tail. This has been attributed to a reversible time-dependent geminate recombination. The reprotonation is an adiabatic process, so that the excited ROH* can undergo a second cycle of deprotonation. Based on the experimental and theoretical studies, Agmon, Pines, and Huppert [154–156] have analyzed the overall kinetics according to a two-step model:

$$\text{ROH*} \underset{k_r}{\overset{k_d}{\rightleftarrows}} [\text{RO}^{-*}...\text{H}^+]_{(r=a)} \overset{\text{DSE}}{\rightleftarrows} \text{RO}^{-*} + \text{H}^+ \qquad (2.19)$$

The first step, which is described by back-reaction boundary condition with intrinsic rate constants k_d and k_r, is followed by a diffusional second step in which the hydrated proton is removed from the parent molecule. Separation of a contact ion pair from the contact radius, a, to infinity is described by the transient numerical solution of the Debye–Smoluchowski equation (DSE). The asymptotic expression (the long time behavior) for the fluorescence of ROH*(t) is derived as [160]

$$[ROH^*] \cong \frac{\pi}{2} a^2 \exp(R_D / a) \frac{k_r}{k_d (\pi D)^{3/2}} t^{-3/2} \qquad (2.20)$$

In Equation (2.20), $D = D_{H^+} + D_{RO^{-*}}$ is the mutual diffusion coefficient of the proton and its conjugate base, and R_D is the Debye radius, given by

$$R_D = \frac{|z_1||z_2|e^2}{\varepsilon k_B T} \qquad (2.21)$$

where z_1 and z_2 are the charges of the proton and anion, and ε is the static dielectric constant of the solvent. Equation (2.20) predicts that the time-dependent population of ROH* conforms to a power law at $t^{-3/2}$. Actually, it has been shown that time-resolved fluorescence data of a variety of hydroxyarenes exhibit a log-time tail that follows the $t^{-3/2}$ time dependence. In Figure 2.8, the fluorescence decay profiles of protonated p-methoxyaniline and 1-naphthol in H_2O are shown as typical examples. It can be

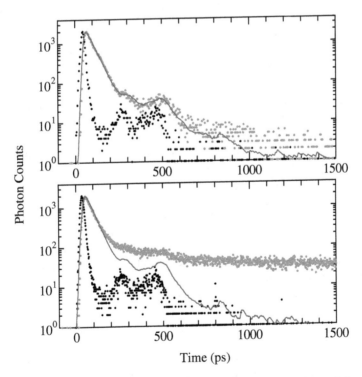

FIGURE 2.8 Fluorescence decay curves of (top) p-methoxyanilinium ion and (bottom) 1-naphthol in aqueous solution. Solid lines represent the least-squared fitted curves with a single exponential function. (From S. Kaneko, S. Shiobara, and S. Tobita, unpublished results.)

found for 1-naphthol that the rapid decay due to proton transfer is followed by a long-time tail resulting from reversible recombination of RO$^-$* with the geminate proton. In protonated p-methoxyaniline such a long-time tail in fluorescence decay profile is not prominent, because the proton dissociation of protonated p-methoxyaniline results in a proton and neutral molecule pair.

In proton dissociation of excited hydroxyarenes, the geminate recombination strongly depends on the coulomb potential between the ejected proton and the parent anion molecule, as well as the mutual diffusion coefficient. In contrast, in proton dissociation of protonated amines, the contribution of geminate recombination is expected to be less important, because the reaction produces a geminate pair between proton and excited neutral amines [92–94].

2.8 APPLICATIONS OF EXCITED-STATE PROTON TRANSFER REACTIONS

Excited-state proton transfer has received much attention because of its importance in fundamental aspects of reaction dynamics and also because of the various possibilities for applications, as in, e.g., effective photoprotecting agents [161–165], laser dyes [166–172], photodynamic therapy [173,174], and fluorescence probes for biological molecules [3,175]. In this section, applications of an excited-state intramolecular proton-transfer (ESIPT) system to photoprotecting agents (UV absorbers) are described briefly.

First, direct measurements of ESIPT have been carried out for 2-hydroxyphenyl-1,3,5-triazines with picosecond pulses [176–179]. The proton transfer rate was estimated by measuring the rise rate of the green fluorescence due to the proton-transferred form. Kramer and coworkers have reported on the application of intramolecular proton transfer of triazine and benzotriazole derivatives to UV absorbers [161–164]. Intramolecular proton transfer of these molecules in the excited state proceeds adiabatically to produce its tautomeric form in the excited state. The excited tautomeric molecule is deactivated by rapid radiationless transition, which probably arises from the presence of intramolecular hydrogen bond. Thus, the overall process results in the transformation of harmful ultraviolet radiation into thermal energy. Photophysical properties of UV absorbers MA-TIN 1 (2-[2-hydroxy-3-*tert*-butyl-5-(O-[2-hydroxy-3-(2-methylpropenoxyloxy)-propyl]-2-carbonyloxyethyl)phenyl]-benzotriazole) and MA-TZ-1 (2,4-bis(2,4-dimethylphenyl)-6-[2-hydroxy-4-(2-hydroxy-3-[2-methylpropenoyloxyl])-propoxyphenyl)-1,3,5-triazine) (see Figure 2.9) in polymer matrixes were investigated by means of time-resolved emission measurements [161,163]. Various copolymers of MA-TIN 1 and MA-TZ-1 with styrene, methyl methacrylate, and methacrylic acid have been synthesized by radical polymerization. Their absorption spectra in the long-wavelength UV region appeared unchanged compared to those of the monomeric UV absorbers, indicating that the intramolecular hydrogen bond, which is essential for the photostability of this type of UV absorbers, is still intact in the copolymer [163]. On the basis of the copolymerization parameters

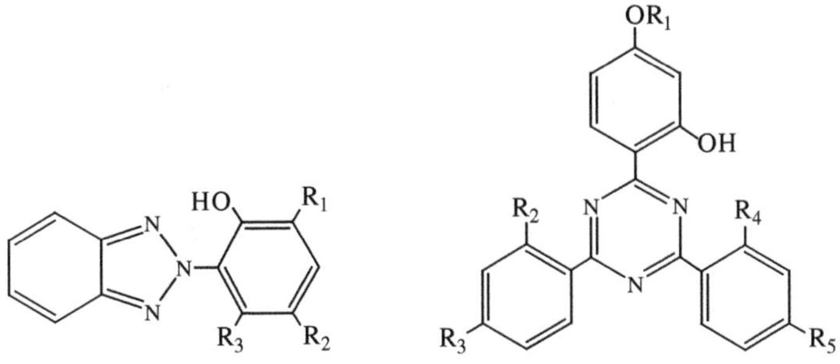

Compound	R_1	R_2	R_3	R_4	R_5
MA-TIN 1	$(CH_3)_3$	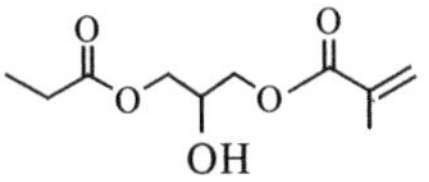	H	–	–
MA-TZ 1	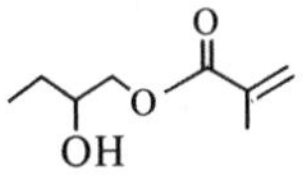	CH_3	CH_3	CH_3	CH_3

FIGURE 2.9 Molecular formulas of MA-TIN 1 and MA-TZ 1. (From Stein, M., Keck, J., Waiblinger, F., Fluegge, A.P., Kramer, H.E.A., Hartschuh, A., Port, H., Leppard, D., Rytz, G., *J. Phys. Chem. A,* 2002, 106, 2055. With permission.)

obtained, it is revealed that in copolymers of MA-TIN 1 and MA-TZ-1 with styrene and methyl methacrylate the UV absorber is present to a higher extent than it is when simply present as a mixture of monomeric UV absorbers in the monomer feed. Radiationless deactivation is more efficient in copolymer than in mixtures at temperatures far below 200 K. However, at room temperature these deactivation steps are more efficient in mixtures than in copolymers. Both MA-TZ 1 in the copolymerized state as well as physically admixed to a polar polymer matrix showed phosphorescence evolution when irradiating samples with radiation ($\lambda_{exc} = 313$ nm) from a 100 W Hg lamp. The phosphorescence evolution is due to open conformers of the UV absorbers, i.e., with intermolecular rather than intramolecular hydrogen bonds, which are formed in polar matrices under the influence of UV radiation (Figure 2.10). As described above, at room temperature the radiationless deactivation is less favored in copolymers than in mixtures. However, in copolymers the UV stabilizer cannot migrate out of the polymer, which might favor copolymers for long-term use. Kramer and coworkers [161] conclude that UV absorbers such as MA-TZ 1 are good light-protecting agents.

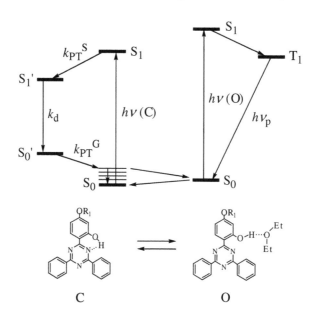

FIGURE 2.10 Intramolecular proton transfer reactions of 2-(2-hydroxy-4-methoxyphenyl)-4,6-diphenyl-1,3,5-triazine in the S_1 state (k_{PT}^S) and ground state (k_{PT}^G) and light-induced opening of the intramolecular hydrogen bond. (From Waiblinger, F., Keck, J., Stein, M., Fluegge, A.P., Kramer, H.E.A., Leppard, D., *J. Phys. Chem. A,* 2000, 104, 1100. With permission.)

2.9 CONCLUSIONS

Fundamental aspects of excited-state proton transfer reactions in solution as well as in gas-phase clusters have been revealed by recent dynamic studies using various experimental techniques and theoretical studies. Proton-induced fluorescence quenching competing with prototropic equilibrium is important to determine the accurate pK_a^* value of aromatic compounds, which have an intramolecular charge transfer character in the excited state. The proton transfer from an aromatic molecule to solvent is closely related to the electronic structure of not only the protonated form, but also of the deprotonated form, and the proton transfer rate is correlated with the reaction free energy change. The electronic rearrangements associated with the proton transfer play a crucial role in proton transfer reactions in solution. Also, the microenvironment around excited acid as well as bulk solvent property influences remarkably the proton transfer rate. The overall proton-dissociation processes are affected by the geminate recombination of the two separated ions. There are some cases that proton and hydrogen atom transfer reactions take place in the triplet state. Excited-state proton transfer reactions are applicable to a wide range of fields.

REFERENCES

1. Alnaut, L.G., Formosinho, S.J., *J. Photochem. Photobiol. A: Chem.*, 1993, 75, 1.
2. Formosinho, S.J., Alnaut, L.G., *J. Photochem. Photobiol. A: Chem.*, 1993, 75, 21.
3. Mishra, A.K., Fluorescence of excited singlet state acids in certain organized media: Applications as molecular probes, in *Understanding and Manipulating Excited State Processes, Molecular and Supramolecular Photochemistry Series*, V. Ramamurthy and K.S. Schanze, Eds., Marcel Dekker, New York, 2001, vol. 8, chap. 10.
4. Caldin, E.F., in *The Mechanisms of Fast Reactions in Solution*, IOS Press, Amsterdam, 2001, chap. 8.
5. Waluk, J., Conformational Aspects of Intra- and Intermolecular Excited-State Proton Transfer, in *Conformational Analysis of Molecules in Excited States*, J. Waluk, Ed., Wiley-VCH, New York, 2000.
6. Valeur, B., *Molecular Fluorescence*, Wiley-VCH, Weinheim, 2002.
7. Sharma, A., Schulman, S.G., Introduction to Fluorescence Spectroscopy, in *Technique in Analytical Chemistry Series*, F.A. Settle, Ed., Wiley, New York, 1999.
8. Shizuka, H., Yamaji, M., *Bull. Chem. Soc. Jpn. Accounts*, 2000, 72, 267.
9. Martynov, I.Y., Demyashkevich, A.B., Uzhinov, B.M., Kuz'min, M.G., *Russ. Chem. Rev. (Usp. Khim.)*, 1977, 46, 3.
10. Shizuka, H., *Acc. Chem. Res.*, 1985, 18, 141.
11. Ireland, J.F., Wyatt, P.A.H., *Adv. Phys. Org. Chem.*, 1976, 12, 131.
12. Klopffer, W., *Adv. Photochem.*, 1977, 10, 311.
13. Kosower, E.M., Huppert, D., *Ann. Rev. Phys. Chem.*, 1986, 27, 127.
14. Tolbert, L.M., Solntsev, K.M., *Acc. Chem. Res.*, 2002, 25, 19.
15. Syage, J.A., *J. Phys. Chem.*, 1995, 99, 5772.
16. Förster, T., *Naturwiss.*, 1949, 26, 186.
17. Förster, T., *Z. Elektrochem.*, 1950, 54, 42.
18. Förster, T., *Pure Appl. Chem. Int. Ed. Engl.*, 1964, 2, 1.
19. Grabowski, Z.R., Grabowska, A.Z., *Z. Phys. Chem. (Wiesbaden)*, 1976, 101, 197.
20. Grabowski, Z.R., Rubaszewska, W., *J. Chem. Soc. Faraday Trans. I*, 1977, 72, 11.
21. Weller, A., *Prog. React. Kinet.*, 1961, 1, 189.
22. Weller, A., *Ber. Bunsenges. Phys. Chem.*, 1956, 66, 1144.
23. Weller, A., *Ber. Bunsenges. Phys. Chem.*, 1952, 56, 662.
24. Vander Donckt, E., *Prog. React. Kinet.*, 1970, 5, 273.
25. Schulman, S.G., in *Physical Methods in Heterocyclic Chemistry*, vol. VI, A.R. Katritzky, Ed., Academic Press, New York, 1974, 147.
26. Klöpffer, W., *Adv. Photochem.*, 1977, 10, 311.
27. Jackson, G., Porter, G., *Proc. R. Soc. London, Ser. A*, 1961, 200.
28. Beens, H., Grellmann, K.H., Gurr, M., Weller, A., *Discuss. Faraday Soc.*, 1965, 29, 183.
29. Wehry, E.L., Rogers, L.B., in *Fluorescence and Phosphorescence Analyses*, Vol. 2, D.M. Hercules, Ed., Plenum Press, New York, 1976.
30. Schulman, S.G., in *Modern Fluorescence Spectroscopy*, E.L. Wehry, Ed., Wiley-Interscience, New York, 1966.
31. Schulman, S.G., *Fluorescence and Phosphorescence Spectroscopy*, Pergamon, Oxford, 1977.
32. Tsutsumi, K., Shizuka, H., *Chem. Phys. Lett.*, 1977, 52, 485.

33. Tsutsumi, K., Shizuka, H., *Z. Phys. Chem. (Wiesbaden)*, 1978, 111, 129.
34. Shizuka, H., Tsutsumi, K., Takeuchi, H., Tanaka, I., *Chem. Phys. Lett.*, 1979, 62, 408.
35. Shizuka, H., Tsutsumi, K., Takeuchi, H., Tanaka, I., *Chem. Phys.*, 1981, 59, 183.
36. Shizuka, H., Tsutsumi, K., *J. Photochem.*, 1978, 9, 334.
37. Tsutsumi, K., Sekiguchi, S., Shizuka, H., *J. Chem. Soc., Faraday Trans. 1*, 1982, 78, 1087.
38. Swaminathan, M., Dogra, S.K., *Can. J. Chem.*, 1983, 61, 1064.
39. Tsutsumi, K., Shizuka, H., *Z. Phys. Chem. (Wiesbaden)*, 1980, 122, 129.
40. Hafner, F., Wörner, J., Steiner, V., Hauser, M., *Chem. Phys. Lett.*, 1980, 72, 139.
41. Harris, C.M., Selinger, B.K., *J. Phys. Chem.*, 1980, 84, 891, 1366.
42. Schulman, S.G., Rosenberg, L.S., Vincent, W.R., Jr., *J. Am. Chem. Soc.*, 1979, 101, 139.
43. Förster, T., *Chem. Phys. Lett.*, 1972, 17, 309.
44. Schulman, S.G., Sturgeon, R.J., *J. Am. Chem. Soc.*, 1977, 99, 7209.
45. Tobita, S., Shizuka, H., *Chem. Phys. Lett.*, 1980, 75, 140.
46. Shizuka, H., Tobita, S., *J. Am. Chem. Soc.*, 1982, 104, 6919.
47. Förster, T., *Z. Elektrochem.*, 1950, 54, 531.
48. Webb, S.P., Philips, L.A., Yeh, S.W., Tolbert, L.M., Clark, J.H., *J. Phys. Chem.*, 1986, 90, 5154.
49. Shizuka, H., Ogiwara, T., Kimura, E., *J. Phys. Chem.*, 1985, 89, 4302.
50. Okada, T., Kawai, M., Ikemachi, T., Mataga, N., Sakata, Y., Misumi, S., Shionoya, S., *J. Phys. Chem.*, 1984, 88, 1976.
51. Shizuka, H., Ishii, Y., Morita, T., *Chem. Phys. Lett.*, 1979, 62, 408.
52. Shizuka, H., Nakamura, M., Morita, T., *J. Phys. Chem.*, 1980, 84, 989.
53. Shizuka, H., Obuchi, H., *J. Phys. Chem.*, 1982, 86, 1297.
54. Suzuki, S., Baba, H., *Bull. Chem. Soc. Jpn.*, 1967, 40, 2199.
55. Suzuki, S., Fujii, T., Sato, K., *Bull. Chem. Soc. Jpn.*, 1972, 45, 1937.
56. Knochenmuss, R., Muino, P.L., Wickleder, C., *J. Phys. Chem.*, 1996, 100, 11218.
57. Magnes, B.-Z., Strashnikova, N.V., Pines, E., *Isr. J. Chem.*, 1999, 29, 361.
58. Coulson, C.A., Longuet-Higgins, H.C. *Proc. R. Soc. London, Ser. A*, 1947, 191, 39.
59. Fukui, K., Yonezawa, T., Shingu, H., *J. Chem. Phys.*, 1952, 20, 722.
60. Fukui, K., Yonezawa, T., Nagata, C., *Bull. Chem. Soc. Jpn.*, 1954, 27, 423.
61. Wheland, G.W., *J. Am. Chem. Soc.*, 1942, 64, 900.
62. Creed, D., *Photochem. Photobiol.*, 1984, 29, 537.
63. Lumry, R., Hershberger, M., *Photochem. Photobiol.*, 1978, 27, 819.
64. Rayner, D.M., Szabo, A.G., *Can. J. Chem.*, 1978, 56, 743.
65. Beddard, G.S., Fleming, G.R., Porter, G., Robbins, R., *Philos. Trans. R. Soc. London, Ser. A*, 1980, 298, 321.
66. Cowgill, R.W., *Biochim. Biophys. Acta*, 1967, 122, 6.
67. Picci, R.W., Nesta, J.M., *J. Phys. Chem.*, 1976, 80, 974.
68. Szabo, A.G., Rayner, D.M., *J. Am. Chem. Soc.* 1980, 102, 554.
69a. Chang, M.C., Perich, J.W., McDonald, D.B., Fleming, G.R., *J. Am. Chem. Soc.*, 1983, 105, 3819.
69b. Perich, J.W., Chang, M.C., McDonald, D.B., Fleming, G.R., *J. Am. Chem. Soc.*, 1983, 105, 3824.
70. Lehrer, S.S., *J. Am. Chem. Soc.*, 1970, 92, 3459.
71. Ricci, R.W., *Photochem. Photobiol.*, 1970, 12, 67.
72. Nakanishi, M., Tsuboi, M., *Chem. Phys. Lett.*, 1978, 57, 262.
73. Robbins, R.J., Gleming, G.R., Beddard, G.S., Robinson, G.W., Thistlethwaite, P.J.,

Woolfe, G.J., *J. Am. Chem. Soc.*, 1980, 102, 6271.

74. Saito, I., Sugiyama, H., Yamamoto, A., Muramatsu, S., Matsuura, T., *J. Am. Chem. Soc.*, 1984, 106, 4286.
75. Saito, I., Muramatsu, S., Sugiyama, H., Yamamoto, A., Matsuura, T., *Tetrahedron Lett.*, 1985, 26, 5891.
76. Hershberger, M.V., Lumry, R., Verrall, R., *Photochem. Photobiol.*, 1981, 22, 609.
77. Lasser, N., Feitelson, J., Lumry, R., *Isr. J. Chem.*, 1977, 16, 330.
78. Gudgin-Templeton, E.F., Ware, W.R., *J. Phys. Chem.*, 1984, 88, 4626.
79a. Santus, R., Bazin, M., Aubailly, M., *Rev. Chem. Intermed.*, 1980, 2, 231.
79b. Grossweiner, L.I., Bredzel, A.M., Blum, A., *Chem. Phys.*, 1981, 57, 147.
79c. Kirby, E.P., Steiner, R.F., *J. Phys. Chem.*, 1983, 87, 189.
80. Gudgin, E., Lopez-Delgado, R., Ware, W.R., *J. Phys. Chem.*, 1983, 87, 1559.
81. Shizuka, H., Serizawa, M., Kobayashi, H., Kameta, K., Sugiyama, H., Matsuura, T., Saito, I., *J. Am. Chem. Soc.*, 1988, 110, 1726.
82. Shizuka, H., Serizawa, M., Shimo, T., Saito, I., Matsuura, T., *J. Am. Chem. Soc.*, 1988, 110, 1930.
83. Pines, E., Fleming, G.R., *Chem. Phys.*, 1994, 182, 393.
84. Pines, E., Tepper, D., Magnes, B-Z., Pines, D., Barak, T., *Ber. Bunsenges. Phys. Chem.*, 1998, 102, 504.
85. Agmon, N., *J. Chem. Phys.*, 1999, 100, 2175.
86. Gross, K.C., Seybold, P.G., *Int. J. Quant. Chem.*, 2000, 80, 1107.
87. Wehry, L., Rogers, L.B., *J. Am. Chem. Soc.*, 1965, 87, 4234.
88. Schulman, S.G., Vincent, W.R., Underberg, W.J.M., *J. Phys. Chem.*, 1981, 85, 4068.
89a. Tolbert, L.M., Haubrich, J.E., *J. Am. Chem. Soc.*, 1990, 112, 8163.
89b. Tolbert, L.M., Haubrich, J.E., *J. Am. Chem. Soc.*, 1994, 126, 10593.
90. Koifman, N., Cohen, B., Huppert, D., *J. Phys. Chem. A*, 2002, 106, 4336.
91. Cohen, B., Huppert, D., *J. Phys. Chem. A*, 2001, 105, 2980.
92. Tajima, S., Shiobara, S., Shizuka, H., Tobita, S., *Phys. Chem. Chem. Phys.*, 2002, 4, 3376.
93. Shiobara, S., Kamiyama, R., Tajima, S., Shizuka, H., Tobita, S., *J. Photochem. Photobiol. A*, 2002, 154, 53.
94. Shiobara, S., Tajima, S., Tobita, S., *Chem. Phys. Lett.*, 2003, 280, 673.
95. Moreira, P.F., Giestas, L., Yihwa, C., Vautier-Giongo, C., Quina, F.H., Macanita, A.L., Lima, J.C., *J. Phys. Chem. A*, 2003, 107, 4203.
96. Tran-Thi, T.H., Prayer, C., Millie, P., Uznanski, P., Hynes, J.T., *J. Phys. Chem. A*, 2002, 106, 2244.
97. Tran-Thi, T.H., Gustavsson, T., Prayer, C., Pommeret, S., Uznanski, P., Hynes, J.T., *Chem. Phys. Lett.*, 2000, 229, 421.
98. Granucci, G., Hynes, J.T., Millie, P., Tran-Thi, T.H., *J. Am. Chem. Soc.*, 2000, 122, 12243.
99. Hynes, J.T., Tran-Thi, T.H., Granucci, G., *J. Photochem. Photobiol. A*, 2000, 154, 3.
100. Domcke, W., Sobolewski, A.L., *Science*, 2003, 202, 1693.
101. Sobolewski, A.L., Domcke, W., Dedonder-Lardeux, C., Jouvet, C., *Phys. Chem. Chem. Phys.*, 2002, 4, 1093.
102. Sobolewski, A.L., Domcke, W., *J. Phys. Chem. A*, 2001, 105, 9275.
103. Barroso, M., Arnaut, L.G., Formosinho, S.J., *J. Photochem. Photobiol. A*, 2002, 154, 13.
104. Giestas, L., Yihwa, C., Lima, J.C., Vautier-Giongo, C., Lopes, A., Macanita, A.L., Quina, F.H., *J. Phys. Chem. A*, 2003, 107, 3263.

105. Mishra, A.K., Shizuka, H., *J. Chem. Soc., Faraday Trans. 1*, 1996, 92, 1481.
106. Shizuka, H., Saito, T., Morita, T., *Chem. Phys. Lett.*, 1978, 56, 519.
107. Shizuka, H., Takada, K., Morita, T., *J. Phys. Chem.*, 1980, 84, 989.
108. Shizuka, H., Ogiwara, T., Narita, A., Sumitani, M., Yoshihara, K., *J. Phys. Chem.*, 1986, 90, 6708.
109. Shizuka, H., Kameta, K., Shinozaki, T., *J. Am. Chem. Soc.*, 1985, 107, 3956.
110. Shizuka, H., Serizawa, M., *J. Phys. Chem.*, 1986, 90, 4573.
111. Rayner, D.M., Wayatt, P.A.H., *J. Chem. Soc. Faraday Trans. 2*, 1974, 70, 945.
112. Shizuka, H., Kimura, E., *Can. J. Chem.*, 1984, 62, 2041.
113. El-Sayed, M.A., *J. Chem. Phys.*, 1962, 26, 573.
114. El-Sayed, M.A., *J. Chem. Phys.*, 1963, 28, 2834.
115. El-Sayed, M.A., *J. Chem. Phys.*, 1964, 41, 2462.
116. Ware, W.R., Watt, D., Holmes, J.D., *J. Am. Chem. Soc.*, 1974, 96, 7853.
117. Bonner, T.G., Phillips, J., *J. Chem. Soc. B*, 1966, 650.
118. Fisher, A., Grigor, B.A., Parker, J., Vaughan, J., *J. Am. Chem. Soc.*, 1961, 82, 4208.
119. Taft, R.W., in *Steric Effects in Organic Chemistry*, M.S. Newman, Ed., Wiley, New York, 1956, 619.
120. Shizuka, H., Fukushima, M., *Chem. Phys. Lett.*, 1983, 101, 598.
121. Shizuka, H., Hagiwara, H., Fukushima, M., *J. Am. Chem. Soc.*, 1985, 107, 7816.
122. Kaneko, S., Yamaji, M., Hoshino, M., Shizuka, H., *J. Phys. Chem.*, 1992, 96, 8028.
123. Kohno, S., Hoshino, M., Shizuka, H., *J. Phys. Chem.*, 1991, 95, 5489.
124. Robinson, G.W., *J. Phys. Chem.*, 1991, 95, 10386.
125. Krishnan, R., Fillingim, T.G., Lee, J., Robinson, G.W., *J. Am. Chem. Soc.*, 1990, 112, 1353.
126. Robinson, G.W., Thistlethwaite, P.J., Lee, J., *J. Phys. Chem.*, 1986, 90, 4224.
127. Lee, J., Robinson, G.W., Webb, S.P., Philips, L.A., Clark, J.H., *J. Am. Chem. Soc.*, 1986, 108, 6538.
128. Lee, J., Robinson, G.W., Bassez, M.-P., *J. Am. Chem. Soc.*, 1986, 108, 7477.
129. Lee, J., Griffin, R.D., Robinson, G.W., *J. Chem. Phys.*, 1985, 82, 4920.
130. Eigen, M., Kustin, K., *J. Am. Chem. Soc.*, 1960, 82, 5952.
131. Clark, J.H., Shapiro, S.L., Campillo, A.J., Winn, K.R., *J. Am. Chem. Soc.*, 1979, 101, 746.
132. Huppet, D., Tolbert, L.M., Linares-Samaniego, S., *J. Phys. Chem. A*, 1997, 101, 4602.
133. Pines, E., Pines, D., Barak, T., Magnes, B.-Z., Tolbert, L.M., Haubrich, J.E., *Ber. Bunsenges. Phys. Chem.*, 1998, 102, 511.
134. Solntsev, K.M., Huppert, D., Agmon, N., Tolbert, L.M., *J. Phys. Chem. A*, 2000, 104, 4658.
135. Agmon, N., Huppert, D., Masad, A., Pines, E., *J. Phys. Chem.*, 1991, 95, 10407.
136. Htun, M.T., Suwaiyan, A., Klein, U.K.A., *Chem. Phys. Lett.*, 1995, 242, 506.
137. Htun, M.T., Suwaiyan, A., Klein, U.K.A., *Chem. Phys. Lett.*, 1995, 242, 512.
138. Lima, J.C., Abreu, I., Brouillard, R., Maçanita, A.L., *Chem. Phys. Lett.*, 1998, 298, 189.
139. Rønne, C.R., Keiding, S.J., *Mol. Liquids*, 2002, 101, 199.
140. Buchner, R., Barthel, J., Stauber, J., *Chem. Phys. Lett.*, 1999, 206, 57.
141. Rønne, C., Åstrand, P.-O., Keiding, S.R., *Phys. Rev. Lett.*, 1999, 82, 2888.
142. Pant, D., Levinger, N. E., *J. Phys. Chem. B*, 1999, 102, 7846.

143. Jimenez, R., Fleming, G.R., Kumar, P.V., Maroncelli, M., *Nature,* 1994, 269, 471.
144. Jarzeba, W., Walker, G.C., Johnson, A.E., Kahlow, M.A., Barbara, P.F., *J. Phys. Chem.,* 1988, 92, 7039.
145. Knochenmuss, R., Fischer, I., Lührs, D., Lin, Q., *Isr. J. Chem.,* 1999, 29, 221.
146. Knochenmuss, R., Cheshnovsky, O., Leutwyler, S., *Chem. Phys. Lett.,* 1988, 144, 317.
147. Steadman, J., Syage, J.A., *J. Chem. Phys.,* 1990, 92, 4630.
148. Syage, J.A., Steadman, J., *J. Chem. Phys.,* 1991, 95, 2497.
149. Knochenmuss, R., Holtom, G.R., Ray, D., *Chem. Phys. Lett.,* 1993, 215, 188.
150. Knochenmuss, R., Smith, D.E., *J. Chem. Phys.,* 1994, 101, 7327.
151. Timoneda, J., Hynes, J.T., *J. Phys. Chem.,* 1991, 95, 10431.
152. Pines, E., Huppert, D., *Chem. Phys. Lett.,* 1986, 126, 88.
153. Pines, E., Huppert, D., *J. Chem. Phys.,* 1986, 84, 3576.
154. Agmon, N., *J. Chem. Phys.,* 1988, 89, 1524.
155. Agmon, N., Pines, E., Huppert, D., *J. Chem. Phys.,* 1988, 88, 5631.
156. Pines, E., Huppert, D., Agmon, N., *J. Chem. Phys.,* 1988, 88, 5620.
157. Gopich, I.V., Solntsev, K.M., Agmon, N., *J. Chem. Phys.,* 1999, 110, 2164.
158. Poles, E., Cohen, B., Huppert, D., *Isr. J. Chem.,* 1999, 29, 347.
159. Cohen, B., Huppert, D., *J. Phys. Chem. A,* 2002, 106, 1946.
160. Agmon, N., Goldberg, S.Y., Huppert, D., *J. Molec. Liquids,* 1995, 64, 161.
161. Stein, M., Keck, J., Waiblinger, F., Fluegge, A.P., Kramer, H.E.A., Hartschuh, A., Port, H., Leppard, D., Rytz, G., *J. Phys. Chem. A,* 2002, 106, 2055.
162. Waiblinger, F., Keck, J., Stein, M., Fluegge, A.P., Kramer, H.E.A., Leppard, D., *J. Phys. Chem. A,* 2000, 104, 1100.
163. Keck, J., Kramer, H.E.A., Port, H., Hirsch, T., Fischer, P., Rytz, G., *J. Phys. Chem.,* 1996, 100, 14468.
164. Stuber, G.J., Kieninger, M., Schettler, H., Busch, W., Goeller, B., Franke, J., Kramer, H.E.A., Hoier, H., Henkel, S., Fischer, P., Port, H., Hirsch, T., Rytz, G., Birbaum, J.-L., *J. Phys. Chem.,* 1995, 99, 10097.
165. O'Connor, D.B., Scott, G.W., Coulter, D.R., Yavrouian, A., *J. Phys. Chem.,* 1991, 95, 10252.
166. Khan, A.U., Kasha, M., *Proc. Natl. Acad. Sci. USA,* 1983, 80, 1767.
167. Brucker, G.A., Swinney, T.C., Kelly, D.F., *J. Phys. Chem.,* 1991, 95, 3190.
168. Chou, P.T., McMorrow, D., Aartsma, T.J., Kasha, M., *J. Phys. Chem.,* 1984, 88, 4596.
169. Parthenopoulos, D.A., McMorrow, D., Kasha, M., *J. Phys. Chem.,* 1991, 95, 2668.
170. Acuna, A.U., Costela, A., Munoz, J.M., *J. Phys. Chem.,* 1986, 90, 2807.
171. Costela, A., Amat, F., Catalan, J., Douhal, A., Figuera, J.M., Munoz, J.M., Acuna, A.U., *Opt. Commun.,* 1987, 64, 457.
172. Costela, A., Munoz, J.M., Douhal, A., Figuera, J.M., Acuna, A.U., *Appl. Phys. B,* 1989, 49, 545.
173. Das, K., Ashby, K.D., Wen, J., Petrich, J.W., *J. Phys. Chem. B,* 1999, 102, 1581.
174. Smirnov, A.V., Das, K., English, D.S., Wan, Z., Kraus, G.A., Petrich, J.W., *J. Phys. Chem. A,* 1999, 102, 7949.
175. Bhattacharyya, K., *Acc. Chem. Res.,* 2003, 26, 95.
176. Shizuka, H., Matsui, K., Okamura, T., Tanaka, I., *J. Phys. Chem.,* 1975, 79, 2731.
177. Shizuka, H., Matsui, K., Hirata Y., Tanaka, I., *J. Phys. Chem.,* 1976, 80, 2070.

178. Shizuka, H., Matsui, K., Okamura, T., Tanaka, I., *J. Phys. Chem.* 1977, 81, 2243.
179. Shizuka, H., Machii, M., Higaki, Y., Tanaka, M., Tanaka, I., *J. Phys. Chem.* 1985, 89, 320.
180. Huppert, D., Tolbert, L.M., and Linares-Samaniego, S., *J. Phys. Chem. A,* 1997, 101, 4602.

3 Photoreactivity of n,π*-Excited Azoalkanes and Ketones

Werner M. Nau and Uwe Pischel

CONTENTS

3.1 INTRODUCTION

Being two of the simplest organic chromophores, the carbonyl and azo groups in ketones and azoalkanes have been of great historical importance for the development of photochemistry [1,2]. Ketones were among the first compounds used in rationally designed, practically useful [3,4], and mechanistically understood photoreactions [5,6]. Azo compounds, on the other hand, while traditionally associated with dye-industrial applications, remain attractive radical initiators as well as precursors in the photochemical synthesis of diradicals [7] and strained polycyclic compounds resulting from denitrogenation [8,9]. Other applications include their use as photochromic systems based on *cis-trans* isomerization [10–12]. In general, simple ketones and azoalkanes posses an n,π* electronic configuration in their first excited state.

3.1.1 A GLIMPSE AT THE DEVELOPMENT OF THE MECHANISTIC PHOTOCHEMISTRY OF KETONES

The reactions of n,π* triplet-excited ketones are classics in photochemistry and have contributed much to the definition of molecular organic photochemistry as a field of its own [13]. For example, it took 60 years to realize that it is the triplet state of benzophenone [6] that is responsible for its photoreduction, one of the first described photoreactions [3]. With the further development of molecular organic photochemistry, the dependence of photochemical reactivity on the electronic configuration of excited states has received more detailed attention. Thus, it has been recognized that n,π*-excited states behave more "radical-like" in their reactions than π,π*-excited states [13–25]. The extensively studied hydrogen abstraction reactions of triplet-excited ketones, which mimic the behavior of alkoxyl radicals [15,16,24–29], have been taken as a test case for the characteristic radical-like n,π* reactivity.

The singlet-excited states of ketones, and in particular their aromatic derivatives, are relatively short-lived and yield on intersystem crossing (ISC) their longer-lived triplet states, whose reactivity generally prevails. Consequently, mechanistic investigations have been generally carried out by employing suitable triplet ketones as prototypal n,π* states, such that some important aspects on photochemical reactivity have received comparably little attention, for example, a comprehensive comparison of singlet and triplet reactivity of n,π*-excited states. Hence, the reactivity of singlet-excited ketones has been subject to debate and experimental reactivity data are limited to a restricted number of aliphatic derivatives, mostly diketones. An illustrative example is the controversy on whether singlet or triplet n,π*-excited states are more reactive in hydrogen abstractions [30], which has only been settled quite recently [31].

3.1.2 A GLIMPSE AT THE DEVELOPMENT OF THE MECHANISTIC PHOTOCHEMISTRY OF AZOALKANES

Following some early reports [32–44], the photoreactivity of azoalkanes has been recently studied in great detail [9,31,45–69]. The availability of comprehensive photoreactivity data for azoalkanes now offers the unique possibility for a comparison with the previously examined photochemical reactivity of ketones, and thus to contrast the reactivity of two fundamentally different n,π* chromophores (N=N vs. C=O) and to allow a better generalization of the proposed reaction characteristics of n,π* chromophores [31,70,71]. In fact, many generalizations made for "n,π* photochemistry" have been proven incorrect in recent studies, and these studies have yielded additional insights and have led to some new viewpoints of photochemical reactivity, demonstrating that the efficiency for singlet oxygen production from n,π* triplet-excited states is not necessarily low [46], that not all charge-transfer-induced quenching processes of n,π*-excited states are accelerated in polar solvents [55,65], that not all n,π*-excited states are electrophilic [71], and that some n,π*-excited states are even sufficiently reactive to oxidize water, chloroform, and acetonitrile [47,52,71] and show deuterium isotope effects with methanol as high as 20 [56,65].

In addition, the reactivity of both singlet- and triplet-excited azoalkanes has been investigated [31,37,41,45,47–49,51,55,57-61,65,71–76]. This allows a comparison of the photoreactivity of singlet- and triplet-excited n,π* states, which has proven difficult and limited to only few additives for ketones [31,70,77].

3.1.3 Scope of Review

Numerous reviews on the photochemistry and photophysics of ketones [18,23,25,78,79] and azoalkanes [38,43,80,81] are already available. This chapter focuses on intermolecular photoreactions of azoalkanes, which are compared with known data for ketones (see Structure 3.1). Unimolecular reactions such as the Norrish type-I α-cleavage reaction of ketones [17,82–87] and their Norrish type-II reactions [20,24,73,83,88–91] as well as denitrogenation [8,9,43,68,92,93] and cis-trans isomerization of azoalkanes [43,92,93] are not discussed. The emphasis lies, besides data compilation, on mechanistic understanding, such that classical applications of azo compounds as dyes [94] or more recent applications of azo compounds in photochromic materials [10–12], or ketones as radical initiators in polymerization [95], are omitted as well.

The primary attention of the following review chapter lies on some new developments, which have largely resulted from investigations on the comparative photochemistry of bicyclic azoalkanes with simple ketones. The prototypal structures are shown above, all of which possess an n,π* configuration in both the lowest singlet- and triplet-excited states. All ketones further undergo ISC, which in the case of acetone and biacetyl is sufficiently slow to result in singlet-excited states with ns lifetimes, which are accessible to study intermolecular reactivity. The azoalkane DBH-T is a member of the exceptional class displaying spontaneous ISC and long triplet lifetimes [96–98], while the azoalkane 2,3-diazabicyclo[2.2.2]oct-2-ene (DBO) is a representative of the small class of fluorescent azoalkanes [34,42,43,56,72,81,99–104]. DBO is known for its extraordinarily long-lived [34,56,72,81,100,103,104] and unusually structured [33,80] fluorescence, which is still not yet understood in detail.

pyrazoline   DBH   DBH–T   DBO   DBN

Me_2CO AP   BP   BA

STRUCTURE 3.1

The photophysical background is described first. We then turn to the principal mechanisms of photoreactions of n,π*-excited states, and finally tabulate and discuss quenching rate constants and photoreaction pathways.

3.2 PHOTOPHYSICAL PROPERTIES

In Table 3.1, the most important general properties of the azo and the carbonyl chromophore are summarized. Table 3.2 shows detailed photophysical data for azoalkanes and ketones, which are discussed in the following sections.

3.2.1 ELECTRONIC TRANSITIONS

Even in the absence of a delocalized π system, simple ketones and azoalkanes can give rise to absorption in the near UV or even visible region. However, the absorbance is low due to the n,π* character of the related electronic transition. The relevant orbitals involved in the transitions for the different chromophores are depicted in Scheme 3.1.

In the case of ketones, the lone pair orbital from which excitation occurs is essentially a p-type atomic orbital on oxygen. The associated n,π* transition has no electric transition dipole moment (only quadrupole, see Scheme 3.1) and is therefore forbidden. In addition, orbital overlap is poor such that typical extinction coefficients (ε) as low as *ca.* 10 M⁻¹cm⁻¹ for aliphatic ketones and *ca.* 200 M⁻¹cm⁻¹ for aromatic ketones result.

For azoalkanes, the orbital from which excitation occurs is the antisymmetric combination of two s-p-hybrid nitrogen lone pairs, *i.e.,* an in-plane n_ molecular orbital [105]. A special characteristic of azoalkanes is that their configuration can be *cis* or *trans* (Scheme 3.1), which has implications for the spatial distribution of lone-pair

TABLE 3.1

Generalized Photophysical and Photochemical Properties of Ketones and *cis*-Azoalkanes with Lowest n,π* Excited-State Configuration[a]

	Ketones	*cis*-Azoalkanes
Absorption, electronic transition	Weak, forbidden, unstructured	Allowed, weak, structured
Fluorescence	Weak, unstructured	Strong, if unreactive, structured
Fluorescence rate constant	10^7–10^8 s^{-1}	10^6–10^7 s^{-1}
Singlet-triplet energy gap	< 10 kcal mol^{-1}	≈ 20 kcal mol^{-1}
Mixing with higher triplet-excited states	With low-lying π,π*	No mixing with π,π*
State-switching in triplet manifold	Possible	Not reported
Intersystem crossing	Fast, efficient	None
Phosphorescence (on direct excitation)	Observable, also in solution	Not observable
Phosphorescence rate constant	30–300 s^{-1}	≈ 3 s^{-1}
State reactivity	Triplet reacts	Singlet reacts
Philicity in hydrogen abstractions	Strongly electrophilic	Also nucleophilic
Efficiency of singlet oxygen formation	Low	High

[a] Exceptions are addressed in text.

electron density [105] and the "allowedness" of the related electronic transitions [93]. The lowest energy transition is allowed for *cis*-azoalkanes, and the electric transition dipole moment points along the composite p-orbitals of the azo π system [106,107]. However, orbital overlap is small, which greatly reduces the intensity of the n_,π* absorption band of *cis*-azoalkanes to result in typical values of *ca.* 100 M^{-1}cm^{-1} for the extinction coefficient [93]. The strength of the absorption band of *cis*-azoalkanes (as well as diketones such as biacetyl) depends in a distinct and pronounced manner on the polarizability of the environment, which can be employed to probe this property in supramolecular cavities and related inhomogeneous media [108,109]. The absorption band (as well as the associated fluorescence band) sometimes displays vibrational fine-structure due to the allowed nature of the related transition [80,93].

For *trans*-azoalkanes, the lowest energy transition is forbidden, in fact symmetry-forbidden, since the electron is promoted from an in-plane n_ orbital with A$_g$ symmetry to an out-of-plane π*-orbital with B$_g$ symmetry [93,105]. The n_,π* absorption band of *trans*-azoalkanes is therefore about a factor of 10 weaker than for *cis* azo compounds [93], unless intensity can be borrowed from a close-lying π,π* transition, *e.g.*, in azobenzene [80,93]. As for ketones, the absorption bands display no structure, which reflects the forbidden nature of the lowest-energy transition. The comparison of the extinction coefficients of *cis* and *trans* isomers is interesting since it implies that the absorption of *cis*-azoalkanes, which is 10 times stronger, is about "90% allowed."

TABLE 3.2
Photochemical Parameters of *cis*-Azoalkanes and Ketones[a]

	DBH	DBH-T	DBO	Me$_2$CO	BP	BA
$^1\tau$/ns	0.15	0.7–3.7[b]	10–1000[c]	1.5–2.1[d]	0.005	ca. 10[e]
$^3\tau$/μs	<0.001	0.10–0.63[f]	ca. 0.03[g]	30[h]	61[i]	ca. 500
Φ_{ISC}	≈0	≈0.5	≈0	1.0	1.0	1.0
Φ_{fl}	<0.001	ca. 0.02	0.56[j]	0.001[d]	0.00	ca. 0.03[k]
Φ_{ph} (solution)	0.00	0.00[l]	0.00	0.001[m]	0.009[h]	ca. 0.1[k]
Φ_{r}	1.0	0.59	0.001–0.03[n]	0.0	0.0	0.0
E_S/kcal mol^{-1}	84.5	78.1	76	85	76	62
E_T/kcal mol^{-1}	64[o]	62.4	54	78[p]	69	55
ΔE_{ST}/kcal mol^{-1}	20.5	15.7	22	7	7	7
k_{ISC}/s^{-1}	<10^6	2 × 10^8	<10^5	5 × 10^8	2 × 10^{11}	1 × 10^8
k_{fl}/s^{-1}	≤10^7	≈10^7	≈10^6	2 × 10^8	—	3 × 10^7
k_{ph}/s^{-1}	—	2.1–4.4	—	ca. 30	ca. 150	ca. 200
IP_v/eV	8.89[q]	8.50[q]	8.32[q]	9.71[r]	9.05[s]	9.55[t]
EA/eV	—	—	−0.50[u]	0.00[v]	0.62[w]	0.69[x]
$E_{p,ox}$/V	1.97[q]	1.83[q]	1.45[q]	3.06	2.37	(2.3)
$E_{p,red}$/V	—	ca. −2.9[y]	−2.9[z]	<−3.0[aa]	−1.83[bb]	−1.28[cc]
pK_a^{dd}	−1.4[ee]	—	0.4[ee]	−7.5[ff]	−6.1[gg]	ca. −10[hh]
S_Δ	0.00	0.96	0.7	0.31	0.35	0.30

[a] Values refer to fluid deaerated solution at ambient temperature from Refs. [98,157] unless stated differently or directly calculated from values contained in this table. [b] Strongly solvent dependent, see Ref. [47]. [c] Strongly solvent dependent, see Ref. [56]. [d] Solvent dependent, see Ref. [31,70]. [e] See also Ref. [31]. [f] Solvent dependent, see Ref. [50]. [g] From Ref. [202]. [h] From Ref. [70,242]. [i] In water from Ref. [243]. [j] Value reported for gas phase from Ref. [33]. [k] See Ref. [244]. [l] Phosphorescence is observed as an exceptional case in solid at 77 K with a quantum yield of 5.2 %, see Ref. [98]. [m] Estimated from the relative spectral areas of fluorescence and phosphorescence from Ref. [242]. [n] Solvent-dependent, from Ref. [60]. [o] From Ref. [147]. [p] See Ref. [31]. [q] See Ref. [198]. [r] From Ref. [245]. [s] From Ref. [246]. [t] From Ref. [247]. [u] See Ref. [65]. [v] From Ref. [248]. [w] From Ref. [249]. [x] From Ref. [250]. [y] From Ref. [49]. [z] From Ref. [55]. [aa] From Ref. [70]. [bb] Half-wave potential, from Ref. [164]. [cc] See Ref. [251]. [dd] pK_a value for conjugate acid. [ee] From Ref. [252].[ff] From Ref. [253].[gg] From Ref. [254].[hh] Estimated from Ref. [255].

The MO energy diagrams are shown in Scheme 3.1. The energetic order of orbitals is derived from photoelectron-spectroscopic assignments [81,105,110]. It should be noted that the assignment of n_ and n$_+$ is sometimes inverted for *trans*-azoalkanes as a consequence of different conventions of orbital nomenclature [93,111]. The LUMO is in all cases the π* orbital. The interaction of the two azo lone pair orbitals places the n_ combination at sufficiently high energy above the π orbital to prevent state switching in different solvents and upon substitution, *i.e.*, the n_ MO remains the HOMO and the n_,π* transition is the lowest energy transition. This is even the case for azoalkanes in which the energy of the π system is greatly raised by direct attachment of donor-substituted aromatic systems [80,93]. In addition, the combination of the lone-pair orbitals in azoalkanes places the antisymmetric molecular orbital at high energy, which accounts for the fact that azoalkanes absorb at longer wavelengths (400 ± 50 nm) than simple ketones (300 ± 50 nm). Diketones,

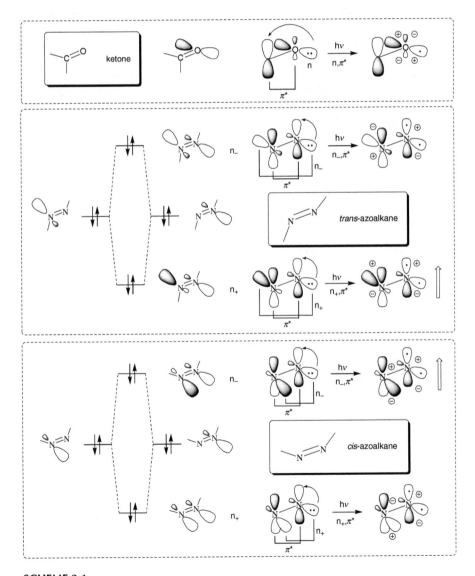

SCHEME 3.1

where the combination of two carbonyl groups produces a similar set of symmetric and antisymmetric orbitals, show also absorption in the visible wavelength range [112,113]. The absorption spectra of azoalkanes display additional interesting trends, for example, the enlargement of the bridge in bicyclic azo compounds from DBH to DBO to DBN causes a bathochromic and hypochromic shift [114].

As a second characteristic of the energetic separation between the n_- and the π orbital in azoalkanes, the n_-,π^* transition and the π,π^* transition are energetically well-separated, which leaves a spectral window in the UV region; see, for example, the UV absorption spectrum of DBO (Figure 3.1). This peculiarity has allowed exact determinations of the oscillator strengths of the n_-,π^* absorption [108]. In addition, the spectral window has been exploited to allow a selective excitation of a second attached chromophore [69,115,116], which is useful, for example, in studies of energy transfer [38–40,44,69,115]. Based on quantum-chemical calculations and assignments of photoelectron spectra, the π MO falls generally in between the n_+ and n_- combinations [93,105]. The n_+,π^* transition is difficult to detect since it is weak and, in the case of *cis*-azoalkanes, forbidden. It is also expected to lie at higher energy than the π,π^* transition, although a tentative assignment of induced circular dichroism spectra of azoalkanes in cyclodextrins suggested in one case a position at slightly lower energy [107].

As depicted in Scheme 3.2 for ketones, polar and in particular protic solvents cause hypsochromic and bathochromic effects on the n,π^* vs. π,π^* transitions, respectively [117]. The lone pairs are stabilized, in particular, through hydrogen bonding. While polar solvents lead to the expected hypsochromic effect on the n_-,π^*-transition in azoalkanes, as well, a full protonation of the azoalkane [118] or

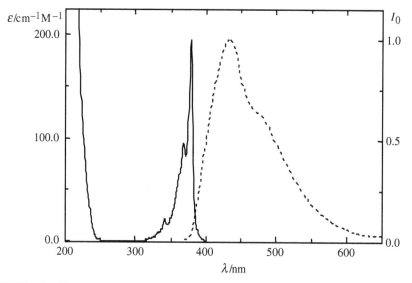

FIGURE 3.1   Absorption spectrum and fluorescence spectrum of DBO in cyclohexane.

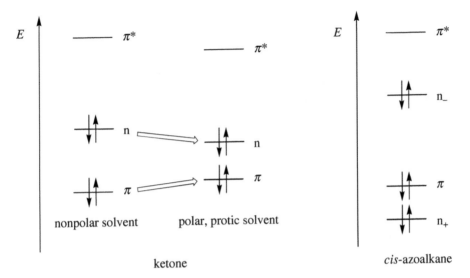

SCHEME 3.2

complexation by a metal ion (*e.g.*, Co^{2+}) [119] "deactivates" one nitrogen lone pair (Scheme 3.3), which results in a more pronounced hypsochromic shift (Figure 3.2). While more basic than ketones, azoalkanes are generally weak bases (Table 3.2), especially in their excited states [58]. The excited state pK_a can be extrapolated from the shift of the ground-state absorption spectra upon protonation (see Figure 3.2) and affords a pK_a value of *ca.* −8 for the singlet-excited state of DBO as an example. The lone pairs are also relatively poor electron donors, much poorer than amines, for example, although transition metal complexes with azoalkanes as ligands [120], adducts with stable cation radical salts [121,122], and charge-transfer complexes with electrophiles such as tetracyanoethylene, carbon tetrabromide, iodine, Ag^+, and Hg^{2+} have been reported [123,124].

Instructive is also a comparison of the state energy diagrams of ketones and azoalkanes (Scheme 3.4, S_2 energy for acetone from Ref. [125]). The lower ^{1}(n,π)* excitation energy of the azoalkane can again be rationalized in terms of the energetically high-lying n_- orbital (Scheme 3.2). For azoalkanes, the ^{1}(π,π)* state lies at *ca.* 135 kcal mol^{-1} according to its absorption near 210 nm (see Figure 3.1), and the ^{3}(π,π)* state has been estimated to lie *ca.* 40 kcal mol^{-1} above the ^{3}(n,π)* state according to EELS measurements for DBH [126]. In the excited carbonyl chromophore, the unpaired electrons prefer to occupy remote orbitals on different atoms (one half-vacant oxygen lone pair orbital and carbon-centered π*), which results in a very small exchange interaction, and (related) a small singlet-triplet energy gap (*ca.* 7 kcal mol^{-1}). For *cis*-azoalkanes, there is a higher probability that the unpaired electrons are found in orbitals on the same atom (due to the symmetric delocalization of the symmetrical π* and n_- orbitals). A sizable exchange interaction

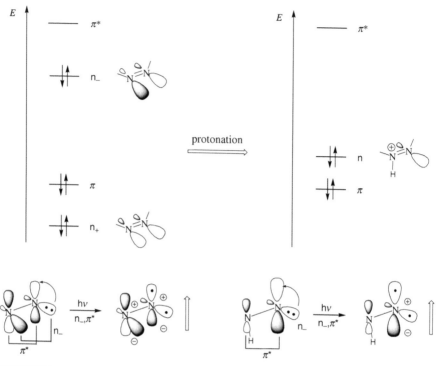

SCHEME 3.3

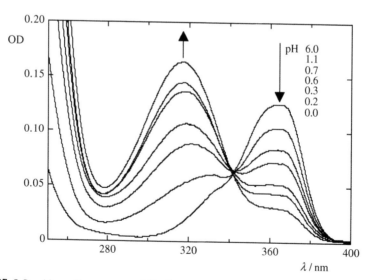

FIGURE 3.2 Absorption spectra of DBO (*ca.* 2.5 mM) in water in dependence on pH, adjusted through addition of H_2SO_4.

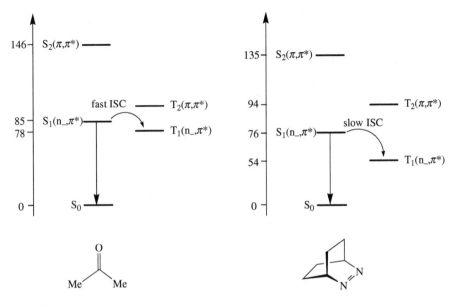

SCHEME 3.4

and a significant singlet-triplet energy gap result (*ca.* 22 kcal mol^{-1}), which results in a more pronounced difference in photoreactivity of singlet- and triplet-excited azoalkanes than that of singlet- and triplet-excited ketones.

The change in dipole moment upon excitation provides another contrast between ketones and azoalkanes [55,70,127]. For ketones, electronic excitation involves the promotion of a lone pair electron from oxygen to the carbonyl carbon (Scheme 3.1), which may be part of a delocalized π system in the case of aromatic ketones. This leads to a marked redistribution of electron density, which can be assessed through the change in dipole moment following excitation. In the case of benzophenone, the dipole moment decreases from 2.95 Debye [128] in the ground state to 2.1 Debye [129] in the triplet-excited state. For *cis*-azoalkanes, although they possess an inherently large dipole moment in the ground state (3.5 Debye for DBO) [55,130], there is only a small change upon excitation to the singlet- or triplet-excited state (*ca.* 3.2 or 3.0 Debye for DBO) [55]. Although the electron is promoted from the n_ composite lone pair MO of the azo group to the π* orbital (Scheme 3.1), the electron density remains symmetrically distributed over the two nitrogens, such that the dipole moment of the molecule remains similar. It should be noted that the change of dipole moment upon excitation dominates solvation effects, knowledge of which is important to understand host-guest photochemistry [127] and to predict solvent-dependent reactivity in charge-transfer-induced quenching [55,70].

Note that the ^{3}(n,π)* and ^{3}(π,π)* states of ketones lie very close in energy. This is a consequence of the large singlet-triplet energy gap for π,π* transitions (*ca.* 30

to 40 kcal mol^{-1}, large exchange interaction), and the small gap for n,π*-transitions in ketones, which brings the two states closer together compared to the spacing in the singlet manifold. The attachment of a large aromatic system [131] or π-donor substituents [132–135] to the carbonyl chromophore, and in some cases even the choice of a polar and hydrogen-bond donating solvent [136] may switch the lowest triplet-excited-state electronic configuration from n,π* to π,π* (Scheme 3.2). Interestingly, cation-induced switching between n,π* and π,π* has been observed in Y zeolites [137]. This is due to the slight increase in energy of the π system HOMO or a stabilization of the lone pair MO, *e.g.*, through the engagement of the lone pairs in dipolar interactions or hydrogen bonds with the solvent.

3.2.2 RATE CONSTANTS OF PHOTOPHYSICAL PROCESSES

The n$_-$,π* transition in *cis*-azoalkanes is allowed, such that fluorescence emission, which in the absence of competitive deactivation processes is proportional to the oscillator strength of the lowest-energy transition, should also be favored in comparison to ketones as well as *trans*-azoalkanes, for which the lowest energy transition is forbidden. However, experimental data (Table 3.2) suggest that the natural fluorescence rate constant (or radiative decay rate constant of the singlet-excited state) for the rigid *cis*-azoalkanes is *lower* than for ketones (no fluorescence from *trans*-azoalkanes has yet been quantified to allow comparison). This is presumably related to a structural change in the case of acetone as well as biacetyl, which produces, for example, a pyramidalized excited-state geometry of singlet-excited acetone [85]. The transition to the ground state becomes allowed from this distorted structure, thereby accounting for a comparably large radiative decay rate of singlet-excited acetone.

Disregarding inconsistencies in the natural fluorescence rate constants, it is important to note that aliphatic ketones (biacetyl and acetone) as well as several *cis*-azoalkanes show fluorescence in solution [81]. This includes not only the examples of DBH-T [47,98] and DBO [56,103,138] (see Structure 3.1), but also the smaller derivatives (see Structure 3.2) 2,3-diazabicyclo[2.1.1]hex-2-ene [139], diazirines, *e.g.*, adamantyldiazirine [140], and even some sterically hindered azobenzenes [141]. In contrast, DBH is not fluorescent with the exception of a report in siliceous zeolites [66]; it undergoes denitrogenation with unit quantum efficiency from the

STRUCTURE 3.2

singlet-excited state instead [33]. The next higher homologue, DBN, on the other hand, exhibits neither fluorescence nor photodecomposition, but deactivates rapidly in a radiationless manner [114,142].

The rate constant for ISC, on the other hand, is much larger for ketones than for cis-azoalkanes (Table 3.2), especially if one disregards the exceptional case of DBH-T [96–98]. This can be directly rationalized in terms of the singlet-triplet energy gap (Table 3.2), which is two to three times larger for cis-azoalkanes and reduces the required overlap between the initial and final states. Intersystem crossing from $^1(n,\pi)^*$ states is known to occur particularly fast to $^3(\pi,\pi)^*$ states, since in this case the change in spin state is coupled with a concomitant change of the orbital quantum number to ensure an overall conservation of angular momentum (El-Sayed rules [143]). The initially produced $^3(\pi,\pi)^*$ state may rapidly convert internally to the $^3(n,\pi)^*$ state to result in an overall fast $^1(n,\pi)^* \rightarrow {}^3(n,\pi)^*$ ISC process. This requires the $^3(\pi,\pi)^*$ state to be accessible, either via an energetic downhill process, a thermally activated transition, or through state mixing, which is commonly the case for aromatic ketones, but not conceivable for aliphatic ketones, diketones, or azoalkanes. Azoalkanes, in particular, possess energetically high-lying $^3(\pi,\pi)^*$ states (Scheme 3.4), which, in combination with the large energy gap, suppress a fast intersystem crossing.

Kasha has sought for phosphorescence from azo compounds in his famous review in 1947 [144], and this quest was later continued by Engel [99,103,115,145,146]. Some phosphorescent azo compounds of the DBH-T type were later identified [96–98] and phosphorescence from azoalkanes can nowadays be deliberately induced by measurement in heavy-ion exchanged zeolites [147]. In contrast to phosphorescence, singlet-triplet absorptions (ε ca. 0.01 M^{-1}cm^{-1}) have been claimed for some azo compounds, either oxygen-induced [148], directly, e.g., for the smallest cyclic derivatives, diazirines [149–151] or, more recently, heavy-atom-assisted in thallium-exchanged zeolites [147]. Some of these reports are debatable, since the assignment of the related absorption features to singlet-triplet absorptions is the opposite of unambiguous [152]. In particular, one must note that the natural phosphorescence rate constant for azoalkanes [66,98,147], which should be related to the intensity of the singlet-triplet absorption, falls one to two orders of magnitude below those of ketones (Table 3.2), for which the occurrence of singlet-triplet absorptions is more rigorously established [13].

The natural phosphorescence rate constants suggest again that the $S_0 \leftrightarrow T_1$ transitions are about "10 to 100 times more allowed" for ketones than for cis-azoalkanes. It should be noted that the phosphorescence rate of azoalkanes is not accelerated by heavy atoms such as in 1,2-dibromoethane as solvent [50] or thallium in zeolites [147]. This is expected since phosphorescence from a $^3(n,\pi)^*$ state to the π ground state is an El-Sayed rule allowed process [143], which is inherently fast (in comparison to phosphorescence from $^3(\pi,\pi)^*$ states) and is less likely to be much assisted by heavy atoms [153].

Since ISC in ketones and azoalkanes is "forbidden," the process should be sensitive to internal or external heavy atom effects, which can accelerate ISC. Nevertheless, most experiments to induce ISC by using heavy-atom substituents [98,103], heavy-atom solvents [56,154], xenon matrices at low temperature [155], or xenon or cesium loaded zeolites [155] have remained unsuccessful, *i.e.*, the fluorescence lifetime of the ketones or azoalkanes remained similar and the phosphorescence intensity remained unchanged, and in the case of most azoalkanes it remained absent. Only recently, there have been reports on enhanced phosphorescence of alkanones and azoalkanes in thallium-exchanged zeolites [147] and even in siliceous zeolites without heavy atoms [66]; the latter experimental effect is not yet fully understood.

In contrast to the absence of heavy-atom effects on both the formation and radiative decay of the triplet state, the ISC process can instead be assisted by molecular triplet oxygen [46,156,157]. The same mechanism is presumed to operate for other paramagnetic additives, *e.g.*, the stable free radical TEMPO as quencher (see discussion of quenching rate constants below).

3.3 FACTORS AFFECTING PHOTOCHEMICAL REACTIVITY

3.3.1 QUENCHING MECHANISMS

The principal photochemical reaction mechanisms of interest are hydrogen atom abstraction from hydrogen donors and exciplex formation with electron donors. For these mechanisms quite interesting effects have been observed by using *cis*-azoalkanes as chromophores. Of course, electron and energy transfer processes involving n,π*-excited states are also relevant, but can be treated with classical models [158–161].

The mechanism of quenching by hydrogen atom abstraction has been recently studied in a series of joint experimental and high-level quantum-chemical investigations [52,54,56,64,67]. In these studies, the singlet-excited azoalkane DBO was used as experimental model for an n,π*-excited state. DBO does not undergo ISC, such that the photoreactivity of the triplet state can be ignored in the experimental analysis. For the quantum-chemical calculations, 1,2-pyrazoline was employed as a smaller cyclic model azoalkane. The mechanism of hydrogen atom abstraction by an n,π*-excited state is schematically shown in Figure 3.3(a) in the form of a modified correlation diagram. Following excitation to the singlet-excited state, hydrogen transfer is initiated by the in-plane lone pair orbital. The process is thermally activated, which accounts for the sizable activation barriers and the observed large deuterium isotope effects [52,54,56]. The abstracting electronegative atom of the n,π*-excited state (X), the active lone-pair orbital, and both atoms of the H–R bond are aligned in a colinear fashion in the transition state. Having surmounted the transition state, the system evolves in the direction of the radical pair state to encounter a surface crossing with the ground-state potential energy surface, which has been characterized as a conical intersection (CI). At the conical intersection, the system proceeds efficiently back to the ground state to yield only minor amounts of radicals as photoproducts. Small photoreaction quantum yields result (<10%), which are a signature of hydrogen abstraction reactions of singlet-excited states, be it intermolecularly or intramolecularly, be it for azoalkanes or for ketones [73]. This mechanism is referred to as an "aborted" hydrogen abstraction, since the process of

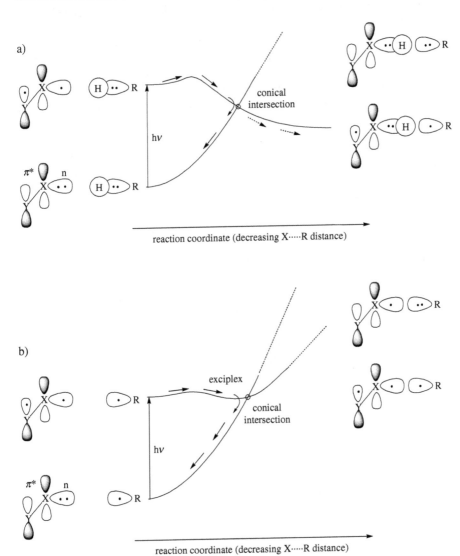

FIGURE 3.3 Modified correlation diagrams for the interaction of n,π*-excited states (X=O,N; Y=C,N) with (a) hydrogen donors H–R and (b) electron donors R reflecting the occurrence of transition states, exciplexes, and conical intersections along the reaction pathway. Dashed lines lead to strongly repulsive states. (From Ref. [62]. With permission from Wiley-VCH.)

hydrogen abstraction is initiated (the transition state is surmounted) but not completed, which can give rise to the striking observation of large deuterium isotope effects in the absence of significant photoproduct formation. The energy hypersurfaces for the triplet reaction should be qualitatively similar, with the exception that the conical intersection is replaced by a singlet-triplet crossing, which provides a somewhat less efficient deactivation pathway.

The second characteristic reaction mechanism of n,π*-excited states is exciplex formation with amines and aromatics [55,59,60,62,70,162–166]. For ketones, which have a stronger electron acceptor potential than azoalkanes, electron transfer is possible as well, and this interplay between electron transfer and exciplex formation as quenching mechanisms of n,π*-excited ketones by electron donors has been intensively investigated [165–167]. For azoalkanes as weaker electron-acceptors, electron transfer can be excluded, leaving exciplex formation as a prominent deactivation pathway [55,59–62]. The structures and properties of the exciplexes of ketones and azoalkanes with amines have recently been calculated (Figure 3.4) [55,59,62,67,70,168]. Although nonemissive, they present real minima on the excited-state energy surfaces in both the singlet and triplet manifold. Structurally, these exciplexes involve a three-electron two-center bond, which results from a collinear overlap of the lone pair orbital of the amine and a singly occupied lone pair of the n,π*-excited state (Figure 3.4). The amount of charge transfer in the n,π* exciplexes ranges typically from 5 to 20% [55,59,60,70].

The exciplexes of n,π*-excited states are critical intermediates, which can lead to three follow-up reactions (1) (exergonic) inner-sphere electron transfer (which competes with a direct outer-sphere electron transfer) [169], (2) exciplex-mediated hydrogen atom abstraction (the occurrence of which in n,π* exciplexes is still in debate) [70,164,170], and (3), exciplex-induced quenching [62]. The last mechanism prevails when the energetics of electron transfer is endergonic. Again, high-level

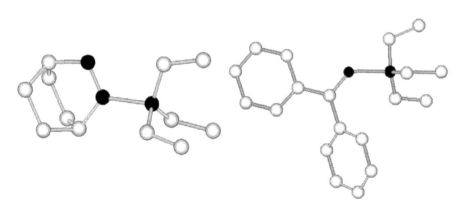

FIGURE 3.4 Calculated geometries (UHF–PM3 method) of the (triplet) exciplexes between DBO and triethylamine (left) and benzophenone and triethylamine (right).

quantum-chemical calculations were employed to reproduce the exact quenching mechanism for the singlet-excited state of azoalkanes. Interestingly, it was found that a conical intersection exists again in close vicinity of the exciplex minimum (Figure 3.3(b)), which accounts for the experimentally observed efficient deactivation (low quantum yields of photoreaction). The mechanism is transferable to ketones, in particular acetone for which the energetics of electron transfer remains endergonic for many amines [70]. In the case of triplet n,π*-excited states, triplet exciplexes are involved, which are structurally similar, see Figure 3.4 [55,70]. The role of conical intersections is then taken over by a singlet-triplet crossing region from which deactivation back to the ground-state molecules occurs [67].

The actual importance of exciplex-mediated hydrogen atom abstraction is unknown. In the case of amines as donors, we have reached the conclusion that such a reaction is unlikely to apply [70]. This means that the interaction of n,π*-excited states with amines can be suitably accounted for by assuming exciplex-induced quenching in competition with direct hydrogen abstraction, where the latter is of the aborted type for singlet-excited states. In the case of exergonic energetics, an electron transfer quenching mechanism may also become operative. On the other hand, exciplex-mediated hydrogen abstraction from alkylated benzenes presents a reasonable reaction pathway, especially for π,π* triplet-excited ketones [164,170].

In summary, a continuum of possible quenching and photoreaction mechanisms exists for n,π*-excited ketones and azoalkanes, which is summarized in Scheme 3.5.

3.3.2 Thermodynamics and Philicity

The reaction enthalpies in Table 3.3 show that hydrogen atom abstraction by excited ketones and azoalkanes from C–H, O–H, and N–H bonds of alcohols and amines is an exothermic process. In general, singlet-excited azoalkanes are more

TABLE 3.3
Enthalpies for Hydrogen Abstraction by Ketones and Azoalkanes for Different Hydrogen Donors

	$\Delta_r H_{298}$ / kcal mol^{-1}				
	CH$_3$OH		(CH$_3$)$_2$NH		H$_2$O
Excited State	**From OH**	**From CH**	**From NH**	**From CH**	**From OH**
^{1}Me$_2$CO*	−9	−17	−20	−21	6
^{3}Me$_2$CO*	−2	−10	−13	−14	13
^{3}BP*	−1	−9	−13	−13	14
^{1}DBO*	−15	−23	−27	−27	0
^{3}DDBH*	−1	−10	−13	−13	13

Note: Data converted from Ref. [71].

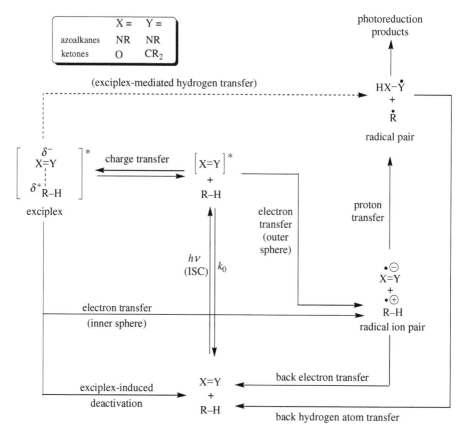

SCHEME 3.5

reactive than singlet-excited ketones, which can be related to the lower N=N π bond energy [71], while triplet-excited azoalkanes display comparable energetics to triplet-excited ketones. The lower excitation energy of azoalkane triplet states (Table 3.2) is generally compensated by the comparably strong N–H bond in the intermediary hydrazinyl radical, which accounts for the similar energetics of the photoinduced hydrogen transfer. This N–H bond is generally stronger than the O–H bond in ketyl radicals, which is also borne out by the fact that hydrogen transfer from ketyl radicals to azoalkanes occurs efficiently [171]. The quenching of excited ketones by secondary amines has been suggested to involve a competition between C–H and N–H abstraction [70,172], which is also observed for azoalkanes [58,60,65]. Hydrogen abstraction from the O–H bonds of aliphatic alcohols is a rather uncommon phenomenon for triplet ketones [173], but this reaction pathway becomes dominant for azoalkanes, which reveals a contrast in selectivity. In fact, singlet-excited azoalkanes are even quenched by hydrogen abstraction from the very strong O–H bonds of water [52,71].

The higher propensity of excited azoalkanes to abstract electrophilic hydrogens is not only manifested in the strong quenching by protic O–H bonds of water and alcohols, but also in the efficient interaction with the C–H bonds of acetonitrile and chloroform (*cf.* Scheme 3.6) [47,54,56,65,71]. The latter are unreactive toward excited ketones and are broadly employed as "photochemically inert" solvents.

The peculiar reactivity pattern of azoalkanes signifies a higher nucleophilicity (and vice versa a lower electrophilicity) of excited azoalkanes compared to ketones. In fact, ketones are commonly considered as electrophilic species, which resemble simple alkoxyl radicals in respect to their reactivity. From Zimmerman's pictorial description of the n,π*-excited state configuration for ketones [14], the electrophilic character and similarity to alkoxyl radicals is readily evident (Scheme 3.7). Excited azoalkanes do not display the same polarization pattern in their excited state (Scheme 3.7), which can account for their lower electrophilicity [71,174].

3.4 QUENCHING RATE CONSTANTS

Quenching rate constants of representative n,π*-excited states are compiled in Table 3.4 through Table 3.12. The focus of the discussion is on the intermolecular photoreactivity of n,π*-excited azoalkanes, using DBO and DBH-T as representative singlet- and triplet-excited states. Different classes of quenchers are discussed first, including olefins and dienes, alcohols and ethers, sulfides, amines, and aromatic compounds. The photoreactivity of azoalkanes is then compared with that of representative n,π*-excited ketones, *i.e.,* acetone, benzophenone and biacetyl. Furthermore, alkoxyl radicals as ground state models of n,π*-excited states are

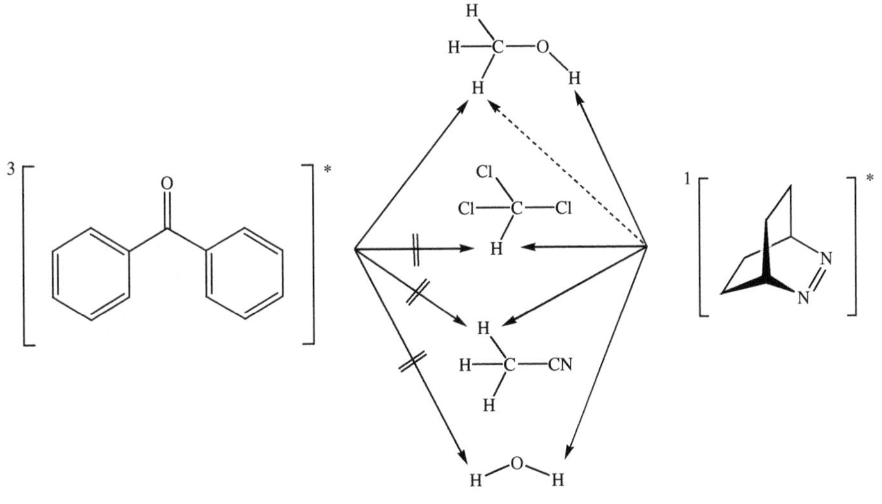

SCHEME 3.6

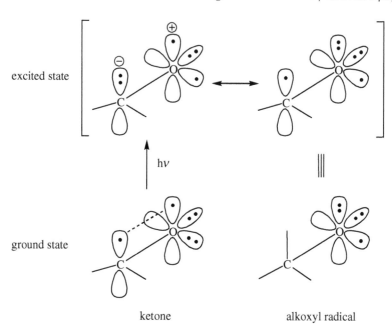

excited state

ground state

ketone   alkoxyl radical

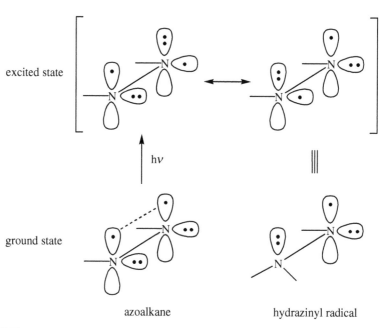

excited state

ground state

azoalkane   hydrazinyl radical

SCHEME 3.7

included. The consideration of the gas phase reactivity of n,π*-excited states and energy transfer between azoalkanes and ketones concludes the discussion.

3.4.1 QUENCHING BY DIENES AND OLEFINS

Dienes and olefins quench the fluorescence of azoalkanes with moderate rate constants on the order of 10^6 to 10^7 $M^{-1}s^{-1}$ (cf. Table 3.4) [33,34,51,72,145]. Quenching of DBO by dienes is supposed to operate mainly via hydrogen atom transfer, with preferred attack at α-CH positions to produce allylic radicals. The favorable energetics results in rate constants which are by a factor of 10 to 100 higher than for CH hydrogen abstractions from alkanes, e.g., cycloalkane. Substantial deuterium isotope effects (ca. 2 to 3) have been noted for cyclohexadienes, which corroborates the involvement of hydrogen transfer in the quenching mechanism (cf. Table 3.5) [72]. Noteworthy, the fluorescence quenching rate constants of DBO with dienes are quite similar to those with other strong hydrogen donors, such as diphenylmethanol, which forms a highly stabilized diphenyl ketyl radical upon hydrogen abstraction.

A detailed study on the fluorescence quenching of DBO by dienes and olefines revealed that the reactivity of these quenchers parallels their electron donor ability and that even poor hydrogen donors such as 1,1-diphenylethylene reacted with noticeable rate constants [72]. Based on these observations the involvement of an exciplex between excited DBO and the quencher was suggested, which might react further either via hydrogen transfer or electron transfer. A recent study [51] revealed that the fluorescence quenching rate constants of a DBO derivative by olefins can be correlated with a combination of electronic and thermodynamic parameters such as ionization potentials and CH bond dissociation energies, which corroborates the parallel involvement of both quenching pathways, i.e., exciplex formation and hydrogen transfer.

Table 3.4 also contains bimolecular quenching rate constants with olefins bearing either electron-withdrawing, e.g., acrylonitrile and tetrachloroethylene, or electron-pushing substituents, such as in tetramethylethylene or 1,2-diethoxyethylene. The following reactivity trend applies: electron-rich olefins > unsubstituted olefins > electron-poor olefins, spanning one order of magnitude difference between the rate constants. Once more, these data confirm the involvement of charge transfer, i.e., exciplex formation, in the overall quenching mechanism.

Generally, triplet-excited DBH-T is quenched by dienes and some olefins with rate constants one to two orders of magnitude higher than singlet-excited DBO (cf. Table 3.6), which is due to a change in quenching mechanism from hydrogen abstraction/ exciplex formation to efficient triplet-triplet energy transfer in the case of DBH-T. Indeed, since the triplet energy of DBH-T lies at ca. 63 kcal mol^{-1}, triplet energy transfer should be thermodynamically feasible for most dienes ($E_T < 60$ kcal mol^{-1}) [175] and some olefins such as fumaronitrile (E_T ca. 59 kcal mol^{-1}) [176]. The fact that the rates fall below the diffusion-controlled limit can presumably be accounted for in terms of steric effects on triplet-triplet energy transfer [40,44,177,178], which

TABLE 3.4

Rate Constants for Singlet and Triplet Quenching of the Azoalkanes DBH-T[a] and DBO

Quencher	k_q / 10^8 M^{-1}s^{-1}		Quencher	k_q / 10^8 M^{-1}s^{-1}	
	^{3}DBH-T*	^{1}DBO*		^{3}DBH-T*	^{1}DBO*
Dienes and Olefins					
1,3-cyclohexadiene	5.2	2.0[b]	tetramethylethylene	0.90	0.12[c]
2,3-dimethylbutadiene	10.0	0.18[d]	1,1-diphenylethylene	9.0	0.18[e]
trans-1,3-pentadiene	25.0	0.12[d]	*cis*-1,2-diethoxyethylene	0.36	0.20[c]
1,4-cyclohexadiene	0.24[f,g]	0.41[e]	ethyl vinyl ether	0.014	0.015[c]
cyclopentadiene		0.56[e,h]	tetrachloroethylene	1.5	0.020[c]
cyclopentene		0.087[e,h]	acrylonitrile	0.062	0.032[c]
cyclohexene		0.084[e,h]	fumaronitrile	64.0[i]	110[c]
Alcohols and Ethers					
methanol	<0.001[j]	0.018[j]	1,4-dioxane	0.0008[f]	0.0043[f]
2-propanol	≤0.0064[f,g]	0.049[f]	diethylether		0.008[k]
diphenylmethanol	0.19[f]	0.26[f]	di-*n*-butylether		0.0064[l]
phenol		9.5[c]	diisopropylether		0.007[k]
tetrahydrofuran	0.009[f]	0.02[f,k]			
Amines					
n-propylamine		0.29[k]	diisopropyl-3-pentylamine		0.076[m]
tert-butylamine		0.082[n]	DABCO		0.42[m]
diethylamine	0.61[j]	12[n]	triphenylamine	6.6[o]	2.1[n]
2,2,6,6-tetramethylpiperdine		4.0[n]	*N,N*-dimethylaniline		2.4[n]
triethylamine	0.19[j,p]	0.72[q,n]	*p*-methyl-*N,N*-dimethylaniline		5.3[n]
tri-*n*-propylamine		0.85[m,n]	*N,N,N',N'*-TMPD		73[n]
N-ethyldicyclohexylamine		2.5[m,n]	*N,N*-diphenylamine	13[o]	52[n]
triisopropylamine	0.11[m]		aniline		31[n]
Sulfides					
dimethylsulfide		0.60[k]	diisopropylsulfide		0.18[q]
diethylsulfide		0.49[k]			
Aromatics					
benzene		<0.0004[h]	methoxybenzene		0.96[c]
toluene		0.0022[c]	1,4-dimethoxybenzene		1.2[c]
1,4-dimethylbenzene		0.0044[c]	*p*-methyl-anisole		0.52[c]
1,3,5-trimethylbenzene		0.027[c]	1,4-dicyanobenzene		25[r,s]
1,2,4,5-tetramethylbenzene		0.028[c]	chlorobenzene		0.001[r,s]
1,2,3,4,5,6-hexamethylbenzene		0.16[c]	naphthalene		0.067[t]
			1-methylnaphthalene	100	0.043[u]

TABLE 3.4 (CONT.)

	k_q / 10^8 M^{-1}s^{-1}			k_q / 10^8 M^{-1}s^{-1}	
Quencher	^{3}DBH-T*	^{1}DBO*	Quencher	^{3}DBH-T*	^{1}DBO*
Miscellaneous					
oxygen	3.1[v]	87[v,w]	TEMPO	3.4	42[s]
methyl viologen	30[x]	88[x]	biacetyl	50[y]	130[c]
tributyltinhydride	0.35[f,g]	0.62[f]	cyclohexane	< 0.0001[j]	0.0037[j]
chloroform	<0.0001[j]	0.061[j]	carbon tetrachloride		0.010[u,z]
1,2-dibromoethane		0.004[u]	carbon tetrabromide	<0.2	100[r]
1,1-dimethylhydrazine	14[o]	10[aa]			

[a] Measured by time-resolved transient spectroscopy of the T-T absorption (λ_{obs} = 450 nm) of DBH-T or its similarly reactive C=C hydrogenated derivative DBH-T-H$_2$; unless stated differently, the absorption was observed in carbon tetrachloride, due to the absence of singlet quenching by this solvent, *cf.* Ref. [47]. [b] Stated as mean value from Refs. [33,34,72,145] in alkane solvents; mean value in acetonitrile from Refs. [72,145] is 3.8 × 10^7 M^{-1}s^{-1}. [c] Measured by time-resolved fluorescence spectroscopy in acetonitrile, this work. Aromatic compounds and phenol see Ref. [174]. [d] From Ref. [34] in *n*-hexane; for 1,3-pentadiene see also Refs. [32,33]. [e] From Ref. [72]; in isooctane; values for 1,4-cyclohexadiene and 1,1-diphenylethylene in acetonitrile are 1.7 × 10^7 M^{-1}s^{-1} and 1.8 × 10^7 M^{-1}s^{-1}, respectively (*cf.* Ref. [72]). [f] From Ref. [48]; in neat quenchers, benzene or acetonitrile. [g] From Refs. [31,45]. [h] From Ref. [33]; in isooctane. [i] Value in acetonitrile: 3.8 × 10^9 M^{-1}s^{-1}. [j] From Ref. [71]; data for methanol, cyclohexane and chloroform are in neat quencher; data for ^{3}DBH-T* quenching by amines in benzene. [k] From Ref. [65]; in neat ethers or benzene. [l] From Ref. [62]; in acetonitrile. [m] From Ref. [59]; in benzene. [n] From Ref. [60]; in benzene. [o] From Refs. [45,49]; in benzene. 1,1-dimethylhydrazine from Ref. [49]. [p] The value reported in Refs. [45,49] is one order of magnitude too high. [q] From Ref. [55]; in benzene. [r] From Ref. [145]; in acetonitrile. [s] From Refs. [72]; in acetonitrile. [t] From Ref. [69]; in cyclohexane. [u] Calculated from fluorescence lifetime in neat quencher. [v] From Refs. [46,157]; in benzene. [w] Value in isooctane is 5.9 × 10^9 M^{-1}s^{-1} (*cf.* Ref. [36]); for other values see also Ref. [13], p. 590, and Ref. [33]. [x] This work; in acetonitrile/water (9/1). [y] Also determined from growth of T–T absorption of biacetyl (at 310 nm) and growth of emission due to biacetyl phosphorescence (λ_{max} = 540 nm). [z] A value of 1.9 × 10^7 M^{-1}s^{-1} has been reported for a DBO derivative in Ref. [72]. [aa] From Ref. [36].

may become important for the very small azo chromophore. Note that triplet-triplet energy transfer is supposed to operate via a Dexter mechanism, which requires orbital overlap between donor and acceptor [161]. Note that olefins with higher triplet energies than DBH-T, *e.g.*, tetramethylethylene, are less efficient quenchers, presumably by hydrogen atom abstraction. The most important mechanistic pathways for excited state quenching of azoalkanes and ketones by dienes and olefins are summarized in Scheme 3.8.

TABLE 3.5
Deuterium Isotope Effects on the Quenching of Azoalkanes and Ketones

Chromophore	Quencher[a]	Solvent	k_H/k_D
^{1}DBO*	d_1-chloroform	neat	9.4[b]
	d_3-acetonitrile	neat	1.9[b]
	d_{12}-cyclohexane	neat	3.7[b]
	d_8-1,3-cyclohexadiene	isooctane	2.1[c]
		acetonitrile	1.7[c]
	d_8-1,4-cyclohexadiene	isooctane	2.7[c]
		acetonitrile	3.0[c]
	d_4-methanol	neat	20.6[b,d]
	d_1-O-methanol	neat	8.6[b,d]
	d_3-C-methanol	neat	1.1[d]
	d_1-O-phenol	acetonitrile/water	3.7[e]
	d_{12}-1,3,5-trimethylbenzene	acetonitrile	5.9[e]
	d_8-tetrahydrofuran	neat	5.7[b,d]
		gas phase	2.4[d]
	d_{10}-diethylether	neat	3.9[d]
		gas phase	2.4[d]
	d_6-dimethylsulfide	benzene	1.3[d]
		gas phase	1.3[d]
	d_2-N-tert-butylamine	water	5.1[f]
	d_1-N-diethylamine	benzene	1.3[d]
		gas phase	1.1[d]
	d_1-N-di-n-propylamine	acetonitrile/water	1.8[f]
	d_{15}-triethylamine	benzene	1.8[f]
^{1}Me$_2$CO*	d_{15}-triethylamine	acetone	1.2[g]
	d_1-Sn-tributyltinhydride	tetramethyltin	1.3[h]
^{3}Me$_2$CO*	d_8-toluene	acetonitrile	1.3[i]
	d_4-methanol	acetonitrile	3.7[i]
	d_{15}-triethylamine	acetone	1.5[g]
^{3}BP*	d_8-toluene	neat	1.8[j]
		acetonitrile	2.4[k]
	d_4-methanol	neat	2.7[j]
	d_8-2-propanol	neat	2.6[j]
	d_2-N-sec-butylamine	benzene	1.4[l]
	d_1-α-C-sec-butylamine	benzene	1.3[l]
	d_2-N-tert-butylamine	benzene	1.9[l]
tert-BuO•	d_8-2-propanol	benzene[m]	3.3[n]
	d_8-tetrahydrofuran	benzene[m]	2.8[n]
	d_{15}-triethylamine	acetonitrile	1.1[g,o]
	d_1-Sn-tributyltinhydride	benzene	1.2[p]

[a] Quencher in its deuterated form. [b] From Ref. [71]. [c] From Ref. [72]. [d] From Ref. [65]. [e] From Ref. [174]. [f] From Ref. [60]. [g] From Ref. [70]. [h] From Ref. [31]. [i] From Ref. [256]. [j] From Ref. [257]. [k] From Ref. [164]. [l] From Ref. [191]. [m] Mixture of benzene/di-tert-butylperoxide (1/2). [n] From Ref. [26]. [o] Value for cumyloxyl radical. [p] From Ref. [28].

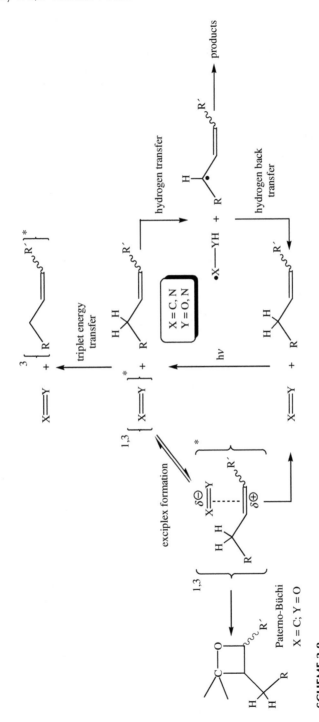

SCHEME 3.8

TABLE 3.6

Quenching Rate Constants for Triplet n,π*-Excited States

Quencher	k_q / 10^8 $M^{-1}s^{-1}$			
	^{3}DBH-T*[a]	^{3}Me$_2$CO*	^{3}BP*	^{3}BA*
Dienes				
1,3-cyclohexadiene	5.2		74[b]	2.9[c]
2,3-dimethylbutadiene	10	39[d,e]	56[b]	
trans-1,3-pentadiene	25	44[d]	55[b]	2.5[f]
1,4-cyclohexadiene	0.24	0.63[g]	2.9[h]	
Olefins				
tetramethylethylene	0.90	0.51[i]	9.0[h,i,j]	0.062[k,l,m]
1,1-diphenylethylene	9.0		60[n]	
cis-1,2-diethoxyethylene	0.36	13[o]	41[o]	2.3[k,p]
ethyl vinyl ether	0.014	0.20[o]	0.8[o]	0.033[k,m]
tetrachloroethylene	1.5	6.9[i]	10[q]	0.0004[r]
acrylonitrile	0.062	6.0[q]	0.34[s]	<0.0006[k,t]
fumaronitrile	64	74[d]	80[u]	
Alcohols				
methanol	<0.001	0.001[v,w]	0.0021[x]	
2-propanol	≤0.0064	0.01[v,w]	0.018[x,y]	0.00003[c]
diphenylmethanol	0.19		0.075[y,z,aa]	0.0007[c]
Ethers				
tetrahydrofuran	0.009	0.026[bb]	0.096[aa,cc]	
1,4-dioxane	0.0008		0.0043[aa,dd]	
Amines				
diethylamine	0.61	2.2[ee]	34[ff]	0.22[c]
triethylamine	0.19	4.0[ee]	28[gg]	0.50[c]
triphenylamine	6.6		7.6[hh]	0.31[c]
N,N-diphenylamine	13		130[ii]	17[c]
Aromatics				
benzene		0.028[jj]	0.0001[jj]	
toluene		0.035[v]	0.005[kk]	
Miscellaneous				
oxygen	3.1	39[ll]	17[ll,mm]	4.1[nn]
methyl viologen	30		27[oo]	
TEMPO	3.4		75[pp]	
biacetyl	50	200[qq]	12[rr]	
tributyltinhydride	0.35	5.4[g,ss]	3.0[b,tt]	0.15[c]

TABLE 3.6 (CONT.)

	k_q / 10^8 $M^{-1}s^{-1}$			
Quencher	³DBH-T*ª	³Me₂CO*	³BP*	³BA*
cyclohexane	<0.0001	0.0034[v]	0.0072[uu]	
chloroform	<0.0001	0.0001[bb]	0.0007[vv]	

[a] See Table 3.4. [b] This work; in chloroform. [c] From Ref. [77]; in benzene. [d] From Ref. [215] in acetonitrile; for fumaronitrile see also Refs. [258–260]. [e] Value for 2-methylbutadiene. [f] From Ref. [261]; in carbon tetrachloride. [g] From Ref. [31]; in acetonitrile. [h] From Ref. [206]; in benzene; for 1,4-cyclohexadiene see also Refs. [262,263]. [i] See also Ref. [258]; in acetonitrile.. [j] See also Ref. [211]; in benzene. [k] From Ref. [214]; in benzene. [l] See also Ref. [216]; in various solvents. [m] See also Ref. [217]; in acetonitrile. [n] From Ref. [264]; in benzene. [o] From Ref. [212]; in acetonitrile. [p] Value for 1,2-dimethoxyethylene. [q] See p. 436 in Ref. [13]. [r] From Ref. [265]; in benzene. [s] From Ref. [266]; in benzene. [t] Value for methacrylonitrile. [u] From Ref. [176]; in benzene. [v] From Ref. [256]; in acetonitrile. [w] See also Ref. [30]; neat quencher. [x] From Ref. [257]; in neat quencher; for 2-propanol see also Refs. [262,263]; in benzene. [y] From Ref. [189]; in benzene. [z] See also Ref. [267]; in benzene. [aa] From Ref. [48]; for diphenylmethanol see also Ref. [45]. [bb] From Ref. [71]; in neat quencher. [cc] From Ref. [268]; in benzene. [dd] From Refs. [269,270]; in neat quencher. [ee] From Ref. [70]; in neat acetone; for triethylamine see also Refs. [172,271] with values of 3.9 and 2.9 × 10⁸ M⁻¹s⁻¹, respectively, in acetonitrile. [ff] From Ref. [191]; in benzene; value for di-n-propylamine. [gg] Mean value from Refs. [29,189-191,223,226]; in benzene. [hh] From Ref. [220]; in benzene. [ii] From Ref. [272]; in isooctane; in Ref. [273] a value of 4.3 × 10⁹ M⁻¹s⁻¹ in isopropanol was reported. [jj] From Ref. [163]; neat benzene. For benzophenone see also Ref. [263]; in acetonitrile. [kk] From Ref. [274]; in benzene; see also Ref. [257]. [ll] From Ref. [157]; in acetonitrile. [mm] From Ref. [275]. [nn] From Ref. [276]; in benzene. [oo] From Ref. [277]; in acetonitrile/water 9:1. [pp] Value for p-methylbenzophenone from Ref. [21]. [qq] Calculated from the data in Table III of Ref. [278] in n-hexane; see also Ref. [260]. [rr] From Ref. [243]; in water. [ss] See also Ref. [279]; in n-hexane. [tt] See also Refs. [28,280]. [uu] From Ref. [281]; in acetonitrile. [vv] From Ref. [47]; in neat chloroform.

3.4.2 Quenching by Alcohols and Ethers

Alcohols and ethers act as classical hydrogen donors in the quenching of excited azoalkanes [45,48]. These quenchers are weak electron donors and do not interact with azoalkanes in a charge transfer mechanism. This has been verified experimentally and theoretically for the fluorescence quenching of DBO [62,65]. Quantum-chemical calculations on the CASPT2 level of theory confirmed the absence of a charge transfer minimum structure in the case of ethers as quenchers [62,67]. However, this does not exclude the participation of polar effects in the transition state [65,179–182], which are signaled by the slight increase of the quenching rate of DBO in their interaction with ethers with lower ionization potentials. Generally, the bimolecular quenching rate constants for singlet-excited DBO by alcohols and ethers lie on the order of 10^5 to 10^7 $M^{-1}s^{-1}$ (cf. Table 3.4), typical for hydrogen transfer. The operation of hydrogen atom transfer is further corroborated by the observation of substantial deuterium isotope effects for ethers (k_H/k_D ca. 4 to 6) and methanol (k_H/k_D ca. 21), which reach the theoretical values for abstraction from CH or OH groups (cf. Table

3.5) [52,56,62,65]. Phenols are known to quench DBO with higher rate constants ($ca.$ 10^9 $M^{-1}s^{-1}$), which can be traced back to their low OH bond dissociation energies [183], resulting from the resonance stabilization of phenoxyl radicals [174]. The observation of substantial deuterium isotope effects for the fluorescence quenching of DBO by phenols corroborates, however, that hydrogen transfer operates for phenols as well ($cf.$ Table 3.5).

In the fluorescence quenching of DBO by methanol not only hydrogen abstraction from CH bonds is observed, but also from OH bonds, which has rarely been reported for ketones [173]. This is manifested in the large deuterium isotope effect for OH abstraction (8.6) compared to CH abstraction (1.1) [52]. From the standpoint of thermodynamics this is quite a peculiar behavior, since the bond dissociation energy of CH in methanol is by $ca.$ 8 kcal mol^{-1} lower than that for OH. DBO seems to possess a preferential reactivity toward "electrophilic" hydrogen atoms, contrasting the reactivity pattern of n,π*-excited ketones [14,184]. This can be rationalized by considering (1) the strongly exergonic thermodynamics for abstraction from both positions, $i.e.$, CH and OH, which results in a lower selectivity; (2) the lower electrophilicity of the excited azo chromophore; and (3) a low antibonding for hydrogen abstractions from heteroatoms such as O, S, or N [71,185,186].

Among alcohols, 2-propanol and diphenylmethanol stand out with respect to their reactivity in hydrogen transfer reactions with azoalkanes [45,48]. The abstraction from secondary α-CH groups with two radical-stabilizing substituents, $i.e.$, methyl or phenyl, proceeds with quite high rate constants, due to strongly exergonic thermodynamics [184].

Quenching rate constants of triplet-excited DBH-T by some alcohols and ethers have also been measured. Generally, the reactivity of this azo chromophore is one order of magnitude lower than for singlet-excited DBO. This is a thermodynamic effect, which relates to the lower excitation energy of the triplet-excited state of DBH-T compared to singlet-excited DBO ($cf.$ Table 3.2). Noteworthy, in the case of DBH-T the preferential attack at OH bonds is lost, $i.e.$, faster quenching by tetrahydrofuran than by methanol is observed. This might be rationalized by the fact that the reaction is almost thermoneutral, which causes a higher selectivity [184].

Interestingly, the quantum yield of hydrogen abstraction from diphenylmethanol by singlet-excited DBO and triplet-excited DBH-T was found to be dependent on the spin multiplicity of the excited state. While DBH-T shows a quantum yield of $ca.$ 0.12, a significantly lower value of $ca.$ 0.02 was determined for DBO. This difference could be related to a slower, because spin forbidden, hydrogen back transfer in the triplet radical pair derived from DBH-T, compared to the singlet radical pair with DBO, resulting in a higher yield of free radicals. Furthermore, as mentioned above, the conical intersection between the S_1 and S_0 surface provides in the case of singlet-excited DBO a more efficient radiationless deactivation pathway than the T_1-S_0 crossing for triplet-excited DBH-T. The photoproducts derived from hydrogen abstraction have been investigated in detail [45,48]. The final reduction product of the azo chromophore is always the corresponding hydrazine, while the products of the hydrogen donor are derived from radical recombination processes, *e.g.*, pinacols in the case of secondary alcohols.

3.4.3 QUENCHING BY AMINES AND SULFIDES

Table 3.4 and Table 3.10 summarize quenching data of singlet-excited DBO by amines and sulfides. These quenchers are stronger electron donors than ethers or alcohols and therefore mainly involved in charge-transfer-induced mechanisms. Sulfides are somewhat weaker electron donors than amines, which is manifested in lower quenching rate constants, except where structural effects such as steric or stereoelectronic hindrance are involved (see below) [59].

At first glance the quenching of DBO fluorescence by tertiary amines follows the electron donor potential of the amine: the stronger electron-donating the amine, the faster the quenching. This is valid for n-electron donating aliphatic as well as for π-electron donating aromatic amines [60]. Multiple Rehm–Weller plots such as for charge-transfer-induced quenching of n,π*-excited ketones by amines were not observed for azoalkanes [165,187,188]. Exclusive variation of electronic parameters without major structural alteration, for instance by changing the substitution patterns of *N,N*-dimethylanilines, led to excellent linear correlations of the quenching rate with the electron donor strength of the amine [60]. This observation corroborates the involvement of charge transfer in the overall quenching mechanism, as has been demonstrated in detail for ketones [37,41,60,70,72,162,165,189]. However, calculations of the free energy of electron transfer under formation of radical ion pairs results in endergonic thermodynamics for most amines, since DBO is a very poor electron acceptor [55,59–61]. Only the strongest electron donors such as TMPD are able to photoreduce DBO by full electron transfer. For those cases quenching rate constants are close to the diffusion-controlled limit [60]. Time-resolved photoconductivity measurements lend support to an unfavorable electron transfer, since no indications for the formation of free radical ions could be obtained for amines such as triethylamine or diphenylamine [61].

To verify the involvement of exciplex formation, solvent effects were investigated. Contrary to what is common for full electron transfer typically involved in amine-induced quenching of triplet-excited benzophenone, *i.e.*, acceleration of the quenching rate with increasing solvent polarity [190,191], an "inverted" solvent effect has been noted for the fluorescence quenching of DBO by various amines and sulfides (*cf.* Table 3.7 and Table 3.8) [55,60,65]. This interesting result can be explained by considering an exciplex with only small charge transfer contribution [55]. Similar observations of an "inverted" solvent effect were also made for the quenching of acetone fluorescence by amines, where exciplexes with weak charge transfer character are involved [70].

These experimental observations were fully in line with quantum-chemical calculations (CASPT2 level) [62,67]. It was demonstrated that charge-transfer complexes, *i.e.*, exciplexes, exist as minimum structures along the reaction

TABLE 3.7
Quenching Rate Constants of Azoalkanes and Ketones by Triethylamine in Different Solvents

Medium	k_q / 10^8 $M^{-1}s^{-1}$			
	[1]DBO*[a]	[1]Me$_2$CO*[b]	[3]BP*[c]	[3]BA*[d]
gas phase	6.4[e]			
perfluorohexane	2.4			
cyclohexane	1.4	25	14	
Freon-113	1.8			
benzene	0.72		170	0.50
ethylacetate	0.56	21		
1,4-dioxane	0.61			
acetone	0.62	21		
acetonitrile	0.44	24	380	2.7

[a] From Ref. [55]; except for gas phase. [b] From Ref. [59]. [c] From Ref. [190]. [d] From Ref. [77]. [e] From Ref. [65].

TABLE 3.8
Quenching Rate Constants of DBO by Different Amines in Different Solvents

Quencher	k_q / 10^8 $M^{-1}s^{-1a}$		
	n-Hexane	Benzene	Acetonitrile
triethylamine[b]	1.4	0.72	0.44
tri-*n*-propylamine		0.85	0.58
N-ethyldicyclohexylamine	5.9	2.5	1.9
N,N-dimethylaniline		2.4	1.2
N,N,N',N'-TMPD	130.0	73	95
triphenylamine[c]	4.3	2.1	1.3

[a] Values from Ref. [60], except for triethylamine (*cf.* Ref. [55]). [b] The respective quenching rate constants for triplet-excited biacetyl are 5.0×10^7 $M^{-1}s^{-1}$ in benzene and 2.7×10^8 $M^{-1}s^{-1}$ in acetonitrile. [c] The respective quenching rate constants for triplet-excited biacetyl are 3.1×10^7 $M^{-1}s^{-1}$ in benzene and 1.6×10^9 $M^{-1}s^{-1}$ in acetonitrile.

coordinate of excited state and electron donor approach. Furthermore, close-lying conical intersections provide efficient radiationless pathways for the deactivation of exciplexes, which accounts for the absence of exciplex emission. Indeed, emission of exciplexes derived from n,π*-excited states and n-electron donors has been observed quite rarely, which emphasizes eventually the importance of the conical intersections in photoreactions of n,π*-excited states. One of the few reports deals with the emission of an intramolecular triplet exciplex of an amino-substituted β-arylpropiophenone [192].

Beside theoretical indications for the existence of exciplexes in the fluorescence quenching of DBO by tertiary amines, further experimental evidence was obtained by observation of steric effects. DBO fluorescence quenching by sterically hindered amines is accompanied by a marked drop of the rate constant, which leads to the straightforward conclusion that chromophore and quencher react at contact distances, which is expected for the formation of compact structures such as exciplexes [59]. Noteworthy, benzophenone as stronger electron acceptor is quenched highly unselectively by sterically hindered amines in a diffusion-controlled reaction [193,194].

For the bicyclic amine DABCO stereoelectronic effects have been observed [59]. DABCO quenches DBO with lower rate constants than tertiary amines with similar electron donor strength, which can be explained by a hindered proton transfer in the initially formed exciplex. The resulting aminoalkyl radical is destabilized due to the unfavorable orientation of the amine lone pair and carbon radical center in a 60° angle. Noteworthy, the observed effects are much smaller than for alkoxyl radicals, which are known to react with amines exclusively via hydrogen atom transfer (see below). Similar observations have been reported for the quenching of triplet-excited benzophenone [29,195]. Also, stereoelectronic effects are deemed responsible for the three times lower quenching rate of DBO by diisopropylsulfide compared to dimethylsulfide.

The conclusions regarding the involvement of exciplexes are only strictly valid for tertiary amines, but seem not to apply for the quenching of DBO by secondary and primary amines. Comparing tertiary and secondary amines based on their electron donor potential, the latter should quench with lower rate constants. As can be seen from the values in Table 3.4 and Table 3.10, this is by far not the case; secondary amines are among the most reactive amines! This observation has been made for aliphatic as well as for aromatic amines and can be explained by invoking hydrogen abstraction from NH bonds as additional or exclusive quenching pathway. Deuterium isotope effects for NH abstraction from di-*n*-propylamine could be observed (*cf.* Table 3.5), which corroborate the involvement of these rather electrophilic hydrogen atoms in the quenching [60]. As already outlined above, hydrogen abstractions of singlet-excited DBO from X–H bonds (X = O, S, N) are not as unusual as for ketones (see below) and have their inherent reason in the strongly exergonic reaction thermodynamics (*cf.* reactivity-selectivity principle), the reduced electrophilicity of the excited azo-chromophore, and transition-state-specific effects such as antibonding [71]. However, comparison of the data for primary and secondary aliphatic amines leads to the conclusion that hydrogen abstraction is less competitive with charge transfer in the case of primary amines. The N–H bond in primary amines is *ca.* 7 kcal mol^{-1} stronger than in secondary amines [191], thus, shifting the reaction thermodynamics to the less exergonic region. However, also for primary amines a very high deuterium isotope effect (5.1 for *tert*-butylamine) was observed, which points to a participation of hydrogen transfer in the quenching

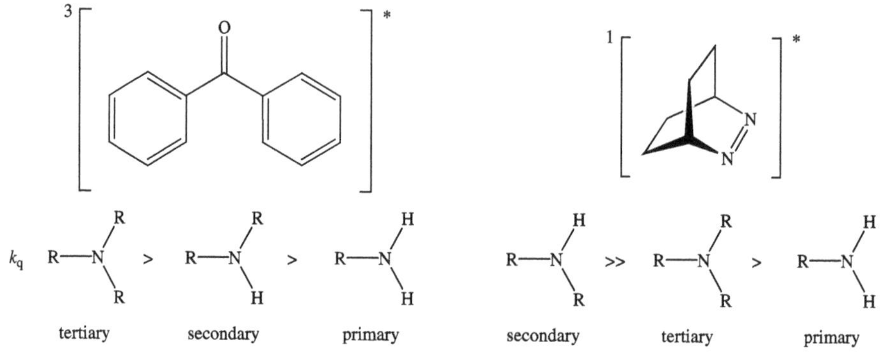

SCHEME 3.9

mechanism [60]. Noteworthy, observations of a preferential reactivity toward NH were made earlier for another singlet-excited polycyclic azoalkane, but no further explanation for this effect was given [41]. For ketones, the expected order of reactivity with amines applies (*cf.* Scheme 3.9), since quenching proceeds exclusively via electron transfer or exciplex formation. Hydrogen transfer from NH is disfavored due to the strongly electrophilic nature of the excited carbonyl chromophore (see above).

For triplet-excited DBH-T relatively few quenching data with amines are known [49]. Generally, the rate constants are up to one order of magnitude lower, which reflects the lower excitation energy of the triplet-excited azo chromophore compared to singlet-excited DBO. With regard to the unusually high reactivity of secondary amines, similar observations are made with DBH-T: secondary amines react two to three times faster than tertiary amines.

Recently, Adam and Nikolaus performed a study on the fluorescence quenching of DBO-analogous polycyclic azoalkanes, which, due to appropriate substitution with electron-withdrawing groups, were found to be better electron acceptors than parent DBO [57]. Calculations of the driving force of electron transfer led to exergonic thermodynamics, which is reflected by one to two orders of magnitude higher quenching rate constants with tertiary amines, compared to DBO. For these cases electron-transfer-induced quenching should be considered as well.

The photoproducts which result from the interaction of azoalkanes with amines have been analyzed [49,57]. The final reduction product of the azoalkane is again the corresponding hydrazine. The oxidation product of the amine depends on its substitution pattern (primary, secondary, or tertiary amine). In the case of primary and secondary amines, imines are formed, resulting formally from a hydrogen transfer from NH *and* α-CH. In the case of tertiary amines no NH are available, thus, the hydrogens in the hydrazine stem exclusively from CH groups, which results in the formation of enamines as oxidation products of tertiary amines. These enamines can be further hydrolyzed to secondary amines and the corresponding aldehyde, giving rise to rather complex product distributions.

3.4.4 QUENCHING BY AROMATIC COMPOUNDS

The electron donor strength of aromatic compounds such as benzene and its methyl derivatives is not sufficient for an interaction with singlet-excited DBO via a charge transfer mechanism. More likely is a hydrogen transfer pathway, especially for methyl-substituted benzene derivatives. The latter posses benzylic CH groups, which are known to facilitate hydrogen abstraction due to a stabilization of the resulting benzylic radical. Based on this mechanism the rate constants listed in Table 3.4 have to be interpreted. The higher the number of methyl substituents, the faster the rate of quenching, which can be traced back to a statistical effect. Significant deuterium isotope effects have been noted for the fluorescence quenching of DBO by aromatics (cf. Table 3.5), which supports hydrogen transfer as main interaction pathway [174].

An increase of the electron-donating or -accepting properties of the aromatics by substitution with methoxy or cyano groups leads to increased quenching rate constants. This is accordance with a charge transfer mechanism, either under photoreduction or photooxidation of DBO [72]. Similarly, for the quenching of triplet-excited benzophenone by aromatic compounds, i.e., alkyl- and alkoxybenzenes, exciplexes have been postulated [164,166,167,196]. Since triplet-excited benzophenone is a stronger electron acceptor than excited azoalkanes, the electron donor strength of methylbenzenes is already sufficient for exciplex formation. In support of exciplex-induced quenching of triplet-excited ketones, steric effects with the hexamethylbenzene/hexaethylbenzene couple and triplet benzophenone have been reported [167].

1-Methylnaphthalene quenches triplet-excited DBH-T via an energy transfer mechanism, which explains the quite high quenching rate constant close to diffusion control. Evidence for energy transfer was obtained from monitoring the concomitant build-up of the typical triplet-triplet absorption band of 1-methylnaphthalene.

3.4.5 QUENCHING BY OTHER COMPOUNDS

Quenchers that do not belong to either one of the previously introduced classes of compounds are discussed in this section. They are identified in Table 3.4 as miscellaneous compounds. Tributyltinhydride, chloroform, and cyclohexane react via hydrogen atom transfer [48,54,56,71]. Noteworthy is the high quenching rate constant for tributyltinhydride, compared to cyclohexane, which reflects the favorable thermodynamics for hydrogen abstraction from the hydride [48].

Carbon tetrabromide and methylviologen show particularly high reactivity toward singlet-excited DBO, with rate constants close to diffusion control. Clearly, photoinduced electron transfer to the quencher under photooxidation of DBO is responsible, which has been described recently for DBO and derivatives [197]. In contrast to its rather unfavorable electrochemical reduction potential, DBO is a good electron donor [198]. In combination with a strong electron acceptor, electron transfer becomes feasible. Blackstock and Kochi reported about ground-state charge-transfer complexes between DBO and carbon tetrabromide [123]. This is in line with the involvement of photoinduced electron transfer in the case of carbon tetrabromide.

Heavy-atom-induced ISC is not indicated, as reflected in the weak quenching of singlet-excited DBO in 1,2-dibromoethane (as solvent). In the case of methylviologen, unambiguous evidence for electron transfer was obtained by detection of the characteristic radical cation of the acceptor (λ_{max} = 610 nm). The quantification of this transient signal allowed for the determination of the electron transfer quantum yield, which was unity in the case of triplet-excited DBH-T and roughly one order of magnitude lower for singlet-excited DBO. This difference can be explained by invoking a fast back-electron transfer within the radical ion pair in the case of DBO, while the same process is spin-forbidden for triplet-excited DBH-T.

Paramagnetic molecules such as the free radical TEMPO also quench excited azoalkanes. Two possible mechanisms have to be considered: (1) spin-catalyzed ISC and (2) triplet energy transfer [199,200]. While singlet-excited DBO is quenched by an accelerated ISC (see also below for oxygen), the triplet state of DBH-T is high-lying enough (cf. Table 3.2) to undergo triplet energy transfer to TEMPO (E_T = 52 kcal mol^{-1}) [199].

Quenching of azoalkanes by oxygen under formation of singlet oxygen is a particularly interesting process. Based on experimental observations it is known that triplet-excited n,π* ketones have a low propensity to form singlet oxygen, while in contrast π,π*-excited ketones generally display high efficiencies of singlet oxygen formation [157,201]. Strikingly, n,π*-excited azoalkanes such as DBH-T and DBO were demonstrated to be quenched by oxygen under efficient formation of singlet oxygen [46,157]. Sensitization by DBH-T proceeds via fast ISC, followed by oxygen quenching of the triplet state. On the other hand, the mechanism for singlet oxygen production by DBO quenching is quite distinct (cf. Scheme 3.10). The singlet-triplet

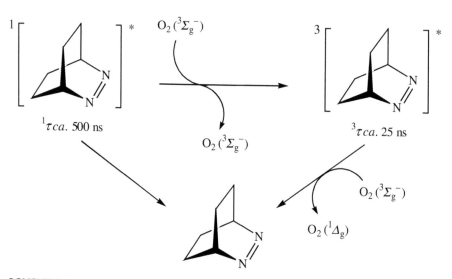

SCHEME 3.10

energy gap of DBO (*ca.* 22 kcal mol^{-1}) falls below that of oxygen (22.5 kcal mol^{-1}), such that direct sensitization of singlet oxygen by fluorescence quenching of the azoalkane is not possible. Moreover, an oxygen-assisted ISC leads to the short-lived (25 ns) triplet state of DBO [202], which nevertheless has a sufficiently long lifetime to sensitize singlet oxygen formation. The entire reason for the rather different quenching rate constants of DBO and DBH-T is not totally clear yet, but could be related to the different electron donor strength of both azoalkanes. It has been pointed out that charge transfer processes play generally an important role in the quenching of excited states by oxygen [203–205].

The importance of a charge transfer pathway in the quenching of azoalkanes by oxygen becomes obvious by comparing the parent DBO with its 1,4-dichloro-substituted derivative. For the latter the quenching rate is lower by a factor of 20 [58], owing to its higher oxidation potential.

3.4.6 AZOALKANE VS. KETONE PHOTOREACTIVITY

Due to a fast and efficient intersystem crossing process (*cf.* Table 3.2) most ketones perform mainly triplet photochemistry [25,73]. In Table 3.6, the photoreactivity of triplet-excited acetone, biacetyl, and benzophenone is compared with the triplet-excited azoalkane DBH-T. The data for both chromophores follow similar trends. Namely, dienes and amines quench with quite high rate constants, while ethers and aromatic compounds react rather inefficiently.

Generally, three competitive pathways for the quenching of triplet n,π*-excited states by dienes and olefins have to be considered: (1) triplet energy transfer, (2) hydrogen transfer, and (3) exciplex formation. For several dienes a triplet energy transfer is responsible for the quite high quenching rate constants. The higher triplet excitation energy of common ketones such as acetone and benzophenone (*cf.* Table 3.2) accounts for the higher rate constants for ketone triplet quenching by dienes compared to DBH-T. Note that the increased reactivity of ketones compared to triplet-excited azoalkanes cannot be related to a more efficient hydrogen transfer. The higher triplet excitation energy of acetone and benzophenone is effectively counter-balanced by a lower gain of energy in the formation of ketyl radicals compared to hydrazinyl radicals in the case of azoalkanes [71], as they would result from a hydrogen abstraction. Biacetyl behaves exceptionally in the sense that its triplet energy is quite low and does not permit triplet energy transfer to the energetically higher-lying triplet states of dienes or olefins. Triplet energy transfer from other ketones to olefins has been documented, *e.g.,* for fumaronitrile [176].

The order of quenching rate constants of triplet-excited azoalkanes and ketones cannot solely be explained on basis of energy transfer processes. Hydrogen abstraction from olefins and dienes plays a role as well, especially for double allylic systems such as 1,4-cyclohexadiene [206]. The resulting bis-allylic radicals are highly stabilized, thus, hydrogen transfer to ketones and azoalkanes is thermodynamically favored. The superior reactivity of the stronger electron acceptor benzophenone can be partly explained by invoking exciplexes, which might act as precursor states for hydrogen abstractions.

Another photoreaction between ketones and olefins or dienes, which has often been connected to the involvement of exciplexes, is the Paterno–Büchi reaction [4,5], *i.e.*, the photocycloaddition of C=C double bonds to carbonyl C=O bonds under formation of oxetanes [17,78,207–217]. Especially for electron-rich olefins such as ethyl vinyl ether or 1,2-diethoxyethylene, intermediary exciplexes have often been postulated [212], with the consequence of a diminished regioselectivity and stereospecificity for oxetane formation. On the other hand, electron-deficient olefins such as α,β-unsaturated nitriles react with a high regioselectivity and stereospecificity due to a well defined transition state, which is based on the electronic requirements of n,π^*-excited ketones [17].

Although common for ketones, the azo-analogous photocycloaddition between C=C and N=N has never been realized intermolecularly for azoalkanes. However, *intramolecular* formation of 1,2-diazetidines resulting from a [2+2] photocycloaddition between an azo chromophore and a C=C double bond has been reported by Hünig and Schmitt [218]. The *intramolecular* [2+2] photocycloaddition of two N=N bonds in azobenzophanes has been postulated as well, but no products could be isolated [219].

Quenchers with heteroatoms such as alcohols, ethers, and amines are either involved in hydrogen transfer or charge transfer reactions (*cf.* Scheme 3.11). The dominance of hydrogen transfer for ethers and alcohols has been corroborated by the observation of substantial deuterium isotope effects, especially for alcohols (see Table 3.5). Absolute quenching rate constants are determined by the involvement of polar effects in the transition state of hydrogen transfer [65,70,179] and the thermodynamics of the process [71]. As mentioned above for the quenching by olefins and dienes, despite rather pronounced differences in triplet excitation energies, ketones and azoalkanes (DBH-T) abstract hydrogen atoms with similar thermodynamics. Therefore, azo chromophores and ketones react with similar quenching rate constants. In the case of the stronger electron acceptor benzophenone, these rate constants are sometimes larger, due to the involvement of polar effects leading to a stabilization of the transition state.

Amines are by far stronger electron donors than ethers, which leads to a switch-over in quenching mechanism to charge transfer, *i.e.*, exciplex formation, for azoalkanes [49,55,59–62] and full electron transfer for stronger electron accepting ketones such as benzophenone [25,29,162,165,189–191,193,194,220–229]. The quenching of acetone by amines has been suggested to involve considerable hydrogen transfer as parallel mechanistic pathway. Depending on the spin multiplicity of the excited ketone state a switch-over in quenching mechanism between hydrogen and charge transfer was discussed [70].

Table 3.9 compiles quenching rate data for singlet-excited ketones such as acetone and biacetyl as well as for singlet-excited DBO. Generally, similar trends as discussed above for triplet state behavior are observed. Noteworthy is the comprehensive set

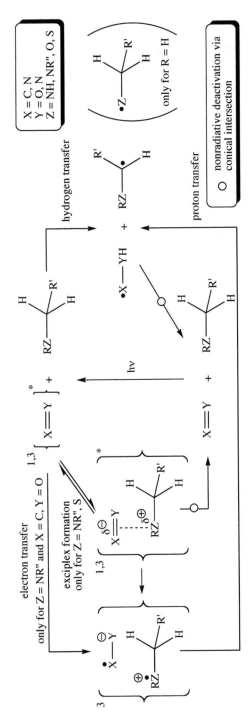

SCHEME 3.11

TABLE 3.9

Quenching Rate Constants for Singlet n,π^*-Excited Chromophores

Quencher	k_q / 10^8 $M^{-1}s^{-1}$		
	^{1}DBO*[a]	^{1}Me$_2$CO*	^{1}BA*
Dienes			
1,3-cyclohexadiene	2.0	10[b]	0.41[c]
1,4-cyclohexadiene	0.41[d]	2.1[e]	
2,3-dimethylbutadiene	0.18	0.65[f]	≤ 0.01[c]
2,5-dimethyl-2,4-hexadiene	1.1[g]	13[b]	3.3[c]
trans-1,3-pentadiene	0.12	0.95[b]	0.02[c]
Olefins			
tetramethylethylene	0.12	1.2[h]	0.04[c]
cis-1,2-diethoxyethylene	0.20	29[i,j]	34[c]
ethyl vinyl ether	0.015	0.25[i,j]	< 0.01[c]
acrylonitrile	0.032	0.12[j,k]	
fumaronitrile	110	25[j,k,l]	
Alcohols			
2-propanol	0.049[d]	0.091[e,m]	< 0.001[n]
phenol	9.5		20[n]
Ethers			
tetrahydrofuran	0.02	0.28[o]	
Amines			
n-propylamine	0.29	2.5[p,q]	
diethylamine	12	18[p,r]	17[n]
diisopropylamine	3.9[s]	9.7[p,q]	13[n]
2,2,6,6-tetramethylpiperidine	4.0	6.1[p,r,u]	
triethylamine	0.72	21[p,q,r]	24[n]
tri-*n*-propylamine	0.85	22[p,q]	31[n]
N-ethyl-diisopropylamine	1.3[s,t]	21[p]	
diisopropyl-3-pentylamine	0.076	1.6[p]	
1,2,2,6,6-pentamethylpiperidine	0.89[s,t]	18[p,u]	
DABCO	0.42	29[p]	76[n]
aniline	31		100[n]
N,N-dimethylaniline	2.4[s]		120[n]
N,N-diphenylamine	52		74[n]
triphenylamine	2.1		49[n]

TABLE 3.9 (CONT.)

	k_q / 10^8 M^{-1}s^{-1}		
Quencher	^{1}DBO*[a]	^{1}Me$_2$CO*	^{1}BA*
Miscellaneous			
tributyltinhydride	0.62[d]	10[e]	0.27[e]
oxygen	87[a]	23[v]	7.0[v]

[a] See Table 3.4. [b] From Ref. [213]; in benzene; see also Ref. [282]. [c] From Ref. [283]; in benzene.
[d] The following quenching rate constants for singlet-excited DBH-T were reported for benzene or neat 2-propanol as solvent in Ref. [31]: 9.5×10^8 M^{-1}s^{-1} for 1,4-cyclohexadiene, 8.9×10^7 M^{-1}s^{-1} for 2-propanol, 9.8×10^8 M^{-1}s^{-1} for tributyltinhydride. [e] From Ref. [31]; in tetramethyltin or neat 2-propanol. [f] This work, obtained according to the experimental procedure described in Ref. [47].
[g] From Ref. [34]; in *n*-hexane. [h] From Ref. [282]; calculated with $\tau_0 = 1.9$ ns; in *n*-hexane. [i] From Ref. [212]; in acetonitrile. [j] From Ref. [209]; in acetonitrile. [k] See also Refs. [210,284]. [l] From Ref. [207]; in acetonitrile. [m] See also Ref. [285]; in acetonitrile. [n] From Ref. [77]; in benzene. [o] From Ref. [71]; in neat quencher. [p] From Ref. [70]; in neat acetone. [q] See also Ref. [286]; in benzene. [r] See also Ref. [287]; in *n*-hexane. [s] From Ref. [60]; in benzene. [t] See also Ref. [59]; in benzene. [u] From Ref. [242]; in neat acetone. [v] From Ref. [157]; in acetonitrile (acetone) and benzene (biacetyl).

of quenching data for electron-donating amines in Table 3.9. Although DBO has a higher singlet excitation energy than biacetyl, the high reduction potential of the diketone overwhelms this "disadvantage" (*cf.* Table 3.2). This fact is translated into quenching rate constants close to diffusion control, clearly pointing to a full electron transfer [77]. The reduction potentials of DBO and acetone are quite low, therefore accounting for the involvement of exciplex formation, due to a less favorable driving force for full electron transfer [55,59–62,70].

Finally, the quenching by oxygen is discussed briefly. Charge transfer contributions to the quenching of excited states by oxygen have been stressed to be of considerable importance [203–205]. While the oxidation potentials of ketones are usually too high to account for exergonic driving forces for electron transfer, singlet-excited DBO can reduce oxygen from an energetic point of view. This accounts partly for the larger quenching rate constant of DBO compared to fluorescence quenching of ketones by oxygen. Another factor is the beforehand mentioned oxygen-catalyzed intersystem crossing as fluorescence quenching mechanism of DBO [46,157].

3.4.7 COMPARISON OF CHROMOPHORE REACTIVITY WITH ALKOXYL RADICALS

In Table 3.10, reactivity data for singlet-excited DBO and triplet-excited carbonyls, *i.e.*, acetone and benzophenone, are compared with those for alkoxyl radicals, *i.e.*, *tert*-butoxyl or cumyloxyl, which are considered as ground state analogues of excited

TABLE 3.10

Quenching Rate Constants for Singlet n,π*-Excited Azoalkanes and Triplet n,π*-Excited Ketones in Comparison with Alkoxyl Radicals

Quencher	k_q / 10^8 M^{-1}s^{-1}			
	^{1}DBO*	^{3}Me$_2$CO*	^{3}BP*	RO•[a]
Dienes				
1,3-cyclohexadiene	2.0[b]		74[c]	0.42[d]
1,4-cyclohexadiene	0.41[b]	0.63[c]	2.9[c]	0.54[d]
1,3-cyclooctadiene	0.12[e]		30[f]	0.036[g]
2,5-dimethyl-2,4-hexadiene	1.1[h]	46[i]		0.026[g]
Olefins				
cyclohexene	0.084[b]	0.092[j]	0.73[f]	0.058[g]
1-octene	0.0055[k]		0.044[f]	0.015[g]
Alcohols				
methanol	0.018[b]	0.001[c]	0.0021[c]	0.0029[d]
ethanol	0.030[l]	0.0054[j]		0.011[d]
2-propanol	0.049[b]	0.01[c]	0.018[c]	0.018[d,m]
diphenylmethanol	0.26[b]		0.075[c]	0.069[d,m]
phenol	9.5[b]		13[n]	3.3[o]
α-tocopherol	53[p]		51[p]	31[q]
Ethers				
diethylether	0.008[b]			0.039[m]
diisopropylether	0.007[b]			0.012[d,m]
tetrahydrofuran	0.02[b]	0.026[c]	0.096[c]	0.083[d,m]
1,4-dioxane	0.0043[b]		0.0043[c]	0.015[m]
Sulfides				
dimethylsulfide	0.60[b]			0.035[r]
diisopropylsulfide	0.18[b]			0.034[r]
di-*n*-butylsulfide	0.49[b,s]		8.3[t]	0.047[r]
Amines				
n-propylamine	0.29[b]	0.22[u]		0.13[u,v]
n-butylamine	0.28[w]	0.13[j,x,y]		0.10[v]
tert-butylamine	0.082[b]	0.019[j]	0.64[z]	0.033[v]
diethylamine	12[b]	2.2[c]	34[c]	0.97[u,v]
diisopropylamine	3.9[h]	1.2[u]		0.42[u]
2,2,6,6-tetramethylpiperidine	4.0[b]	0.49[u,aa]	21[bb]	0.028[u]
triethylamine	0.72[b]	4.0[c]	28[c]	1.6[u,v]

TABLE 3.10 (CONT.)

	k_q / 10^8 M^{-1}s^{-1}			
Quencher	^{1}DBO*	^{3}Me$_2$CO*	^{3}BP*	RO•[a]
tri-*n*-propylamine	0.85[b]	4.5[u]		2.1[u]
triisopropylamine	0.11[b]	0.26[u]	77[cc]	0.059[u]
N-ethyl-diisopropylamine	1.3[h]	7.5[u]	47[cc]	2.5[u]
ethyldicyclohexylamine	2.5[b]	6.8[u]	45[cc]	3.7[u]
diisopropyl-3-pentylamine	0.076[b]	0.17[u]	72[cc]	0.034[u]
1,2,2,6,6-pentamethylpiperidine	0.89[c]	3.5[u,aa]	34[bb]	1.7[u]
ABCO		0.044[u]	1.8[dd]	0.037[u,v]
DABCO	0.42[b]	0.83[u]	24[dd]	0.096[u,v]
aniline	31[b]		51[ee]	1.9[ff,gg]
N,N-diphenylamine	52[b]		130[c]	11[ff,gg]
carbazole	48[w]			5.5[ff]
Aromatics				
toluene	0.0022[b]	0.035[c]	0.005[c]	0.0023[d]
1,3,5-trimethylbenzene	0.027[b]			0.0083[d]
Miscellaneous				
cyclohexane	0.0037[b]	0.0034[c]	0.0072[c]	0.016[g]
tri-*n*-butyltinhydride	0.62[b]	5.4[c]	3.0[c]	2.2[hh]

[a] For *tert*-butoxyl radicals, except otherwise indicated. [b] From Table 3.4. [c] From Table 3.6. [d] From Ref. [26]; in benzene/di-*tert*-butoxyperoxide (1/2). [e] From Ref. [34]; in *n*-hexane. [f] From Ref. [206]; in benzene; for cyclohexene see also Ref. [211]. [g] From Ref. [288]; in benzene/di-*tert*-butoxyperoxide (1/2). Rate constants in benzene for reaction with cumyloxyl radicals are: 4.6×10^6 for 1-octene, 1.0×10^7 for cyclohexene, 2.3×10^6 for 1,3-cyclooctadiene, 2.0×10^6 for cyclohexane; all in M^{-1}s^{-1}; from Ref. [288]. Rate constant for 2,5-dimethyl-2,4-hexadiene refers to cumyloxyl; in benzene. [h] From Table 3.9. [i] From Ref. [215]; in acetonitrile. [j] From Ref. [256]; in acetonitrile; for cyclohexene see also Refs. [258,259]. [k] From Ref. [33]; see also Ref. [72]; in isooctane. [l] Calculated from fluorescence lifetime in neat ethanol; based on a lifetime in gas phase of 1030 ns. [m] From Ref. [230]; in benzene/di-*tert*-butoxyperoxide (1/2). [n] From Ref. [289]; in benzene. [o] From Ref. [288]; in chlorobenzene; Ref. [290]; in benzene/di-*tert*-butoxyperoxide (1/2). Values for cumyloxyl radicals can be found in Refs. [288,291,292]. [p] From Ref. [74]; in benzene. [q] From Ref. [291]; in benzene; see also Refs. [74,293]. [r] From Ref. [232]; in benzene at 37°C. [s] Value for diethylsulfide. [t] From Ref. [294]; in benzene. [u] From Ref. [70]; in neat acetone. For quenching of triplet acetone by *n*-propylamine and triethylamine in acetonitrile see also Refs. [172,271]. In the case of alkoxyl radicals the values for cumyloxyl radicals are reported. [v] From Refs. [29,231]; reaction with *tert*-butoxyl radicals in benzene/di-*tert*-butoxyperoxide (1/2). Values not reported in table: *n*-propylamine: 1.6×10^7 M^{-1}s^{-1}, diethylamine: 7.6×10^7 M^{-1}s^{-1}, triethylamine: 1.8×10^8 M^{-1}s^{-1}, ABCO: 6.0×10^6 M^{-1}s^{-1}, DABCO: 2.8×10^7 M^{-1}s^{-1}. [w] From Ref. [60]; in benzene. [x] From Ref. [271]; in acetonitrile. [y] Value for *n*-hexylamine. [z] From Ref. [191]; in benzene. [aa] See also Ref. [242]; in neat acetone. [bb] From Ref. [229]; in benzene. [cc] From Ref. [194]; in acetonitrile. [dd] From Ref. [223]; in benzene. [ee] From Ref. [273]; in 2-propanol. [ff] From Ref. [295]; in benzene. [gg] See also Ref. [292] for aniline (2.8×10^8 M^{-1}s^{-1}) and diphenylamine (1.5×10^9 M^{-1}s^{-1}); in benzene. [hh] From Ref. [28]; benzene/di-*tert*-butoxyperoxide (1/2).

n,π*-chromophores. The radicals undergo hydrogen atom abstractions [15,16,26,29, 70,180,181,230–232]. Therefore, a drop in reactivity due to stereoelectronic effects, typical for hydrogen abstractions, has been noticed for ethers and amines such as diisopropylether, DABCO or triisopropylamine [29,59,70,230].

The comparison of excited state reactivity with alkoxyl radical reactivity is a valuable mechanistic tool for the understanding of photochemical quenching pathways. For example, the quenching of triplet acetone by an extended set of amines parallels the reactivity of cumyloxyl radicals, while a less significant correlation applies to singlet-excited acetone [70]. Consequently, a switch-over of quenching mechanism of excited acetone by amines has been invoked to explain these observations.

In the case of the fluorescence quenching of DBO by alcohols and ethers, the obtained data correlate very well with those for alkoxyl radicals. At first glance the same appears to apply for the quenching of DBO by amines, especially those, which are involved in hydrogen abstractions (primary and secondary amines). However, significant differences have been noted for tertiary amines, which have been demonstrated to quench singlet-excited DBO via intermediary exciplexes [55,59–62]. Amines such as triisopropylamine and *N,N*-diisopropyl-3-pentylamine show a low reactivity in the fluorescence quenching of DBO due to steric hindrance in exciplex formation [59]. On the other hand with cumyloxyl radicals a drop of the hydrogen transfer rate constant has been noted for these amines as well [70], due to stereoelectronic effects. A comparison of hindered and nonhindered amines shows that stereoelectronic effects are more pronounced for hydrogen transfer to alkoxyl radicals and triplet-excited acetone, than steric hindrance in exciplex formation with singlet-excited DBO.

3.4.8 PHOTOREACTIVITY IN THE GAS PHASE

Measurements in the gas phase offer several advantages, among them the direct comparison to quantum-chemical calculation and the broad availability of thermodynamic and electronic parameters for the gas phase. However, data on the gas phase reactivity of n,π*-excited ketones are rare [172,233–239]. This is mainly caused by experimental difficulties since most ketones such as benzophenone are not sufficiently volatile and therefore require elevated temperatures to be studied experimentally, which limits the comparison with the ubiquitous solution measurements at room temperature [236,237]. In contrast, a comprehensive data set on gas-phase photoreactivity of azoalkanes has recently become available [65], motivated by the observation of Steel and coworkers that the fluorescence of DBO can be easily monitored in the gas phase [33].

Table 3.11 summarizes data on DBO fluorescence quenching by an extended set of compounds including ethers, sulfides, and amines in the gas phase [65]. One major result of this study is the unambiguous confirmation of the "inverted" solvent effect [55], since the quenching by triethylamine is expectedly fastest for the gas phase as most nonpolar medium. Furthermore, it was found that the gas phase quenching rate constants correlate very well with those in solution, which rules out major changes in quenching mechanisms. In other words, ethers, alcohols, secondary and primary amines quench via hydrogen transfer, while tertiary amines and sulfides are mainly involved in charge transfer interactions.

TABLE 3.11
Fluorescence Quenching Rate Constants of DBO in the Gas Phase and in Solution

Quencher	IP_a / eV[a]	$k_q / 10^8$ M^{-1}s^{-1}		k_{gas}/k_{soln}
		Gas Phase (k_{gas})[b]	Benzene (k_{soln})	
Tertiary Amines				
tri-*n*-propylamine	7.23	8.9	0.85[c]	10
triethylamine	7.53	6.4	0.72[c]	9
trimethylamine	7.85	4.3		
Secondary Amines				
2,2,6,6-tetramethylpiperidine	7.59	84	4.0[c]	21
di-*n*-butylamine	7.69	190	12[d]	16
diisopropylamine	7.73	68	3.9[e]	17
diethylamine	8.01	190	12[c]	16
dimethylamine	8.24	70		
Primary Amines				
n-propylamine	8.78	2.2	0.29[c]	8
methylamine	8.90	0.73		
ammonia	10.07	0.064		
Sulfides				
diethylsulfide	8.42	26	0.49[c]	53
dimethylsulfide	8.69	13	0.60[c]	22
Ethers				
diisopropylether	9.20	0.31	0.007[c]	44
tetrahydrofuran	9.40	0.26	0.02[c]	13
diethylether	9.51	0.16	0.008[c]	20
dimethylether	10.03	0.15		
Alcohols				
methanol	10.84	1.1	0.018[c]	61
Miscellaneous				
chloroform	11.37	1.7	0.061[c]	28

[a] Adiabatic ionization potentials, see Ref. [65]. [b] From Ref. [65]. [c] From Table 3.4. [d] From Ref. [60]. [e] From Table 3.9.

Interestingly, a change of the reactivity of singlet-excited DBO with sulfides upon going to the gas phase has been documented. In solution, amines quench DBO fluorescence faster than sulfides as expected, based on electron donor strength. This order of reactivity is inverted in the gas phase, where sulfides quench faster than amines [55,65].

3.4.9 ENERGY TRANSFER BETWEEN KETONES AND BICYCLIC AZOALKANES

In Table 3.12, rate constants for energy transfer between ketones and azoalkanes are compiled. In addition, rate constants for quenching of fluorescence of aromatic compounds by n,π*-excited chromophores are presented. Azoalkanes show quite peculiar UV absorption spectra with large spectral windows between higher energetic π,π*-transitions and the energetically lowest n,π* absorption bands [81]. This fact can be used to perform selective excitation of antenna groups capable of energy transfer to the azo chromophore [69]. In the case of sufficient spectral overlap between donor fluorescence and acceptor absorption, an efficient singlet-singlet energy transfer can be observed [38–40,44,69,240]. Due to the low oscillator strength of the azo n,π*-transition, energy transfer in nonviscous media is presumed to involve the Dexter

TABLE 3.12
Singlet and Triplet Energy Transfer between Cyclic Azoalkanes and Ketones

	Acceptor (k_q / 10^8 M^{-1}s^{-1})			
Donor	Me$_2$CO	BA	DBO	DBN
Singlet-Excited				
benzene		110[a]	1.3[b]	
naphthalene		150[c]	150[b]	
DBO	0.0001[d]	130[e]		50[f]
Triplet-Excited				
DBH-T		50[g]	24[g]	
Me$_2$CO		200[h]	30[i]	
BA			20[j]	
BP		12[k]	67[l]	

[a] From Ref. [240]; in isooctane. [b] From Ref. [39]; in isooctane. [c] From Ref. [296]; in benzene. [d] This work, calculated from lifetime in neat acetone. [e] This work; in acetonitrile. [f] A. Koner, W. M. Nau, unpublished result; in water. [g] This work; in carbon tetrachloride. [h] Calculated from the data in Table III of Ref. [278]; see also Ref. [260]. [i] See Ref. [40]; value for azoisopropane. [j] See Refs. [36,40,103,115,202]. [k] See Ref. [243]; in water. [l] See Ref. [36]; in isooctane.

mechanism [39,40,44,69]. This is experimentally supported by the observation of steric hindrance effects [40], which have been reported for triplet energy transfer involving ketones as well [178].

The rate constants for singlet and triplet energy transfer between the various n,π*-excited chromophores can generally be rationalized in terms of their excited-state energies (see Table 3.2). Strikingly, for systems containing an azo chromophore a so-called "inverted" effect on energy transfer has been observed for the benzene/naphthalene-DBO data pairs in Table 3.12. Although the thermodynamics favors energy transfer from benzene, the spectral overlap integral is larger for naphthalene/DBO, resulting in a larger quenching rate constant for the energetically disfavored energy transfer process. This peculiarity can be rationalized with the narrow n,π*-absorption bands of azoalkanes and the large UV window (Figure 3.1) between the first and higher electronic transitions [39,69,241].

The triplet excitation energy of DBO itself is too low (see Table 3.2) to allow energy transfer from DBO to most chromophores. However, the reverse process has been used very often to sensitize the triplet state of azoalkanes, which is not formed in a direct intersystem crossing process from the singlet-excited state (*cf.* Table 3.2) [35,36,42,103,145]. Triplet-excited ketones and azoalkanes (DBH-T) are able to transfer triplet energy to DBO with reasonably high rate constants.

ACKNOWLEDGMENTS

The authors are grateful for fruitful collaborations and intensive discussions with Prof. W. Adam (Würzburg), Prof. C. Bohne (Victoria), Prof. R.A. Caldwell (Texas), Prof. H. García (Valencia), Prof. E. Haselbach (Fribourg), Prof. D. Klapstein (St. F.X.), Prof. M.A. Miranda (Valencia), Prof. M. Olivucci (Siena), Prof. H. Rau (Stuttgart-Hohenheim), Prof. J.C. Scaiano (Ottawa), Prof. J. Wirz (Basel), and Prof. H.E. Zimmerman (Madison).

ABBREVIATIONS

ΔE_{ST}	S_1–T_1 energy gap
ε	molar extinction coefficient
Φ_{fl}	fluorescence quantum yield
Φ_{ISC}	intersystem crossing quantum yield
Φ_{ph}	phosphorescence quantum yield in solution
Φ_r	photodecomposition quantum yield
λ_{obs}	observation wavelength
τ_0	lifetime without added quencher
$^1\tau$	lifetime of singlet-excited state
$^3\tau$	lifetime of triplet-excited state
ABCO	1-azabicyclo[2.2.2]octane
AP	acetophenone
BA	biacetyl
BP	benzophenone
CI	conical intersection
DABCO	1,4-diazabicyclo[2.2.2]octane
DBH	2,3-diazabicyclo[2.2.1]hept-2-ene
DBH-T	10,10-dimethyl-1,4,4a,5,8,8a-hexahydro-1,4;5,8-dimethano-phthalazine
DBN	2,3-diazabicyclo[2.2.3]non-2-ene
DBO	2,3-diazabicyclo[2.2.2]oct-2-ene
EA	electron affinity
EELS	electron energy-loss spectroscopy
$E_{p,ox}$	electrochemical oxidation peak potential
$E_{p,red}$	electrochemical reduction peak potential
E_S	singlet excitation energy
E_T	triplet excitation energy
HOMO	highest occupied molecular orbital
IP_a	adiabatic ionization potential
IP_v	vertical ionization potential
k_{fl}	natural fluorescence (or radiative) rate constant
k_H/k_D	deuterium isotope effect
k_{ISC}	rate constant for intersystem crossing $S_1 \rightarrow T_1$
k_{ph}	phosphorescence rate constant
k_q	bimolecular quenching rate constant
LUMO	lowest unoccupied molecular orbital
MO	molecular orbital
pK_a	acidity constant in water
S_Δ	efficiency of singlet oxygen formation
TMPD	tetramethylphenylen-1,4-diamine
TEMPO	tetramethylpiperidinoxyl
UV	ultraviolet

REFERENCES

1. Neckers, D.C., Ed. *Selected papers on photochemistry*, SPIE Optical Engineering Press: Bellingham, 1993, Vol. MS65.
2. Roth, H.D. *Pure Appl. Chem.* 2001, 73, 395–403.
3. Ciamician, G., Silber, P. *Ber.* 1900, 33, 2911–2913.
4. Paterno, E., Chieffi, G. *Gazz. Chim. Ital.* 1911, 39, 341–361.
5. Büchi, G., Inman, C.G., Lipinsky, E.S. *J. Am. Chem. Soc.* 1954, 76, 4327–4331.
6. Hammond, G.S., Moore, W.M. *J. Am. Chem. Soc.* 1959, 81, 6334–6334.
7. Adam, W., Grabowski, S., Wilson, R.M. *Acc. Chem. Res.* 1990, 23, 165–172.
8. Adam, W., DeLucchi, O. *Angew. Chem. Int. Ed. Engl.* 1980, 19, 762–779.
9. Adam, W., Trofimov, A.V. *Acc. Chem. Res.* 2003, 36, 571–579.
10. Rau, H. In *Photochromism*, Dürr, H., Bouas-Laurent, H., Eds., Elsevier Science: Amsterdam, 2003, pp. 165–192.
11. Rau, H. In *Photoreactive Organic Thin Films*, Sekkat, Z., Knoll, W., Eds., Academic Press: San Diego, 2002, pp. 3–47.
12. Knoll, H. In *CRC Handbook of Organic Photochemistry and Photobiology (2nd Edition)*, Horspool, W.M., Lenci, F., Eds., CRC Press: Boca Raton, FL, 2004, pp. 1–16 (Chapter 89).
13. Turro, N.J. *Modern Molecular Photochemistry*, University Science Books: Mill Valley, 1991.
14. Zimmerman, H.E. *Adv. Photochem.* 1963, 1, 183–208.
15. Padwa, A. *Tetrahedron Lett.* 1964, 5, 3465–3469.
16. Walling, C., Gibian, M.J. *J. Am. Chem. Soc.* 1965, 87, 3361–3364.
17. Turro, N.J., Dalton, J.C., Dawes, K., Farrington, G., Hautala, R., Morton, D., Niemczyk, M., Schore, N. *Acc. Chem. Res.* 1972, 5, 92–101.
18. Wagner, P.J. *Top. Curr. Chem.* 1976, 66, 1–52.
19. Formosinho, S.J. *Mol. Photochem.* 1977, 8, 459–475.
20. Wagner, P.J. In *Rearrangements in Ground and Excited States*, deMayo, P., Ed., Academic Press: New York, 1980, Vol. 3, pp. 381–444.
21. Lissi, E.A., Encinas, M.V. In *Handbook of Organic Photochemistry*, Scaiano, J.C., Ed., CRC Press: Boca Raton, FL, 1989, Vol. 2, pp. 111–176.
22. Formosinho, S.J., Arnaut, L.G. *Adv. Photochem.* 1991, 16, 67–117.
23. Wagner, P., Park, B.-S. *Org. Photochem.* 1991, 11, 227–366.
24. Wagner, P.J. *Acc. Chem. Res.* 1971, 4, 168–177.
25. Scaiano, J.C. *J. Photochem.* 1973, 2, 81–118.
26. Paul, H., Small, R.D., Scaiano, J.C. *J. Am. Chem. Soc.* 1978, 100, 4520–4527.
27. Dorigo, A.E., McCarrick, M.A., Loncharich, R.J., Houk, K.N. *J. Am. Chem. Soc.* 1990, 112, 7508–7514.
28. Scaiano, J.C. *J. Am. Chem. Soc.* 1980, 102, 5399–5400.
29. Griller, D., Howard, J.A., Marriott, P.R., Scaiano, J.C. *J. Am. Chem. Soc.* 1981, 103, 619–623.
30. Charney, D.R., Dalton, J.C., Hautala, R.R., Snyder, J.J., Turro, N.J. *J. Am. Chem. Soc.* 1974, 96, 1407–1410.
31. Nau, W.M., Cozens, F.L., Scaiano, J.C. *J. Am. Chem. Soc.* 1996, 118, 2275–2282.
32. Andrews, S.D., Day, A.C. *Chem. Commun.* 1967, 477–478.
33. Solomon, B.S., Thomas, T.F., Steel, C. *J. Am. Chem. Soc.* 1968, 90, 2249–2258.
34. Day, A.C., Wright, T.R. *Tetrahedron Lett.* 1969, 10, 1067–1070.

35. Engel, P.S. *J. Am. Chem. Soc.* 1969, 91, 6903–6907.
36. Clark, W.D.K., Steel, C. *J. Am. Chem. Soc.* 1971, 93, 6347–6355.
37. Evans, T.R. *J. Am. Chem. Soc.* 1971, 93, 2081–2082.
38. Engel, P.S., Steel, C. *Acc. Chem. Res.* 1973, 6, 275–281.
39. Engel, P.S., Fogel, L.D., Steel, C. *J. Am. Chem. Soc.* 1974, 96, 327–332.
40. Wamser, C.C., Medary, R.T., Kochevar, I.E., Turro, N.J., Chang, P.L. *J. Am. Chem. Soc.* 1975, 97, 4864–4869.
41. Ramamurthy, V., Turro, N.J. *Ind. J. Chem. B* 1979, 18, 72–74.
42. Engel, P.S. *Chem. Rev.* 1980, 80, 99–150.
43. Engel, P.S., Nalepa, C.J. *Pure Appl. Chem.* 1980, 52, 2621–2632.
44. Wamser, C.C., Lou, L., Mendoza, J., Olson, E. *J. Am. Chem. Soc.* 1981, 103, 7228–7232.
45. Adam, W., Moorthy, J.N., Nau, W.M., Scaiano, J.C. *J. Org. Chem.* 1996, 61, 8722–8723.
46. Nau, W.M., Adam, W., Scaiano, J.C. *J. Am. Chem. Soc.* 1996, 118, 2742–2743.
47. Nau, W.M., Adam, W., Scaiano, J.C. *Chem. Phys. Lett.* 1996, 253, 92–96.
48. Adam, W., Moorthy, J.N., Nau, W.M., Scaiano, J.C. *J. Org. Chem.* 1997, 62, 8082–8090.
49. Adam, W., Moorthy, J.N., Nau, W.M., Scaiano, J.C. *J. Am. Chem. Soc.* 1997, 119, 6749–6756.
50. Adam, W., Moorthy, J.N., Nau, W.M., Scaiano, J.C. *J. Am. Chem. Soc.* 1997, 119, 5550–5555.
51. Adam, W., Nikolaus, A. *Eur. J. Org. Chem.* 1998, 2177–2179.
52. Nau, W.M., Greiner, G., Rau, H., Olivucci, M., Robb, M.A. *Ber. Bunsenges. Phys. Chem.* 1998, 102, 486–492.
53. Nau, W.M. *J. Inf. Recording* 1998, 24, 105–114.
54. Nau, W.M., Greiner, G., Wall, J., Rau, H., Olivucci, M., Robb, M.A. *Angew. Chem. Int. Ed.* 1998, 37, 98–101.
55. Nau, W.M., Pischel, U. *Angew. Chem. Int. Ed.* 1999, 38, 2885–2888.
56. Nau, W.M., Greiner, G., Rau, H., Wall, J., Olivucci, M., Scaiano, J.C. *J. Phys. Chem. A* 1999, 103, 1579–1584.
57. Adam, W., Nikolaus, A. *J. Am. Chem. Soc.* 2000, 122, 884–888.
58. Nau, W.M. *EPA Newsl.* 2000, 70, 6–29.
59. Pischel, U., Zhang, X., Hellrung, B., Haselbach, E., Muller, P.-A., Nau, W.M. *J. Am. Chem. Soc.* 2000, 122, 2027–2034.
60. Pischel, U., Nau, W.M. *J. Phys. Org. Chem.* 2000, 13, 640–647.
61. Pischel, U., Allonas, X., Nau, W.M. *J. Inf. Recording* 2000, 25, 311–321.
62. Sinicropi, A., Pischel, U., Basosi, R., Nau, W.M., Olivucci, M. *Angew. Chem. Int. Ed.* 2000, 39, 4582–4586.
63. Zhang, X., Nau, W.M. *J. Inf. Recording* 2000, 25, 323–330.
64. Sinicropi, A., Pogni, R., Basosi, R., Robb, M.A., Gramlich, G., Nau, W.M., Olivucci, M. *Angew. Chem. Int. Ed.* 2001, 40, 4185–4189.
65. Klapstein, D., Pischel, U., Nau, W.M. *J. Am. Chem. Soc.* 2002, 124, 11349–11357.
66. Márquez, F., Martí, V., Palomares, E., García, H., Adam, W. *J. Am. Chem. Soc.* 2002, 124, 7264–7265.
67. Sinicropi, A., Nau, W.M., Olivucci, M. *Photochem. Photobiol. Sci.* 2002, 1, 537–546.

68. Adam, W., Trofimov, A.V. In *CRC Handbook of Organic Photochemistry and Photobiology (2nd edition)*, Horspool, W.M., Lenci, F., Eds., CRC Press: Boca Raton, FL, 2004, pp. 1–16 (Chapter 93).
69. Pischel, U., Huang, F., Nau, W.M. *Photochem. Photobiol. Sci.* 2004, 3, 305–310.
70. Pischel, U., Nau, W.M. *J. Am. Chem. Soc.* 2001, 123, 9727–9737.
71. Pischel, U., Nau, W.M. *Photochem. Photobiol. Sci.* 2002, 1, 141–147.
72. Engel, P.S., Kitamura, A., Keys, D.E. *J. Org. Chem.* 1987, 52, 5015–5021.
73. Nau, W.M. *Ber. Bunsenges. Phys. Chem.* 1998, 102, 476–485.
74. Nau, W.M. *J. Am. Chem. Soc.* 1998, 120, 12614–12618.
75. Zhang, X., Erb, C., Flammer, J., Nau, W.M. *Photochem. Photobiol.* 2000, 71, 524–533.
76. Marquez, C., Pischel, U., Nau, W.M. *Org. Lett.* 2003, 5, 3911–3914.
77. Turro, N.J., Engel, R. *J. Am. Chem. Soc.* 1969, 91, 7113–7121.
78. Arnold, D.R. *Adv. Photochem.* 1968, 6, 301–423.
79. Horspool, W.M. *Photochemistry* 2002, 33, 53–73 (and previous accounts with the same title within this series).
80. Rau, H. *Angew. Chem. Int. Ed. Engl.* 1973, 12, 224–235.
81. Mirbach, M.J., Liu, K.-C., Mirbach, M.F., Cherry, W.R., Turro, N.J., Engel, P.S. *J. Am. Chem. Soc.* 1978, 100, 5122–5129.
82. Mirbach, M.F., Mirbach, M.J., Liu, K.-C., Turro, N.J. *J. Photochem.* 1978, 8, 299–306.
83. Encinas, M.V., Lissi, E.A. *J. Photochem.* 1978, 8, 131–143.
84. Weiss, D.S. In *Organic Photochemistry*, Padwa, A., Ed., Marcel Dekker: New York, 1981, Vol. 5, p 347.
85. Diau, E.W.G., Kötting, C., Zewail, A.H. *ChemPhysChem* 2001, 2, 273–293.
86. Haas, Y. *Photochem. Photobiol. Sci.* 2004, 3, 6–16.
87. Garcia-Garibay, M.A., Campos, L.M. In *CRC Handbook of Organic Photochemistry and Photobiology (2nd Edition)*, Horspool, W.M., Lenci, F., Eds., CRC Press: Boca Raton, FL, 2004, pp. 1–41 (Chapter 48).
88. Scaiano, J.C., Lissi, E.A., Encina, M.V. *Rev. Chem. Intermediates* 1978, 2, 139–196.
89. De Feyter, S., Diau, E.W.-G., Zewail, A.H. *Angew. Chem. Int. Ed.* 2000, 39, 260–263.
90. Tadashi, H. In *CRC Handbook of Organic Photochemistry and Photobiology (Second Edition)*, Horspool, W.M., Lenci, F., Eds., CRC Press: Boca Raton, FL, 2004, pp. 1–14 (Chapter 55).
91. Scheffer, J.R., Scott, C. In *CRC Handbook of Organic Photochemistry and Photobiology (Second Edition)*, Horspool, W.M., Lenci, F., Eds., CRC Press: Boca Raton, FL, 2004, pp. 1–25 (Chapter 54).
92. Givens, R.S. In *Organic Photochemistry*, Padwa, A., Ed., Marcel Dekker: New York, 1981, Vol. 5, p 227.
93. Dürr, H., Ruge, B. *Top. Curr. Chem.* 1976, 66, 53–87.
94. Zollinger, H. *Colour Chemistry: Syntheses, Properties and Applications of Organic Dyes and Pigments*, 2nd ed., VCH: Weinheim, 1991.
95. Allen, N.S. *Photopolymerization and Photoimaging Science and Technology*, Elsevier: London, 1989.
96. Adam, W., Nau, W.M., Sendelbach, J., Wirz, J. *J. Am. Chem. Soc.* 1993, 115, 12571–12572.
97. Adam, W., Nau, W.M., Sendelbach, J. *J. Am. Chem. Soc.* 1994, 116, 7049–7054.
98. Adam, W., Fragale, G., Klapstein, D., Nau, W.M., Wirz, J. *J. Am. Chem. Soc.* 1995, 117, 12578–12592.

99. Engel, P.S., Soltero, L.R., Baughman, S.A., Nalepa, C.J., Cahill, P.A., Weisman, R.B. *J. Am. Chem. Soc.* 1982, 104, 1698–1700.

100. Steel, C., Thomas, T.F. *Chem. Commun.* 1966, 900–902.

101. Aikawa, M., Yekta, A., Liu, J.-M., Turro, N.J. *Chem. Phys. Lett.* 1979, 68, 285–290.

102. Aikawa, M., Yekta, A., Liu, J.-M., Turro, N.J. *Photochem. Photobiol.* 1980, 32, 297–303.

103. Engel, P.S., Horsey, D.W., Keys, D.E., Nalepa, C.J., Soltero, L.R. *J. Am. Chem. Soc.* 1983, 105, 7108–7114.

104. Mirbach, M.J., Mirbach, M.F., Cherry, W.R., Turro, N.J., Engel, P. *Chem. Phys. Lett.* 1978, 53, 266–269.

105. Houk, K.N., Chang, Y.-M., Engel, P.S. *J. Am. Chem. Soc.* 1975, 97, 1824–1832.

106. Zhang, X., Nau, W.M. *Angew. Chem. Int. Ed.* 2000, 39, 544–547.

107. Mayer, B., Zhang, X., Nau, W.M., Marconi, G. *J. Am. Chem. Soc.* 2001, 123, 5240–5248.

108. Marquez, C., Nau, W.M. *Angew. Chem. Int. Ed.* 2001, 40, 4387–4390.

109. Mohanty, J., Nau, W.M., *Photochem. Photobiol. Sci.* 2004, 3, 1026–1031.

110. Engel, P.S., Gerth, D.B., Keys, D.E., Scholz, J.N., Houk, K.N., Rozeboom, M.D., Eaton, T.A., Glass, R.S., Broeker, J.L. *Tetrahedron* 1988, 44, 6811–6814.

111. Brogli, F., Eberbach, W., Haselbach, E., Heilbronner, E., Hornung, V., Lemal, D.M. *Helv. Chim. Acta* 1973, 56, 1933–1944.

112. Forster, L.S. *J. Am. Chem. Soc.* 1955, 77, 1417–1421.

113. Toledano, E., Rubin, M.B., Speiser, S. *J. Photchem. Photobiol. A* 1996, 94, 93-100.

114. Boyd, R.J., Bünzli, J.C., Snyder, J.P., Heyman, M.L. *J. Am. Chem. Soc.* 1973, 95, 6478–6480.

115. Engel, P.S., Horsey, D.W., Scholz, J.N., Karatsu, T., Kitamura, A. *J. Phys. Chem.* 1992, 96, 7524–7535.

116. Huang, F., Wang, X., Nau, W.M., unpublished results.

117. Reichardt, C. *Solvents and Solvent Effects in Organic Chemistry*, Wiley-VCH: Weinheim, 2003.

118. Haselbach, E. *Helv. Chim. Acta* 1970, 53, 1526–1543.

119. Marquez, C., Nau, W.M., unpublished results.

120. Ackermann, M.N., Kou, L.J., Richter, J.M., Willett, R.M. *Inorg. Chem.* 1977, 16, 1298–1301.

121. Engel, P.S., Hoque, A.K.M.M., Scholz, J.N., Shine, H.J., Whitmire, K.H. *J. Am. Chem. Soc.* 1988, 110, 7880–7882.

122. Engel, P.S., Robertson, D.M., Scholz, J.N., Shine, H.J. *J. Org. Chem.* 1992, 57, 6178–6187.

123. Blackstock, S.C., Kochi, J.K. *J. Am. Chem. Soc.* 1987, 109, 2484–2496.

124. Engel, P.S., Duan, S., Whitmire, K.H. *J. Org. Chem.* 1998, 63, 5666–5667.

125. North, S.W., Blank, D.A., Gezelter, J.D., Longfellow, C.A., Lee, Y.T. *J. Chem. Phys.* 1995, 102, 4447–4460.

126. Allan, M. *J. Electr. Spectr.* 1989, 48, 219–351.

127. Nau, W.M., Zhang, X. *J. Am. Chem. Soc.* 1999, 121, 8022–8032.

128. Gilani, A.G., Mamaghani, M., Anbir, L. *J. Solution Chem.* 2003, 32, 625–636.

129. Shimamori, H., Uegaito, H., Houdo, K. *J. Phys. Chem.* 1991, 95, 7664–7667.

130. Harmony, M.D., Talkington, T.L., Nandi, R.N. *J. Mol. Struct.* 1984, 125, 125–130.

131. Jovanovic, S.V., Morris, D.G., Pliva, C.N., Scaiano, J.C. *J. Photochem. Photobiol. A: Chem.* 1997, 107, 153–158.

132. Yang, N.C., McClure, D.S., Murov, S.L., Houser, J.J., Dusenbery, R. *J. Am. Chem. Soc.* 1967, 89, 5466–5468.

133. Wagner, P.J., Kemppainen, A.E., Schott, H.N. *J. Am. Chem. Soc.* 1973, 95, 5604–5614.

134. Wagner, P.J., Siebert, E.J. *J. Am. Chem. Soc.* 1981, 103, 7329–7335.

135. Srivastava, S., Yourd, E., Toscano, J.P. *J. Am. Chem. Soc.* 1998, 120, 6173–6174.

136. Suter, G.W., Wild, U.P., Schaffner, K. *J. Phys. Chem.* 1986, 90, 2358–2361.

137. Shailaja, J., Lakshminarasimhan, P.H., Pradhan, A.R., Sunoj, R.B., Jockusch, S., Karthikeyan, S., Uppili, S., Chandrasekhar, J., Turro, N.J., Ramamurthy, V. *J. Phys. Chem. A* 2003, 107, 3187–3198.

138. Adam, W., Nikolaus, A., Sauer, J. *J. Org. Chem.* 1999, 64, 3695–3698.

139. Chang, M.H., Dougherty, D.A. *J. Am. Chem. Soc.* 1982, 104, 2333–2334.

140. Buterbaugh, J.S., Toscano, J.P., Weaver, W.L., Gord, J.R., Hadad, C.M., Gustafson, T.L., Platz, M.S. *J. Am. Chem. Soc.* 1997, 119, 3580–3591.

141. Bisle, H., Rau, H. *Chem. Phys. Lett.* 1975, 31, 264–266.

142. Zhang, X., Nau, W.M., unpublished results.

143. Lower, S.K., El-Sayed, M.A. *Chem. Rev.* 1966, 66, 199–241.

144. Kasha, M. *Chem. Rev.* 1947, 41, 401–419.

145. Engel, P.S., Keys, D.E., Kitamura, A. *J. Am. Chem. Soc.* 1985, 107, 4964–4975.

146. Engel, P.S., Culotta, A.M. *J. Am. Chem. Soc.* 1991, 113, 2686–2696.

147. Uppili, S., Marti, V., Nikolaus, A., Jockusch, S., Adam, W., Engel, P.S., Turro, N.J., Ramamurthy, V. *J. Am. Chem. Soc.* 2000, 122, 11025–11026.

148. Metcalfe, J., Chervinsky, S., Oref, I. *Chem. Phys. Lett.* 1976, 42, 190–192.

149. Robertson, L.C., Merritt, J.A. *J. Chem. Phys.* 1972, 56, 2919–2924.

150. Turro, N.J., Cherry, W.R., Mirbach, M.J., Mirbach, M.F., Ramamurthy, V. *Mol. Photochem.* 1978, 9, 111–118.

151. Turro, N.J., Cha, Y., Gould, I.R., Moss, R.A. *J. Photochem.* 1987, 37, 81–86.

152. Adam, W., Dörr, M., Hössel, P. *Angew. Chem. Int. Ed. Engl.* 1986, 25, 818–819.

153. El-Sayed, M.A. *J. Chem. Phys.* 1964, 41, 2462–2467.

154. Wagner, P.J. *J. Chem. Phys.* 1966, 45, 2335–2336.

155. Anderson, M.A., Grissom, C.B. *J. Am. Chem. Soc.* 1996, 118, 9552–9556.

156. Turro, N.J., Renner, C.A., Waddell, W.H., Katz, T.J. *J. Am. Chem. Soc.* 1976, 98, 4320–4322.

157. Nau, W.M., Scaiano, J.C. *J. Phys. Chem.* 1996, 100, 11360–11367.

158. Rehm, D., Weller, A. *Isr. J. Chem.* 1970, 8, 259–271.

159. Marcus, R.A. *Angew. Chem. Int. Ed. Engl.* 1993, 32, 1111–1121.

160. Kavarnos, G.J., Turro, N.J. *Chem. Rev.* 1986, 86, 401–449.

161. Speiser, S. *Chem. Rev.* 1996, 96, 1953–1976.

162. Guttenplan, J.B., Cohen, S.G. *Tetrahedron Lett.* 1972, 13, 2163–2166.

163. Loutfy, R.O., Yip, R.W. *Can. J. Chem.* 1973, 51, 1881–1884.

164. Wagner, P.J., Truman, R.J., Puchalski, A.E., Wake, R. *J. Am. Chem. Soc.* 1986, 108, 7727–7738.

165. Jacques, P. *J. Photochem. Photobiol. A: Chem.* 1991, 56, 159–163.

166. Jacques, P., Allonas, X., von Raumer, M., Suppan, P., Haselbach, E. *J. Photochem. Photobiol. A: Chem.* 1997, 111, 41–45.

167. Jacques, P., Allonas, X., Suppan, P., von Raumer, M. *J. Photochem. Photobiol. A: Chem.* 1996, 101, 183–184.

168. Zimmerman, H.E., Alabugin, I.V. *J. Am. Chem. Soc.* 2001, 123, 2265–2270.
169. Hubig, S.M., Rathore, R., Kochi, J.K. *J. Am. Chem. Soc.* 1999, 121, 617–626.
170. Coenjarts, C., Scaiano, J.C. *J. Am. Chem. Soc.* 2000, 122, 3635–3641.
171. Engel, P.S., Wu, W.-X. *J. Am. Chem. Soc.* 1989, 111, 1830–1835.
172. Abuin, E.B., Encina, M.V., Lissi, E.A., Scaiano, J.C. *J. Chem. Soc. Faraday Trans.1* 1975, 71, 1221–1229.
173. Wagner, P.J., Puchalski, A.E. *J. Am. Chem. Soc.* 1980, 102, 7138–7140.
174. Pischel, U., Patra, D., Koner, A.L., Nau, W.M., *Photochem. Photobiol.*, in press
175. Kellogg, R.E., Simpson, W.T. *J. Am. Chem. Soc.* 1965, 87, 4230–4234.
176. Wong, P.C. *Can. J. Chem.* 1982, 60, 339–341.
177. Wamser, C.C., Chang, P.L. *J. Am. Chem. Soc.* 1973, 95, 2044–2045.
178. Scaiano, J.C., Leigh, W.J., Meador, M.A., Wagner, P.J. *J. Am. Chem. Soc.* 1985, 107, 5806–5807.
179. Zavitsas, A.A., Pinto, J.A. *J. Am. Chem. Soc.* 1972, 94, 7390–7396.
180. Encina, M.V., Lissi, E.A. *Int. J. Chem. Kinet.* 1978, 10, 653–656.
181. Encina, M.V., Diaz, S., Lissi, E. *Int. J. Chem. Kinet.* 1981, 13, 119–123.
182. Naguib, Y.M.A., Steel, C., Cohen, S.G., Young, M.A. *J. Phys. Chem.* 1987, 91, 3033–3036.
183. Bordwell, F.G., Cheng, J.-P. *J. Am. Chem. Soc.* 1991, 113, 1736–1743.
184. Pischel, U., Galletero, M.S., Garcia, H., Miranda, M.A., Nau, W.M. *Chem. Phys. Lett.* 2002, 359, 289–294.
185. Zavitsas, A.A., Melikian, A.A. *J. Am. Chem. Soc.* 1975, 97, 2757–2763.
186. Zavitsas, A.A., Chatgilialoglu, C. *J. Am. Chem. Soc.* 1995, 117, 10645–10654.
187. Jacques, P., Burget, D. *J. Photochem. Photobiol. A: Chem.* 1992, 68, 165–168.
188. Jacques, P., Allonas, X., Burget, D., Haselbach, E., Muller, P.-A., Sergenton, A.-C., Galliker, H. *Phys. Chem. Chem. Phys.* 1999, 1, 1867–1871.
189. Cohen, S.G., Litt, A.D. *Tetrahedron Lett.* 1970, 11, 837–840.
190. Gorman, A.A., Parekh, C.T., Rodgers, M.A.J., Smith, P.G. *J. Photochem.* 1978, 9, 11–17.
191. Inbar, S., Linschitz, H., Cohen, S.G. *J. Am. Chem. Soc.* 1981, 103, 1048–1054.
192. Weigel, W., Wagner, P.J. *J. Am. Chem. Soc.* 1996, 118, 12858–12859.
193. von Raumer, M., Suppan, P., Haselbach, E. *Chem. Phys. Lett.* 1996, 252, 263–266.
194. von Raumer, M., Suppan, P., Haselbach, E. *Helv. Chim. Acta* 1997, 80, 719–724.
195. Scaiano, J.C., Stewart, L.C., Livant, P., Majors, A.W. *Can. J. Chem.* 1984, 62, 1339–1343.
196. Okada, K., Yamaji, M., Shizuka, H. *J. Chem. Soc., Faraday Trans.* 1998, 94, 861–866.
197. Ikeda, H., Minegishi, T., Takahashi, Y., Miyashi, T. *Tetrahedron Lett.* 1996, 37, 4377–4380.
198. Nau, W.M., Adam, W., Klapstein, D., Sahin, C., Walter, H. *J. Org. Chem.* 1997, 62, 5128–5132.
199. Kuzmin, V.A., Tatikolov, A.S. *Chem. Phys. Lett.* 1978, 53, 606–610.
200. Scaiano, J.C. *Chem. Phys. Lett.* 1981, 79, 441–443.
201. Darmanyan, A.P., Foote, C.S. *J. Phys. Chem.* 1993, 97, 5032–5035.
202. Caldwell, R.A., Helms, A.M., Engel, P.S., Wu, A. *J. Phys. Chem.* 1996, 100, 17716–17717.
203. McGarvey, D.J., Szekeres, P.G., Wilkinson, F. *Chem. Phys. Lett.* 1992, 199, 314–319.
204. Grewer, C., Brauer, H.-D. *J. Phys. Chem.* 1994, 98, 4230–4235.

205. Wilkinson, F., McGarvey, D.J., Olea, A.F. *J. Phys. Chem.* 1994, 98, 3762–3769.
206. Encinas, M.V., Scaiano, J.C. *J. Am. Chem. Soc.* 1981, 103, 6393–6397.
207. Dalton, J.C., Wriede, P.A., Turro, N.J. *J. Am. Chem. Soc.* 1970, 92, 1318–1326.
208. Hautala, R.R., Turro, N.J. *J. Am. Chem. Soc.* 1971, 93, 5595–5597.
209. Turro, N.J., Lee, C., Schore, N., Barltrop, J., Carless, H.A.J. *J. Am. Chem. Soc.* 1971, 93, 3079–3080.
210. Barltrop, J.A., Carless, H.A.J. *J. Am. Chem. Soc.* 1972, 94, 1951–1959.
211. Kochevar, I.E., Wagner, P.J. *J. Am. Chem. Soc.* 1972, 94, 3859–3865.
212. Schore, N.E., Turro, N.J. *J. Am. Chem. Soc.* 1975, 97, 2482–2488.
213. Yang, N.C., Hui, M.H., Shold, D.M., Turro, N.J., Hautala, R.R., Dawes, K., Dalton, J.C. *J. Am. Chem. Soc.* 1977, 99, 3023–3033.
214. Jones II, G., Santhanam, M., Chiang, S.-H. *J. Am. Chem. Soc.* 1980, 102, 6088–6095.
215. Turro, N.J., Tanimoto, Y. *J. Photochem.* 1980, 14, 199–203.
216. Turro, N.J., Shima, K., Chung, C.-J., Tanielian, C., Kanfer, S. *Tetrahedron Lett.* 1980, 21, 2775–2778.
217. Gersdorf, J., Mattay, J., Görner, H. *J. Am. Chem. Soc.* 1987, 109, 1203–1209.
218. Hünig, S., Schmitt, M. *Tetrahedron Lett.* 1987, 28, 4521–4524.
219. Röttger, D., Rau, H. *J. Photchem. Photobiol. A: Chem.* 1996, 101, 205–214.
220. Bartholomew, R.F., Davidson, R.S., Lambeth, P.F., McKellar, J.F., Turner, P.H. *J. Chem. Soc., Perkin Trans. 2* 1972, 577–582.
221. Davidson, R.S., Lambeth, P.F., Santhanam, M. *J. Chem. Soc., Perkin Trans. 2* 1972, 2351–2355.
222. Arimitsu, S., Masuhara, H., Mataga, N., Tsubomura, H. *J. Phys. Chem.* 1975, 79, 1255–1259.
223. Parola, A.H., Cohen, S.G. *J. Photochem.* 1980, 12, 41–50.
224. Simon, J.D., Peters, K.S. *J. Am. Chem. Soc.* 1981, 103, 6403–6406.
225. Jacques, P. *Chem. Phys. Lett.* 1987, 142, 96–98.
226. Devadoss, C., Fessenden, R.W. *J. Phys. Chem.* 1991, 95, 7253–7260.
227. Miyasaka, H., Morita, K., Kamada, K., Mataga, N. *Chem. Phys. Lett.* 1991, 178, 504–510.
228. Zhu, Q.Q., Schnabel, W., Jacques, P. *J. Chem. Soc., Faraday Trans.* 1991, 87, 1531–1535.
229. Kluge, T., Brede, O. *Chem. Phys. Lett.* 1998, 289, 319–323.
230. Malatesta, V., Scaiano, J.C. *J. Org. Chem.* 1982, 47, 1455–1459.
231. Nazran, A.S., Griller, D. *J. Am. Chem. Soc.* 1983, 105, 1970–1971.
232. Encinas, M.V., Lissi, E.A., Majmud, C., Olea, A.F. *Int. J. Chem. Kinet.* 1989, 21, 245–250.
233. Rebbert, R.E., Ausloos, P. *J. Am. Chem. Soc.* 1965, 87, 5569–5572.
234. Berger, M., Camp, R.N., Demetrescu, I., Giering, L., Steel, C. *Isr. J. Chem.* 1977, 16, 311–317.
235. Lendvay, G., Bèrces, T. *J. Photochem. Photobiol. A: Chem.* 1987, 40, 31–45.
236. Borisevich, N.A., Kazberuk, D.V., Lysak, N.A., Tolstorozhev, G.B. *Zh. Prik. Spektr.* 1989, 50, 745–749.
237. Matsushita, Y., Kajii, Y., Obi, K. *J. Phys. Chem.* 1992, 96, 4455–4458.
238. Matsushita, Y., Yamaguchi, Y., Hikida, T. *Chem. Phys.* 1996, 213, 413–419.
239. Zalesskaya, G.A., Baranovskii, D.I., Sambor, E.G. *J. Appl. Spectrosc.* 1999, 66, 76–80.

240. Razi Naqvi, K., Steel, C. *Chem. Phys. Lett.* 1970, 6, 29–32.
241. Razi Naqvi, K., Steel, C. *Spectrosc. Lett.* 1993, 26, 1761–1769.
242. Bortolus, P., Dellonte, S., Faucitano, A., Gratani, F. *Macromolecules* 1986, 19, 2916–2922.
243. Almgren, M. *Mol. Photochem.* 1972, 4, 213–216.
244. Almgren, M. *Photochem. Photobiol.* 1967, 6, 829–840.
245. Aue, D.H., Webb, H.M., Bowers, M.T. *J. Am. Chem. Soc.* 1975, 97, 4137–4139.
246. Centineo, G., Fragala, I., Bruno, G., Spampinato, S. *J. Mol. Struct.* 1978, 44, 203–210.
247. Cowan, D.O., Gleiter, R., Hashmall, J.A., Heilbronner, E., Hornung, V. *Angew. Chem. Int. Ed. Engl.* 1971, 10, 401–402.
248. Desfrancois, C., Abdoul-Carime, H., Khelifa, N., Schermann, J.P. *Phys. Rev. Lett.* 1994, 73, 2436–2439.
249. Grimsrud, E.P., Caldwell, G., Chowdhury, S., Kebarle, P. *J. Am. Chem. Soc.* 1985, 107, 4627–4634.
250. Kebarle, P., Chowdhury, S. *Chem. Rev.* 1987, 87, 513–534.
251. Jones II, G., Huang, B., Griffin, S.F. *J. Org. Chem.* 1993, 58, 2035–2042.
252. Nelsen, S.F., Petillo, P.A., Chang, H., Frigo, T.B., Dougherty, D.A., Kaftory, M. *J. Org. Chem.* 1991, 56, 613–618.
253. Levy, G.C., Cargioli, J.D., Racela, W. *J. Am. Chem. Soc.* 1970, 92, 6238–6246.
254. Wells, D.K., Trahanovsky, W.S. *J. Am. Chem. Soc.* 1969, 91, 5871–5872.
255. Meot-Ner, M. *J. Am. Chem. Soc.* 1983, 105, 4906–4911.
256. Porter, G., Dogra, S.K., Loutfy, R.O., Sugamori, S.E., Yip, R.W. *J. Chem. Soc., Faraday Trans. 1* 1973, 69, 1462–1474.
257. Okamoto, M., Teranishi, H. *J. Am. Chem. Soc.* 1986, 108, 6378–6380.
258. Loutfy, R.O., Yip, R.W., Dogra, S.K. *Tetrahedron Lett.* 1977, 18, 2843–2846.
259. Loutfy, R.O., Dogbra, S.K., Yip, R.W. *Can. J. Chem.* 1979, 57, 342–347.
260. Mirbach, M.F., Ramamurthy, V., Mirbach, M.J., Turro, N.J., Wagner, P.J. *Nouv. J. Chim.* 1980, 4, 471–474.
261. Rosenfeld, T., Alchalel, A., Ottolenghi, M. *J. Phys. Chem.* 1974, 78, 336–341.
262. Scaiano, J.C., Wintgens, V., Netto-Ferreira, J.C. *Photochem. Photobiol.* 1989, 50, 707–710.
263. Boate, D.R., Johnston, L.J., Scaiano, J.C. *Can. J. Chem.* 1989, 67, 927–932.
264. Görner, H. *J. Phys. Chem.* 1982, 86, 2028–2035.
265. Bäckström, H.L.J., Sandros, K. *Acta Chem. Scand.* 1958, 12, 823–832.
266. Kuhlmann, R., Schnabel, W. *J. Photochem.* 1977, 7, 287–296.
267. Wagner, P.J. *Mol. Photochem.* 1969, 1, 71–87.
268. Urruti, E.H., Kilp, T. *Macromolecules* 1984, 17, 50–54.
269. Porter, G., Topp, M.R. *Proc. Roy. Soc. A* 1970, 315, 163–184.
270. Previtali, C.M., Scaiano, J.C. *J. Chem. Soc., Perkin Trans. 2* 1972, 1672–1676.
271. Yip, R.W., Loutfy, R.O., Chow, Y.L., Magdzinski, L.K. *Can. J. Chem.* 1972, 50, 3426–3431.
272. Miyasaka, H., Mataga, N. *Bull. Chem. Soc. Jpn.* 1990, 63, 131–137.
273. Santhanam, M., Ramakrishnan, V. *J. Chem. Soc. D, Chem. Commun.* 1970, 344–345.
274. Olea, A.F., Encinas, M.V., Lissi, E.A. *Macromolecules* 1982, 15, 1111–1115.
275. Chattopadhyay, S.K., Kumar, C.V., Das, P.K. *J. Photochem.* 1984, 24, 1–9.
276. Darmanyan, A.P., Foote, C.S., Jardon, P. *J. Phys. Chem.* 1995, 99, 11854–11859.

277. Baral-Tosh, S., Chattopadhyay, S.K., Das, P.K. *J. Phys. Chem.* 1984, 88, 1404–1408.
278. Wilkinson, F., Dubois, J.T. *J. Chem. Phys.* 1963, 39, 377–383.
279. Wagner, P.J. *J. Am. Chem. Soc.* 1967, 89, 2503–2505.
280. Brimage, D.R.G., Davidson, R.S., Lambeth, P.F. *J. Chem. Soc. (C)* 1971, 1241–1244.
281. Giering, L., Berger, M., Steel, C. *J. Am. Chem. Soc.* 1974, 96, 953–958.
282. Yang, N.C., Hui, M.H., Ballard, S.A. *J. Am. Chem. Soc.* 1971, 93, 4056–4058.
283. Monroe, B.M., Lee, C.-G., Turro, N.J. *Mol. Photochem.* 1974, 6, 271–289.
284. Encina, M.V., Lissi, E.A. *J. Polym. Sci. Pol. Chem.* 1979, 17, 1645–1653.
285. Henne, A., Fischer, H. *J. Am. Chem. Soc.* 1977, 99, 300–302.
286. Tominaga, K., Yamauchi, S., Hirota, N. *J. Phys. Chem.* 1988, 92, 5160–5165.
287. Dalton, J.C., Snyder, J.J. *J. Am. Chem. Soc.* 1975, 97, 5192–5197.
288. Baignée, A., Howard, J.A., Scaiano, J.C., Stewart, L.C. *J. Am. Chem. Soc.* 1983, 105, 6120–6123.
289. Das, P.K., Encinas, M.V., Scaiano, J.C. *J. Am. Chem. Soc.* 1981, 103, 4154–4162.
290. Das, P.K., Encinas, M.V., Steenken, S., Scaiano, J.C. *J. Am. Chem. Soc.* 1981, 103, 4162–4166.
291. Valgimigli, L., Banks, J.T., Ingold, K.U., Lusztyk, J. *J. Am. Chem. Soc.* 1995, 117, 9966–9971.
292. MacFaul, P.A., Ingold, K.U., Lusztyk, J. *J. Org. Chem.* 1996, 61, 1316–1321.
293. Evans, C., Scaiano, J.C., Ingold, K.U. *J. Am. Chem. Soc.* 1992, 114, 4589–4593.
294. Guttenplan, J.B., Cohen, S.G. *J. Org. Chem.* 1973, 38, 2001–2007.
295. Leyva, E., Platz, M.S., Niu, B., Wirz, J. *J. Phys. Chem.* 1987, 91, 2293–2298.
296. Birks, J.B., Salete, M., Leite, S.C.P. *J. Phys. B* 1970, 3, 417–424.

4 Photonucleophilic Substitution Reactions

Maurizio Fagnoni and Angelo Albini

CONTENTS

4.1   INTRODUCTION

Nucleophilic substitution reactions form a significant chapter in the photochemistry of aromatics. Interest in the topic started with the observation in Havinga laboratory in 1956 that the hydrolysis of nitrophenyl phosphates and sulfates, not occurring to a significant extent at room temperature, was much accelerated by (room) light and actually occurred smoothly under irradiation, in particular in the case of the *meta* isomers [1]. A striking point in this reaction is the "*meta* activation" by the nitro group, in contrast with the *ortho-para* activation characteristic of thermal nucleophilic substitutions. Along with the opposite stereochemical course of thermal and photochemical cyclization reactions in Vitamin D derivatives, reported shortly afterwards by the same group [2], this was one of the key observations evidencing the different behavior of excited states vs. ground states and gave impetus to the development of organic photochemistry and its rationalization in the following years (see, e.g., Refs. 3–5).

Research in the field has continued actively over the following decades [6–13], though perhaps not as much as in other fields of photochemistry. The main target has been in most cases defining the mechanism and rationalizing the selectivity of the process. The peculiar orientation rules of excited state reactions have been confirmed, albeit in a more complex frame that takes into account the particular mechanism occurring in each case and the role of each step of the reaction.

Since the initial report, it has been speculated that aromatic photosubstitution reactions may have a synthetic significance, but it cannot be said that this point has been addressed in depth. Nevertheless, many reactions have been reported that occur cleanly and selectively, giving a high yield of products otherwise difficult to obtain in a simple way, though usually only in diluted solutions, a limitation that is common to most photochemical reactions. Perhaps even more interesting are the mild conditions of such reactions. This point may lead to a new interest in the field in the frame of the increased drive toward short synthetic sequences occurring under mild conditions, according to the postulates of green chemistry.

In this sense, aromatic nucleophilic photosubstitution reactions can be viewed as an important alternative to aromatic functionalizations promoted by catalysis with metal complexes. It is not surprising that there is some analogy between the two methods. In metal catalysis, the covalent bond between the aromatic ring and the leaving group is either weakened by donation from the HOMO (centered on the ring) to the metal center, accompanied by back donation to the LUMO of the aromatic, or broken by oxidative addition to the metal center (see Scheme 4.1). Upon photoexcitation, the HOMO is again deprived of charge, because of the promotion

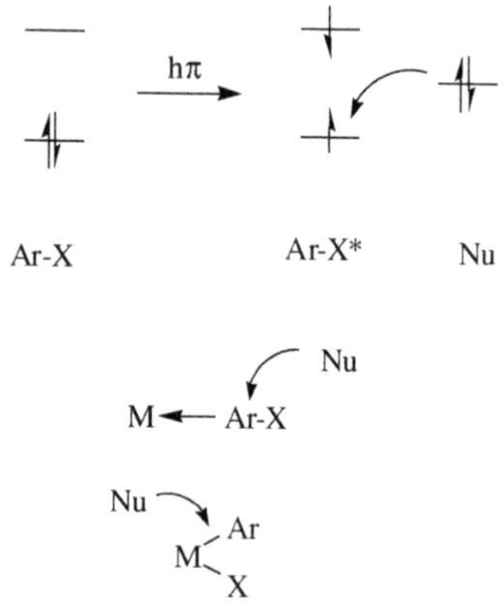

SCHEME 4.1

of an electron to the LUMO, and this allows a much easier interaction with the nucleophile. Metal catalysis has been more extensively applied than photochemical initiation in recent years and such catalyzed reactions have been considered as typical examples of "green" synthetic paths. However, photochemical reactions retain a basic advantage, avoiding the use of the often expensive, unstable, and toxic metal complexes active as catalysts.

In the present report, nucleophilic photosubstitution reactions are reviewed with attention to the possible synthetic significance. The review is more illustrative than exhaustive, minimizing duplication with recent reviews of specific topics, such as $S_{RN}1$ reactions, as indicated in the text. Likewise, early results extensively reviewed in accounts in the 1980s are also not covered. Also not covered are the related reactions with heteroaryl derivatives; there are many examples in this class, often with a remarkable synthetic significance, but the classification remains the same as with homoaromatic molecules. Therefore, for the sake of brevity, these have been completely omitted.

The scope of aromatic photo S_N reaction is extensive and encloses a variety of nucleofugal groups, such as halide, O-leaving groups such as alkoxy, sulfate, phosphate, as well as nitrito anion, nitrogen from diazonium salts, cyanide, and others. There is likewise a large palette of entering nucleophiles, and in view of the synthetic aim of the review, the survey will be according to the bond formed, following the order carbon–carbon and carbon–other IV group elements, carbon–nitrogen (and other V group elements), carbon–sulfur (and other VI group elements, except oxygen), carbon–halogen. The formation of carbon–oxygen bond is reviewed in the last section, since most of these photochemical reactions are carried out in alcohols, water, or in mixed solvents containing them, and thus products resulting from solvation are ubiquitary and are often formed along with the desired substitution product in the presence of other nucleophiles. No special emphasis is given to the mechanism when presenting the reactions, except when this is relevant for the synthetic result. However, it is all-important that the basic mechanistic alternatives are first discussed.

4.2 MECHANISM

The term *photosubstitution* reactions has been applied to a large variety of reactions, which involve quite unrelated mechanisms. In the present context, it will be referred to every photoinitiated reaction between an aromatic and a nucleophile, which occurs via a polar mechanism. The classification is based on the key intermediates involved. A first distinction depends on whether the initiating step is unimolecular or bimolecular. In the first case, path *a* involving heterolytic fragmentation and path *b* involving ionization can be considered (see Scheme 4.2).

Photoheterolysis (path *a*) has been considered a rare event, but evidence has been recently accumulating that electron-donating substituted phenyl halides undergo photochemical cleavage, quite efficiently in some cases, thus giving access to a

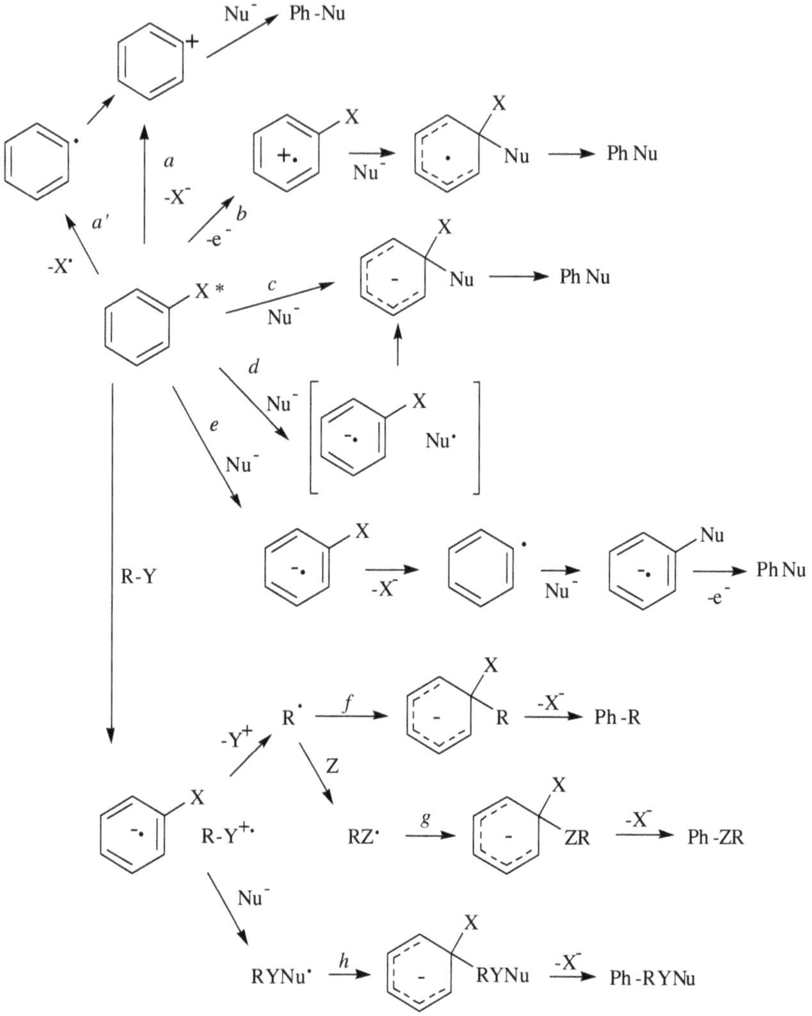

SCHEME 4.2

little known intermediate, the phenyl cation. Another entry to the same species is N$_2$ loss from phenyldiazonium salts, which occurs in a practical way only under photochemical conditions, since thermal dediazoniation usually occurs under reductive catalysis and gives the phenyl radical rather than the cation. A variation to path *a* is homolytic photofragmentation followed by electron transfer between the fragments leading again to an aryl cation (path *a'*; on the contrary, reactions proceeding via aryl radicals do not fall within the scope of the present discussion and are not considered).

On the other hand, photoionization (path *b*) is a common occurrence from electron-donating substituted aromatics in water or other ionic solvents (unless a good nucleofugal group is present, see above). The resulting radical cation may react with a nucleophile and the resulting radical rearomatize by some mechanism.

The other paths are initiated by a bimolecular interaction of the excited state of the reagent with the nucleophile. The S_N2 Ar* reaction (path *c*) matches the usual addition-elimination mechanism of the thermal S_N2 Ar reaction, with the difference that is the electron distribution in the excited state, singlet or triplet according to the case, that now governs the process and thus dictates the orientation rules. When the nucleophile is a good donor, this process shifts toward electron transfer (path *d*), which may lead to the same type of products, but possibly with a different regiochemistry, since here is the electron distribution in the radical anion, not in the excited state, that matters.

The remaining processes are initiated, as in the case of path *d*, by reductive electron transfer, a quite common occurrence with electron-withdrawing substituted aromatics. Indeed, the reduction potential of the electronically excited states jumps upwards by an amount correspondent to the excitation energy (typically 1.5 to 3.5 eV) and electron transfer from good donors such as enolates, but also from weak donors such as amines, aromatics, or alkenes, may become exothermic. A chemical reaction occurs when cleavage of either one of the odd electron transients formed competes efficiently with a chemically unproductive back electron transfer. Different paths involve either the aromatic radical anion or the radical cation of the donor. Fragmentation of the radical anion (path *e*) leads to a phenyl radical, which may combine with the nucleophile. The resulting radical anion gives the final products, e.g., by transferring an electron to the original reagent, in this case giving rise to a chain process ($S_{RN}1$ reaction; this may occur thermally with good donors, but the scope is widened when photoinitiation is used).

Contrary to path *e*, the other paths are based upon the stability of the electron-withdrawing substituted aromatic radical anions. Indeed such extensively delocalized species are often weak bases and nucleophiles and may be quite persistent (in the case of polycyanoaromatics, indefinitely persistent in solution under appropriate conditions). In this case, the following reaction depends on the formation of a reactive intermediate from the cation radical of the donor. Typical examples are as follows. A first possibility is fragmentation of the radical cation, yielding a radical that couples with the aromatic radical anion (path *f*). In a variation of this mechanism, the radical is first trapped by a radicophile and it is the radical adduct that couples with the radical anion (path *g*). A further possibility is addition of a nucleophile to the radical cation, and coupling of the resulting radical with the aromatic radical anion (path *h*).

The mere listing of the various paths above gives a hint of the scope of photochemical S_NAr reactions, which is much wider than that of their thermal counterpart. One may consider, for example, how many weak nucleophiles will interact with the partially filled HOMO of an aromatic for a S_N2Ar* reaction, while interaction with the LUMO of the ground state would be impossible (see Scheme

4.1), or how ubiquitous is single electron transfer in the photochemistry of aromatics, even with weak donors, while thermal reduction occurs only with aggressive reagents such as alkaline metals.

On the other hand, the varied panorama also illustrates the fact that different paths will in general be potentially available and the actual path followed, as well as the efficiency of the overall reaction, will depend on a host of factors such as the lifetime of the singlet and triplet excited states, their redox properties and chemical reactivity, the nature of the nucleophile/electron donor, the medium, etc. This also means that a demonstration of the mechanism is not necessarily a simple job. Steady-state kinetics is usually not sufficient for a complete picture and the contribution by spectroscopic techniques (e.g., epr) or by fast kinetic experiments (e.g., flash photolysis with UV or IR detection) depend on the convenience in detecting the intermediates. As a result, only in a small number of cases has the mechanism been worked out in detail, although the general pattern of most reported reaction has been assigned with reasonable confidence, sometimes on the basis of analogy.

Viewing the thing from the positive side, the high reactivity and a variety of patterns give an opportunity for directing the reaction, for example, by the appropriate choice of direct vs. sensitized irradiation, choice of the medium, etc. Just to mention an issue of growing interest, the course of photo S_NAr reactions may be directed by shifting from solution to organized media, for example, by adding cyclodextrins or surfactants that generate micelles.

4.3 FORMATION OF A C-C OR OTHER C-IV GROUP ELEMENTS BOND

4.3.1 ALKYLATION OF ARYL HALIDES VIA THE S_N1 PATH

Formation of a carbon–carbon bond emerged as one of the most useful classes of photo S_NAr reactions. The most significant reactions pertaining to this class can be attributed to two groups: alkylation and (hetero)arylation on one hand, cyanation and carboxylation on the other. Alkylation occurs starting from aryl halides, either via the $S_{RN}1$ mechanism (path *e*) when good donors such as enolates are used, or via unimolecular fragmentation (path *a*) when this is sufficiently efficient, as with electron-donating substituted aryl halides. Alternatively, alkylation occurs even with moderate donors via one of the electron transfer paths (paths *f*–*h*) with aromatics that are strongly oxidizing agents in the excited state, such as aromatic nitriles.

Photoheterolysis of aryl halides occurs efficiently when a strong electron-donating group is present and irradiation is carried out in a polar medium [14]. The reaction has been developed for chloro and fluoro anilines and the preparatively interesting side is that cleavage occurs from the triplet state of the aromatic and gives a phenyl cation in the triplet state (see Scheme 4.3) [15–17]. This otherwise inaccessible species is characterized by a selective reactivity toward π, not *n*, nucleophiles, contrary to the unselective reactivity observed with the singlet phenyl cation. Thus,

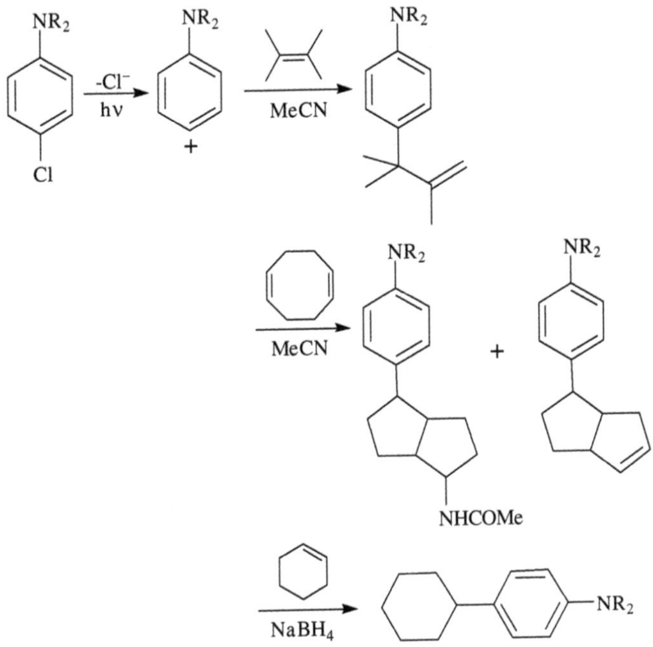

SCHEME 4.3

phenylation of alkenes, aromatics and electron-rich heterocycles takes place from fluoro and chloroaniline. When the irradiation is performed in alcohols, there is no competitive aryl ether formation. The new cation resulting from addition to the π nucleophile presumably undergoes ISC and the ensuing chemistry is that expected from an alkyl cation (deprotonation or elimination of other electrofugal groups, e.g., trialkylsilyl cation, nucleophile addition, Wagner–Meerwein rearrangement). Thus from 4-chloroaniline and alkenes 4-allyl-, 4-alkyl (in the presence of NaBH$_4$) [18], 4-(β-alkoxyalkyl)- (in the presence of alcohols), and 4-(β-aminoalkyl)anilines (in the presence of amines) [19] are formed. Reaction with cyclic dienes leads in some cases to intra-annular carbon-carbon bond formation [20]. Reactions with (silyl) enol ethers or ketene acetals gives α-arylated ketones and esters (see Scheme 4.4) [21]. 4-Phenylanilines are obtained by reaction with benzene and the 2-(4-aminophenyl) derivatives are obtained from the reaction with five-membered heterocycles with complete regioselectivity [22].

The requirement of an activating electron-donating group can be offset by the use of a better electrofugal group, the most typical case being that of diazonium salts. Photolysis here affords the convenient way for obtaining heterolysis of the C-X bond, free from the reductive path leading to phenyl radicals which is most often involved in thermal reactions (usually catalyzed by reducing salts). The photofragmentation can occur either from the singlet or from the triplet state, leading to the corresponding state of the phenyl cation. However, for substrates for which direct irradiation leads to the unselective singlet cation, photosensitization offers a way to arrive at the triplet state, which again shows the characteristic high selectivity for alkenes and arenes (see Scheme 4.5) [23].

Some phenyliodonium phenolates have been found to react photochemically with alkenes, alkynes, and arenes, via a formal nucleophilic substitution, though the reaction may involve the intermediacy of iodanes with subsequent phenyl iodide elimination [24,25].

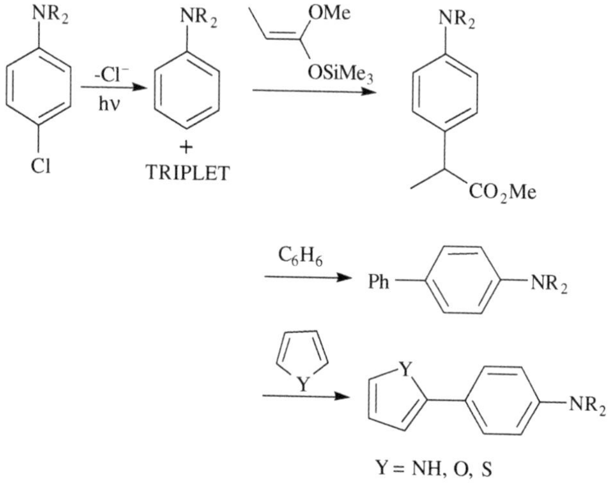

SCHEME 4.4

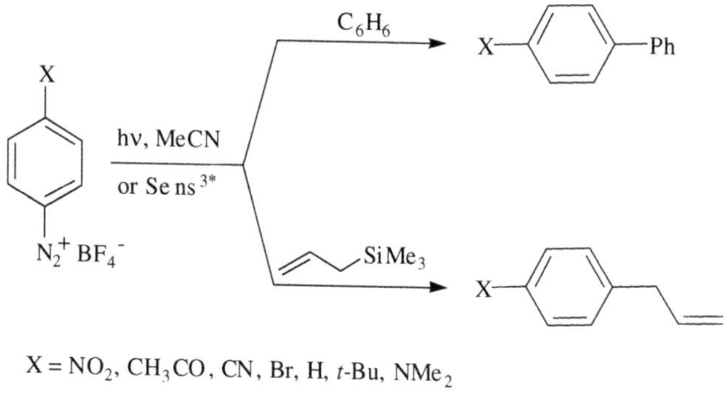

$X = NO_2, CH_3CO, CN, Br, H, t\text{-}Bu, NMe_2$

SCHEME 4.5

4.3.2 ALKYLATION OF ARYL HALIDES VIA THE $S_{RN}1$ PATH

Carbanions and enolates are good electron donors. Excitation of such anions in the presence of aromatics (in some cases, excitation of a preformed ground state complex between the two species) smoothly leads to monoelectronic reduction of the aromatic (path *e*). When this contains a good nucleofugal group fragmentation generates an aryl radical. This is the key step of the highly useful $S_{RN}1$ reaction, a chain process that is propagated via addition of the radical to the enolate and electron transfer from the latter to the starting aromatic compound reinitiating the cycle (see Scheme 4.6). It is important that addition to the enolate is fast (typical rate constant for ketone enolates, 10^8 to 10^9 M^{-1}s^{-1}) allowing to overcome hydrogen abstraction from the solvent that would lead to reductive dehalogenation (as an example, hydrogen transfer rate constant from dimethylsulfoxide to the phenyl radical, k_H 3×10^6 M^{-1}s^{-1}) [26–28]. The process, first reported in 1970 [29], has been extensively investigated and found to offer a significant path for carbon–carbon bond formation. Since the topic is adequately illustrated in several excellent reviews [30–34], only a brief discussion will be presented here.

Photoinitiation is not the only access to this chemistry, e.g., cathodic induced reduction or the use of alkali metals or other inorganic reducing reagents are also possible, but irradiation often is advantageous for preparative purposes. Since this is a chain process, the use of low-power lamps or a low quantum yield initiation step are not necessarily a limitation. Due to the requirement of a fast cleavage at the radical anion stage, aryl halides are by far the most used reagents, in particular iodides and, to a lower extent, bromides. Nucleophiles are carbanions from sufficiently acidic hydrocarbons, e.g., 1, 3-diphenylindane, fluorene or triphenylmethane [35–37] or, more commonly enolates from ketones [38], esters [39], *N,N*-dialkylamides [40], nitriles [41]. C-C bond formation is obtained also with phenoxide or naphthoxide anions [42,43]. A few representative examples of synthetic applications of the $S_{RN}1$

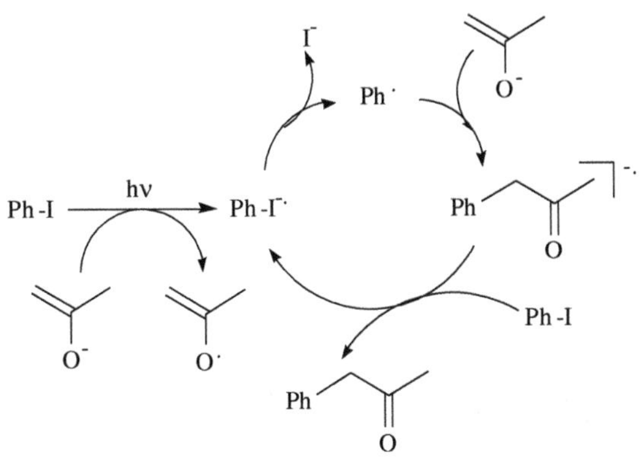

SCHEME 4.6

reaction are reported below (see Scheme 4.7) [44–47] including a tandem radicalic ring-closure – $S_{RN}1$ reaction (see Scheme 4.8) [48].

Nonchain reactions initiated by electron transfer to the aromatic and fragmentation of the radical anion have been reported for good acceptors, e.g., for some polyfluorinated

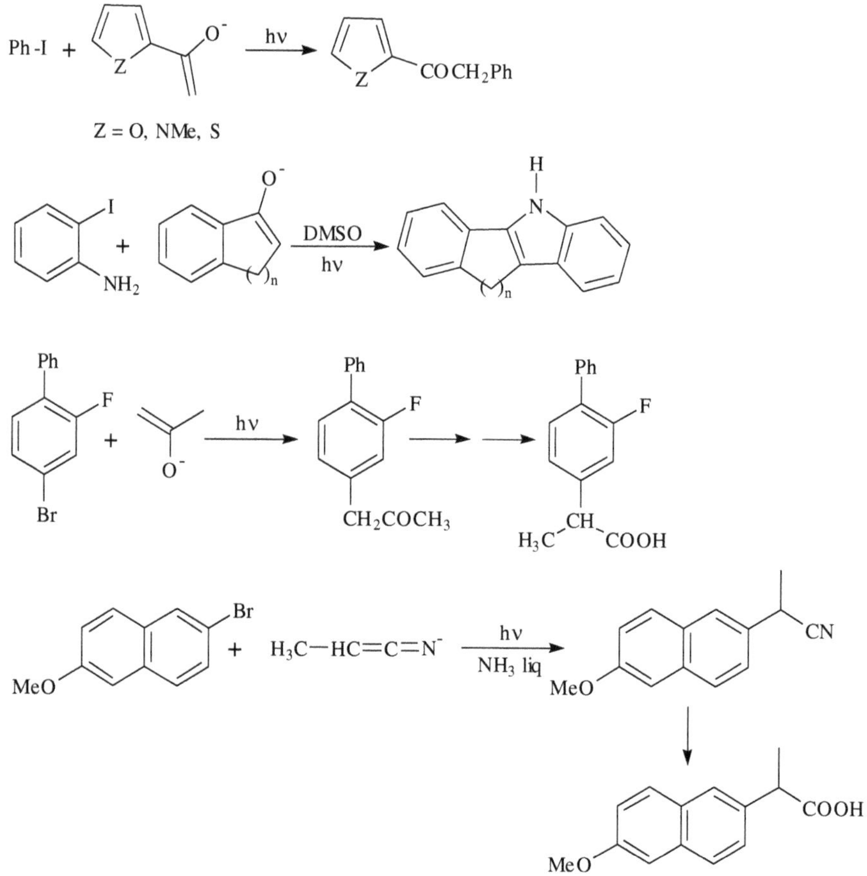

$Z = O, NMe, S$

SCHEME 4.7

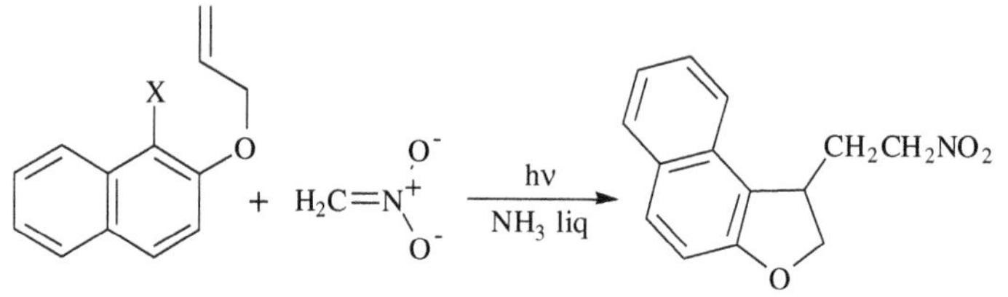

SCHEME 4.8

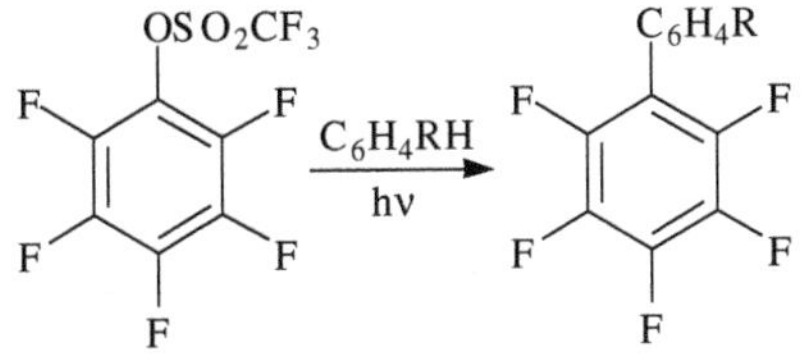

SCHEME 4.9

derivatives. Thus, pentafluorophenylsulfonates and iodide photochemically react with alkylbenzenes, phenyl ethers, and anilines giving the corresponding *ortho* and *para* pentafluorophenyl derivatives, possibly via electron transfer from such donors, fragmentation of the radical anion and coupling of the pentafluorophenyl radical with the cation radical of the donor (see Scheme 4.9) [49,50].

4.3.3 ALKYLATION OF ARYL NITRILES

As an alternative to the reactions presented in the previous sections, formation of a carbon–carbon bond may be based on a persistent aromatic radical anion and a fragmentable radical cation (paths *f–h*). The high energy of the singlet excited state of aromatic nitriles makes these excellent candidates for photoinduced electron transfer processes, greatly extending the range of oxidizable substrates. In this way the relatively stable, nonbasic radical anion of the nitrile is formed (these may be a persistent species observable for months under appropriate conditions) [51]. On the other hand, the simultaneously formed radical cation is a high energy species, since it arises from a weak donor, and this often opens a path to fragmentation. In this way a neutral radical is generated and coupling of the two odd-electron intermediates can result to form an anion according to path *f*. In turn, the anion may undergo either cyanide loss or protonation and elimination of HCN, so that the overall process is substitution of a cyano group.

As discussed in detail elsewhere, fragmentation of a σ bond (C-H, C-C, C-Si, C-Sn, etc., sometimes dubbed as *mesolytic fragmentation*) is a common process among radical cations and is a mild method for the generation of carbon-centered radicals [52–54]. Feasibility has been predicted through thermochemical cycles and the kinetics and mechanism have been studied in detail in a number of significant cases [55].

The reaction may be of some preparative interest for obtaining alkylbenzonitriles and various α-functionalized alkylbenzonitriles starting from polynitriles (see Scheme 4.10 and Scheme 4.11) [56,57]. Donors that can be conveniently used as the precursors of the radicals include π donors, such as alkenes [58,59] and alkyl aromatics [60–63], heteroatom-centered donors, such as carboxylic acids [64] and *tert*-butyl esters [65], ethers [66], ketals [67] (as well as cyclopropanone silyl ketals) [68] and amines, organometallic donors such as silanes, silyl ethers, and silyl amines [69–71] as well as germanes, stannanes, and borates [72].

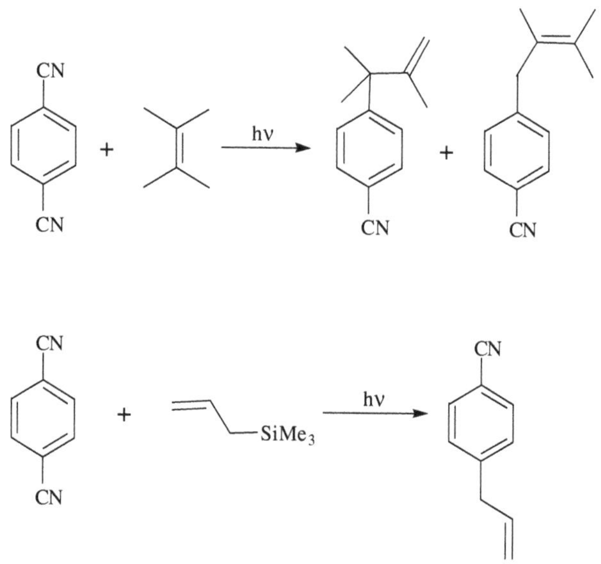

SCHEME 4.10

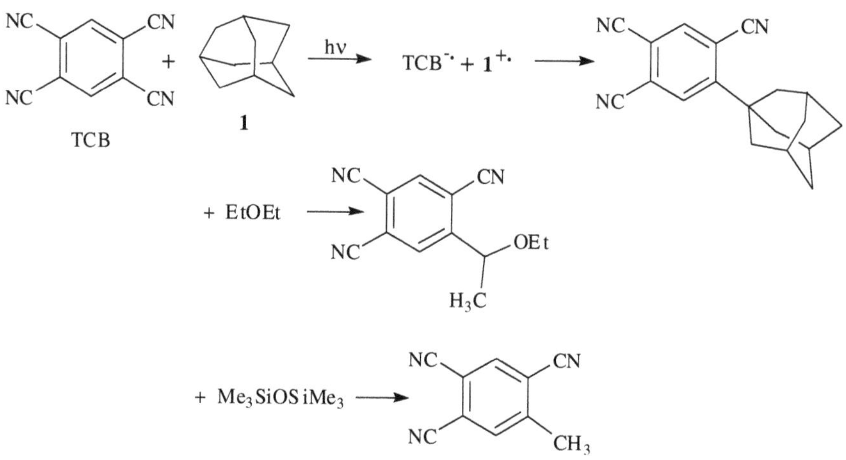

SCHEME 4.11

In particular, allyl derivatives of Si, Ge, and Sn give allylation efficiently [73–77]. The corresponding 5-hexenyl derivatives have been used as radical clock in order to obtain some information about the timing of the steps [69,78].

Even alkanes can be used for the alkylation reaction of strong acceptors such as 1,2,4,5-benzenetetracarbonitrile (TCB) and yield the corresponding alkylbenze netricarbonitriles, where the alkyl radical incorporated is formed with remarkable regioselectivity (e.g., the 1-adamantyl derivative is exclusively obtained from adamantane **1**, see Scheme 4.11) [79,80]. Likewise, acetonitrile can function as a donor, and tricyanophenylacetonitrile is formed by irradiation of tetracyano derivative TCB in this solvent [81].

The process may be more complicated in naphthalene(poly)nitriles and larger aromatics, since addition leading to dihydroaromatics or attack to a different position often accompanies substitution (see Scheme 4.12) [82–86].

When a good radical trap such as an electrophilic alkene is present, the radical may be captured and a more stable radical adduct be formed. At this stage, interaction with the persistent radical anion of the aromatic nitrile may lead either to electron transfer with regeneration of the nitrile, which in this way is not consumed and promotes the alkylation of the alkene functioning as electron accepting sensitizer, or to coupling [87]. The latter process predominates when electron transfer is endothermic and leads to a new photosubstitution process, where the cyano group is substituted by the adduct radical, leading to a highly functionalized product via a three component synthesis. This process is indicated as ROCAS, radical olefin combination aromatic substitution reaction, see path g in Scheme 4.2 and Scheme 4.13, where the alkyl radical arises from the fragmentation of the radical cation of ketal **2** [88].

Another three components synthesis is involved in the extensively investigated NOCAS process (nucleophile olefin combination aromatic substitution, path *h*) [89,90]. In this case, a nucleophile adds to an alkene radical cation and again the interaction between the resulting radical and the radical anion of the aromatic nitrile may follow two paths. The first is electron transfer, which results in sensitized anti-Markovnikov addition onto the alkene, and is favored with stabilized, reducible radicals such as the benzyl radicals obtained from aryl olefins. The latter one is

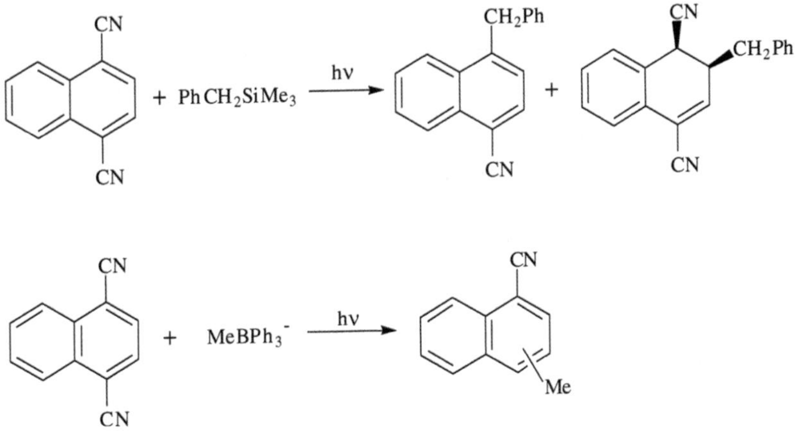

SCHEME 4.12

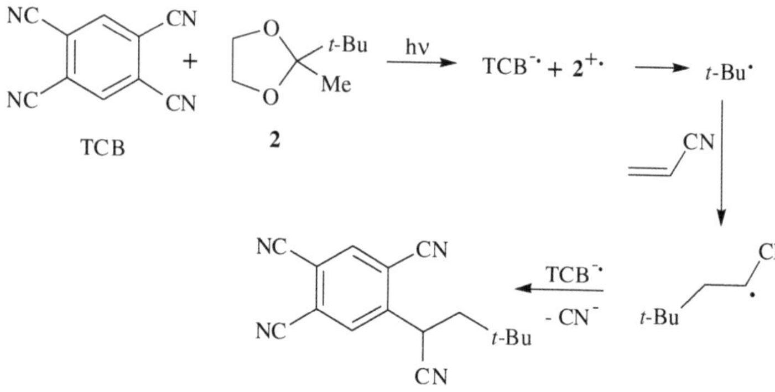

SCHEME 4.13

coupling and is favored with aliphatic alkenes. The two reactions are exemplified by the NOCAS process with dicyanobenzene, 2-methylpropene and methanol [58,91] and by the 1,4-naphthalenedicarbonitrile sensitized anti-Markovnikov methanol addition onto 1,1-diphenylethylene [92,93]. A further possibility is that the radical formed by methanol addition is trapped by an electrophilic alkene, resulting in photosensitized β–alkoxyalkylation of the trap, as also demonstrated) [94].

The detailed study of the reaction makes it useful since it allows predicting its course. Thus the regiochemistry of nucleophile addition onto the alkene radical anion depends on the nature of the former reagent. An illustrative example is offered by the above quoted reaction of 1,4-dicyanobenzene with 2-methylpropene (**3**). When the reaction is carried out in methanol, the alcohol forms a bridged adduct and offers a mechanism for the equilibration of the regioisomeric radicals [91,95]. Thus, the main product results from the attack by the most stable radical (see Scheme 4.14). On the other hand, when a charged nucleophile is used, e.g., fluoride [96] or cyanide [97], no bridged intermediate is involved and the regiochemistry is controlled by the rate of addition of the anion, so that the main product arises from the least substituted radical. Yet a different course is followed in neat acetonitrile. In most cases, the alkene radical cation deprotonates in this solvent, but with 2-methylpropene the solvent acts as neutral, nonprotic nucleophile and traps it. Coupling of the resulting distonic radical cation with the dicyanobenzene radical anion is followed by electrophilic attack by the nitrilium moiety leading to a dihydroisoquinoline (a dihydroisoquinoline has been similarly obtained from tetracyanobenzene and 1-hexene in acetonitrile) [59,98].

The NOCAS process has been extended to conjugated [99] and nonconjugated dienes (in the latter case a ring may be formed, see Scheme 4.15) [100,101], to alkenols (where the intramolecular trapping by the hydroxyl group replaces intermolecular alcohol addition) [102,103] and to allenes [104].

Strained alkenes have been observed to undergo carbon–carbon bond cleavage at the radical cation stage, as shown in the case of α-carene where the four-membered ring is cleaved in the NOCAS products obtained (see Scheme 4.16). Cleavage of a

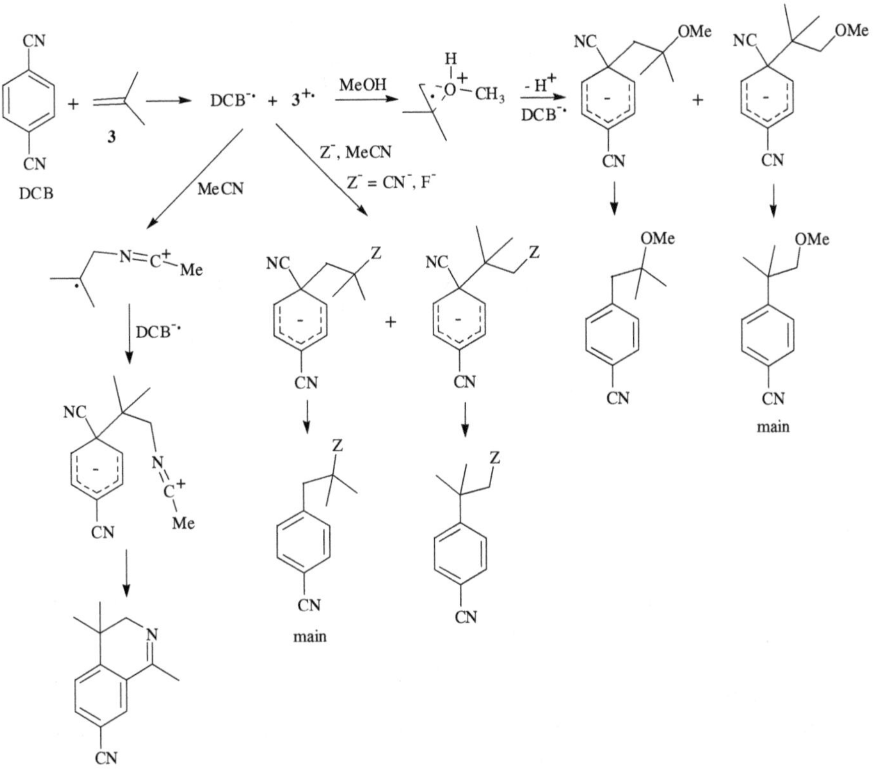

SCHEME 4.14

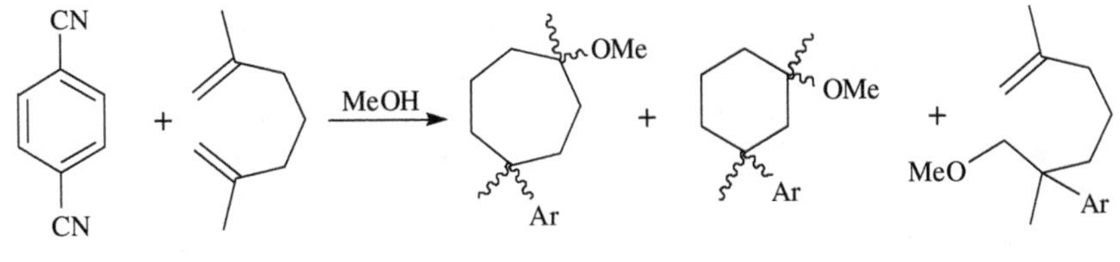

Ar = p-CNC$_6$H$_4$

SCHEME 4.15

strained donor and a NOCAS-type photosubstitution process has been obtained also from a saturated donor such as tricyclene (Scheme 4.16) [105,106].

As it is apparent from the foregoing, introducing C-centered substituents in aryl derivatives via photosubstitution reactions really offers a large choice of synthetically appealing methods. In many respects, the S_N1 reaction is complementary to metal

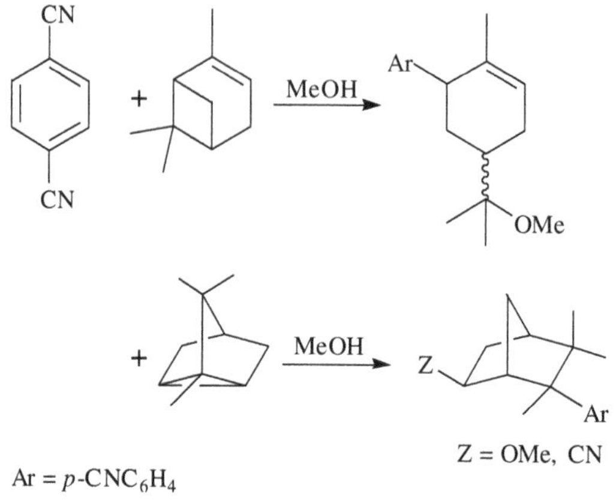

Ar = p-CNC$_6$H$_4$

SCHEME 4.16

catalyzed reactions, e.g., in terms of resistance of the substituents to the reaction conditions or of their activating/deactivating effect. The S$_N$1 alkylation/arylation of haloanilines or of diazonium salts and the various alkylations of aromatic nitriles have no parallel in thermal chemistry and should elicit the attention of synthetic chemists, taking also into account that the mild conditions and relative independence on the medium of the photochemical production of highly reactive intermediates is not only per se an obvious advantage, but further allows for a quite free choice of conditions in view of directing the subsequent chemistry of such intermediates.

4.3.4 CYANATION AND CARBOXYLATION

Benzonitriles are of considerable interest for organic chemistry as a structural motif present in dyes, herbicides, agrochemical, pharmaceuticals, and natural products. Recently there has been an increased interest in their synthesis [107]. The most employed thermal methods for the synthesis of aryl cyanides involve the use of organometallic species as in the case of the palladium catalysed cyanation of aromatic halides [107] or the well known copper mediated Rosenmund–von Braun [108] and Sandmeyer reactions respectively from aryl halides and aryldiazonium salts. Photosubstitution reactions have a role too and in some cases afford an easy way for preparing aromatic nitriles. Although fluorobenzene and anisole irradiated in methanol in the presence of cyanide yielded only 3% of benzonitrile [109], highly substituted aromatics gave better yield. As an example, 3,4-dimethoxy-1-nitrobenzene underwent substitution of the methoxy at the meta position with respect to the nitro group when photolyzed in a t-BuOH-water solution in the presence of KCN. The quantum yield dramatically grew when increasing the t-BuOH content in the solvent mixture (from 0.04 to 0.68 in passing from neat water to 88% t-BuOH) [110]. Water was again detrimental in the photoreaction of 4-methoxy-1-nitronaphthalene and the quantum yield of photosubstitution affording 4-methoxy-1-cyanonaphthalene decreased at a large water content. The substitution products were obtained in 70% yield in a 95:5 MeCN/water mixture [111]. Similar yields were found with 4-nitrobiphenyl, 4,4′-dinitrobiphenyl, 2-nitro- and 2,7-dinitrofluorene. All of these underwent substitution of one (or both) of the nitro groups in t-BuOH-water solutions of KCN [112]. 2-Cyano-4-nitrobiphenyl was likewise synthesized from 2-methoxy-4-nitrobiphenyl in 68.5% yield [112]. Photolysis of 3,4-dimethoxybenzonitrile and 3,4-dimethoxyacetophenone in the presence of cyanide ion gave a marked preference for the substitution in the 4-position in the former case, whereas in the latter a substitution ratio C-3/C-4 3:2 was observed (see Scheme 4.17) [113].

Cyanoanisoles were formed in a quite high yield from the corresponding chloro and fluoroanisoles in t-BuOH-water and in lower yields from the bromides and iodides where reduction competed and actually took place with the same yield as substitution. In halodimethoxybenzenes, the methoxy group exerted an activating and ortho/para orienting effect on the substitution [114]. The reaction was thought to involve a radical cation arising from electron ejection from the triplet excited state. Furthermore, the photocyanation of 2-fluoroanisoles was greatly reduced by the interaction with cyclodextrins (especially by α-cyclodextrin) [115] but replacing such complexing agents by t-BuOH photosubstitution of the fluoride on 2- and 4-fluoroanisoles occurred, though substitution to yield cyanoanisoles was accompanied by photohydroxylation [116].

2-Cyano-4-nitroanisole (5) was obtained in 46% yield from nitroveratrole in a MeCN-water mixture in the presence of sodium cyanide. Surprisingly, the same product was formed when the reaction was carried out in the absence of cyanide but in the presence of tetrabutylammonium fluoride (TBAF) in air equilibrated solution;

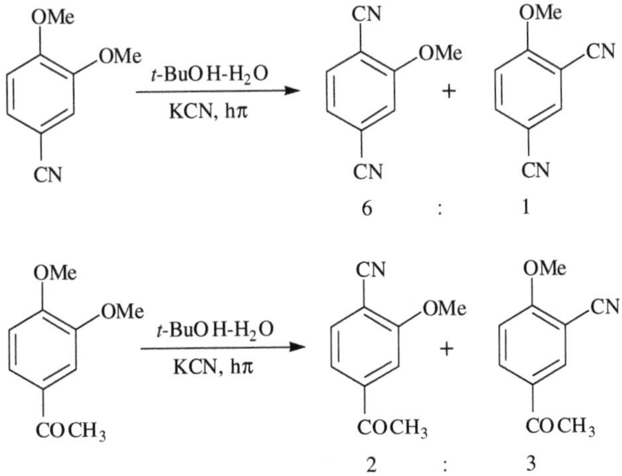

SCHEME 4.17

the yield was largely increased (from 17 to 67%) when the solution was air-bubbled [117]. The Authors suggested a mechanism involving the superoxide anion as depicted in Scheme 4.18.

In aromatic sulfones, sulfonamides, and sulfoxides with electron-donating substituents in the *ortho*, *meta*, and *para* position, the sulfur-containing group was displaced by a cyano group on irradiation in a *t*-BuOH/aqueous solution of KCN (see Scheme 4.19). The yields of photosubstitution products decreased on passing from aryl sulfones to aryl sulfoxides and followed the order *para* > *ortho* > *meta*. [118–120] Analogously, when dichloroanilines were used as the substrates, the corresponding dicyano derivatives were obtained [120].

Cyanoaromatics were formed in more than 70% yield under phase transfer catalysis conditions when starting from nitroaromatics in a 2M KCN biphasic water-methylene chloride system in the presence of a catalytic amount of $Bu_4N^+CN^-$; this

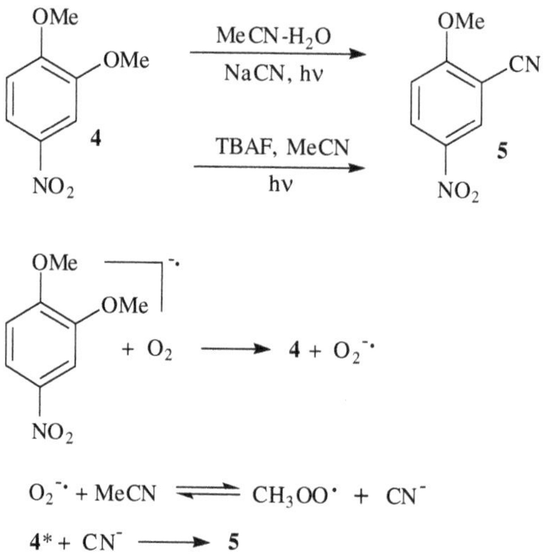

SCHEME 4.18

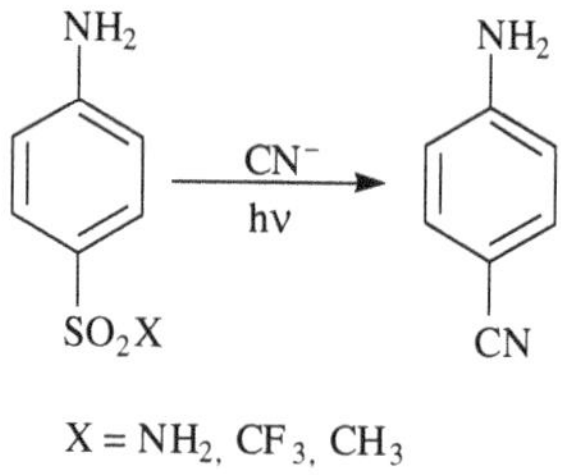

$X = NH_2, CF_3, CH_3$

SCHEME 4.19

procedure was found very convenient thanks to the easy work up [121]. The use of polyethylene glycol as co-solvent in CH_2Cl_2 rather than adding the usual crown ether as the phase transfer catalyst, afforded a successful photocyanation of o- and p-dimethoxybenzenes in aprotic solvent yielding cyanoanisoles [122].

Sequential photocyanation has been achieved in aqueous MeCN for a variety of polyhaloaromatic compounds, including polychlorinated benzenes, biphenyls, naphthalenes, and dioxins. These reactions involve the triplet state and show a low degree of regioselectivity. The mechanism was rationalized as involving a S_N2Ar* path with an autoacceleration of the reaction due to the higher susceptibility to further nucleophilic substitution of the products formed in each step (see Scheme 4.20) [123,124].

Although the cyanide ion is one of the less reactive nucleophiles toward the $S_{RN}1$ process some examples of this class have been reported starting from diazosulfide derivatives [31,125–128]. Tetrabutylammonium cyanide has been used in excess to give photosubstitution in satisfactory yields even if some arylthioethers were obtained along with the nitriles (see Scheme 4.21). Noteworthy, when the aryldiazo precursor bore a chlorine or bromine atom on the aromatic ring dicyano derivatives were formed [126,127].

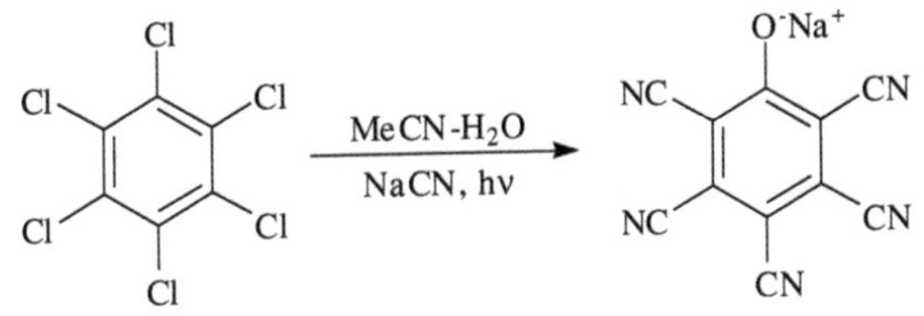

SCHEME 4.20

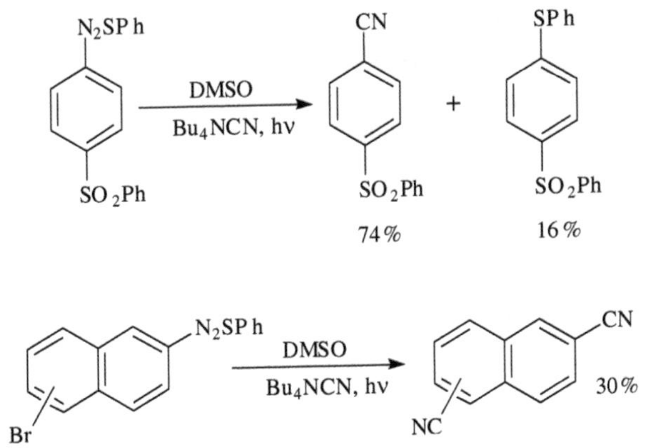

SCHEME 4.21

On the basis of the above presentation, photocyanation appears a scarcely explored method for the introduction of a cyano group in an aromatic ring, despite the fact that the yields were in some cases good and there was an interesting variety of leaving groups that could be displaced in the reaction. The mildness of the conditions employed (no high temperature was required) added to the use of water as the solvent or cosolvent suggests that the process is worth reexamination. A better rationalization of the selectivity would allow a wider application of the method.

Carboxylation is another reaction that can be obtained via the $S_{RN}1$ path. Thus phenyl or naphthyl iodides or bromides, when irradiated in the presence of cobalt diacetate or dicobalt octacarbonyl, carbon monoxide and bases yield carboxylic acids in quite useful yields [129] (the reaction could be conveniently carried out also under phase transfer conditions) [130]. In the presence of alcohols or amine, esters or amides are formed [131a]. Aryl chlorides also react, more favorably by increasing the CO pressure to 2 atm [131b].

4.3.5 FORMATION OF A C-Si BOND

Very few examples were described in this class and the formation of the aryl-silicon bond was obtained analogously to the alkylation of aryl nitriles presented in Section 4.3.3 by ipso-substitution reaction induced by an initial electron transfer process. Thus, tetracyanobenzene in the presence of hexamethyldisilane was in part silylated as illustrated in Scheme 4.22 [132].

4.3.6 FORMATION OF A C-Sn BOND

Photostimulated $S_{RN}1$ reaction is an important tool for the obtainment of aryl-organometallic species (see also Sections 4.4.4 and 4.5.2) [31]. Trimethylstannyl and triphenylstannyl groups were easily linked to an aromatic moiety starting from the corresponding anions and the aromatic chlorides; high yields were obtained [133–137]. Noteworthy, when the reaction conditions were changed and aryl bromides and iodides rather than chlorides were used, a halogen-metal exchange reaction (HME) took place and dehalogenated products were formed in the reaction [133–137]. The last path was attributed to the formation of Ph_3SnX as the primary step, followed by protonation of the anion thus formed by ammonia (see Scheme 4.23) [133].

Interesting is the case of dichlorobenzene in liquid ammonia, since this compound readily afforded the disubstitution product in 88% yield in the presence of Me_3Sn^-.

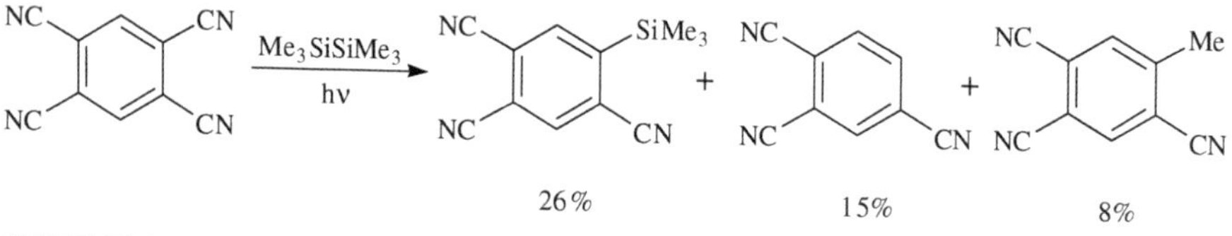

26%   15%   8%

SCHEME 4.22

The product obtained was then treated in situ by Na and t-BuOH generating dianion (6) that, in turn, could undergo a second photostimulated arylation yielding end product (7) in a one-pot process (see Scheme 4.24) [135]. Furthermore, in HMPA as the solvent, the $S_{RN}1$ reaction took place only when photostimulated [136].

Recently, arylstannanes were prepared both from phenols via the corresponding aryl diethyl phosphate intermediates [138,139] and from the trimethylammonium salts of arylamines [140]. Phosphate anion and trimethylamine were used in the place of halide anions as the leaving groups in the reaction. Interestingly, when haloanilines or halophenol derivatives were used as the precursors, disubstitution products were obtained as illustrated in Scheme 4.25 starting from a phosphate or an anilinium salt [138–140].

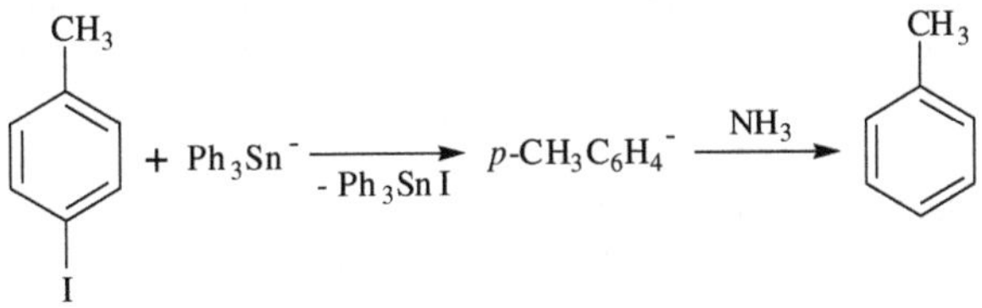

SCHEME 4.23

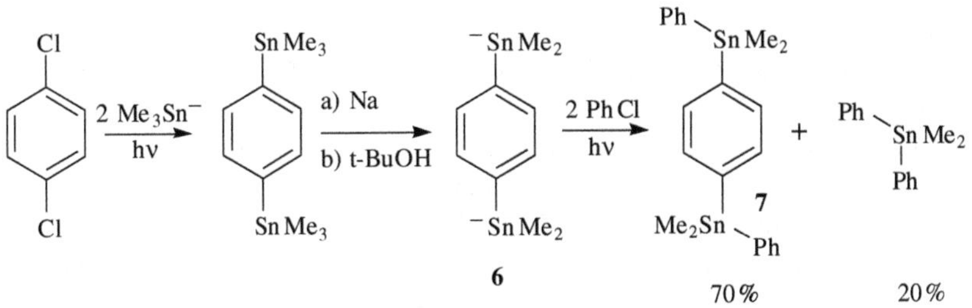

SCHEME 4.24

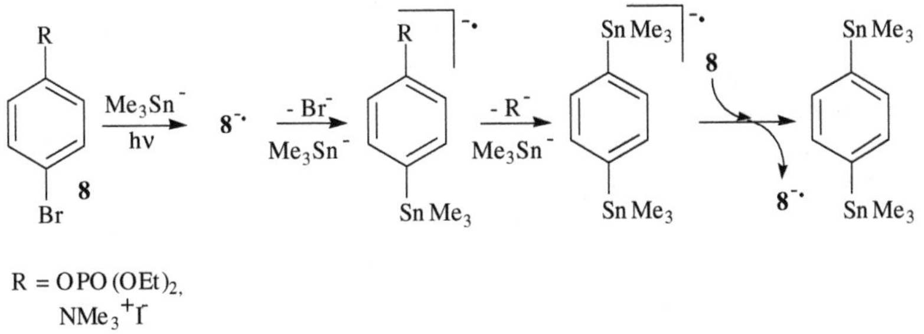

$R = OPO(OEt)_2,$
$NMe_3{}^+I^-$

SCHEME 4.25

4.4 FORMATION OF A C-V GROUP ELEMENTS BOND

4.4.1 FORMATION OF AMINES AND IMINES

An important part of the intramolecular nucleophilic photosubstitutions concerns the photo-Smiles rearrangements [141]. The irradiation of various aryl-ω-aminoalkylethers gave the corresponding substituted anilines. The reaction was largely studied especially from the mechanistic point of view. As an example, 2-(3-nitrophenoxy)ethylamine photorearranged to 2-(3-nitrophenylamino)ethanol in aqueous medium [142]. The authors suggested the involvement of the triplet as the reactive excited state and the ensuing intermediacy of a spiro σ complex. Furthermore, it was apparent that the presence of a base catalyzed the reaction [142]. Unfortunately, the process lacked generality since the corresponding 2-nitro and 4-nitro derivatives were in practice unreactive under the same reaction conditions and the low yield of the rearranged product actually observed were attributed to a thermal contribution [143,144]. This behavior was explained on the basis of the energy gap model [145]. In this model, the size of the energy gap between the excited state nucleophile encounter complex and the ground-state σ complex controlled the process. Thus, the smaller was the gap, the more favored was the reaction and this held in this case for the meta precursor rather than the other two isomers. A different mechanism was envisaged in the case of photolysis of 1-(*p*-nitrophenoxy)-ω-anilinoalkanes (**9**), which yielded *N*-(*p*-nitrophenyl)-ω-anilinoalkanols (**10**) [146–147]. The relative reaction rates in acetonitrile showed an unexpected order, n = 4 > 5 > 3 > 2, which was reversed in the presence of a base. The mechanism proposed is shown in Scheme 4.26 and involves the formation of a radical ion pair followed by a spiro Meisenheimer type complex. The effect of the base on the reaction and the fact that in apolar solvent the photorearrangement did not occur supported the partially ionic character of the mechanism [146–147]. The presence of a methoxy group, a good leaving group in nucleophilic substitution reactions, in position 2 did not alter the course of the reaction, in contrast of the *meta*-directing effect by the nitro group on photosubstitution reactions observed in nitroanisoles and nitroveratroles (see Section 4.7) [148–149].

In any case the presence of the nitro group played a crucial role on the reaction since *N*-[2-(2-methoxy-5-nitrophenoxy)ethyl]aniline (**11**) underwent photosubstitution of the methoxy group *para* to it (see Scheme 4.26) ([149–150]; the orientation rules in the substitution reactions of nitro containing aromatics were discussed in ref. 150). The effect of a base on the reaction was investigated by using ten amines as bases. It was proposed that the quenching mechanism by which these acted involved electron transfer in the case of tertiary amines and a hydrogen abstraction-electron transfer sequence in the case of primary amines [151]. Moreover, the effect of the chain length on the photochemistry of compounds (**9**) (n = 2 to 10, 12, 16) was studied by Mutai et al. These authors supported a change of the mechanism between the lower homologues (n = 2 to 6) in which a rearrangement via nucleophilic attack

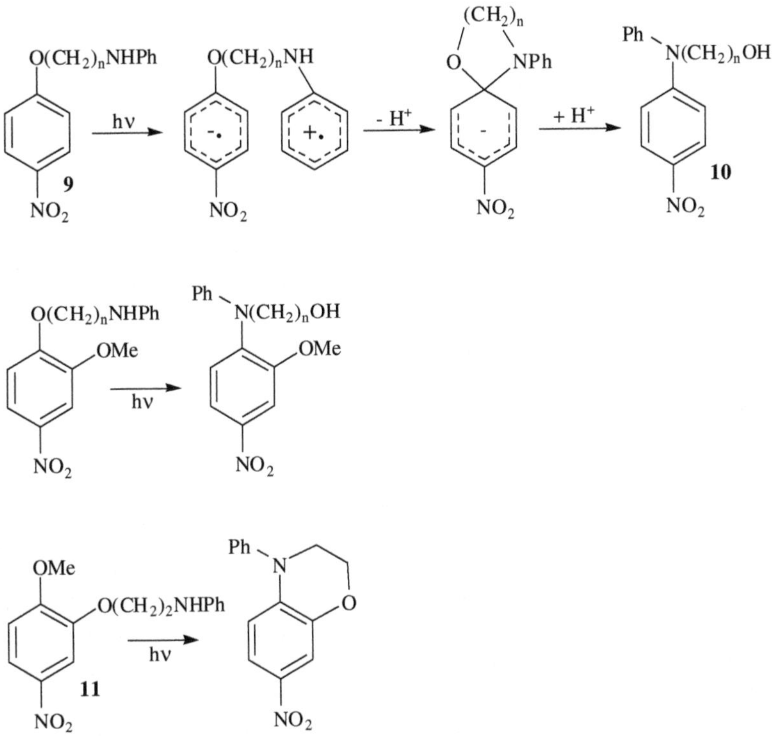

SCHEME 4.26

took place and the higher ones (n = 8) where an intramolecular photoredox reaction accompanied by C-N bond cleavage occurred [152].

Most of the work concerning the formation of aryl-nitrogen bond in photosubstitution reactions has been based on the use of amines as nucleophiles. Interest in this topic arises also from the search of suitable substrates that could behave as photoaffinity labelling agents in order to analyze receptor sites on biological macromolecules. Usually arylazides and diazo compounds have been employed for this purpose but Marquet and coworkers made a systematic work on nitrophenyl ethers in order to evaluate the possibility that such substrates undergo selective photosubstitution reactions. The first precursor considered was 4-nitroanisole [153]; this molecule in the presence of various primary and secondary amines underwent nucleophilic photosubstitution both of the nitro and of the methoxy group. Generally, replacement of the methoxy group rather than of the nitro was preferred and the latter group was substituted only when ammonia and ethylglicinate were used as nucleophiles [153]. Further experiments evidenced that when the nitro group was replaced a S_N2Ar^{3*} mechanism operates, whereas the methoxy group was substituted through a radical ion pair intermediate [154–155], see Scheme 4.27 (compare Scheme 4.2, paths c, d).

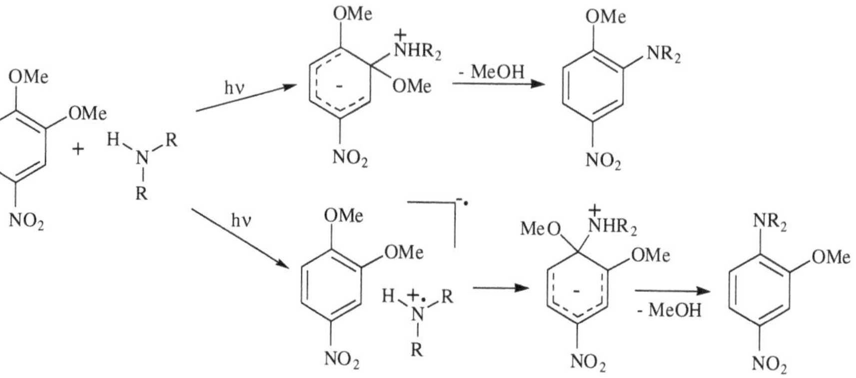

SCHEME 4.27

Irradiation of 4-nitroveratrole under the same conditions caused no substitution of the nitro group. However, this meant no improvement of the selectivity, due to the competitive replacement by an amino group of either of the methoxy groups [153]. Nevertheless, a modified nitroveratrole derivative of the antibiotic cycloheximide (**12**) gave, among further minor products, compounds (**13**) in a regioselective fashion and in satisfactory yield when **12** was photolyzed in the presence of methylamine (see Scheme 4.28) [156].

Methylamine showed an analogous selectivity in the reaction both of various 3,4-dialkyloxy derivatives [156] and of *O*- and *N*-(2-methoxy-4-nitro)phenoxyalkyl derivatives of estrone [157]. Higher aliphatic amines exhibited poor regioselectivity as reported for the reaction of 4-nitroveratrole in aqueous butylamine. The selectivity of substitution was highly dependent on the pH used and increased in highly basic solution in favour of the *meta*-nitroaniline, although a nonnegligible amount of photohydrolysis products was also detected (see Scheme 4.29) [158]. These results led to the conclusion that such precursors were not suitable as a lysine-directed photoaffinity probe. The low selectivity was also confirmed in the reaction with hexylamine where a different mechanism of formation was proposed for each one of the possible *N*-hexylanilines.

Thus, in the case of 4-nitroveratrole a S_N2Ar^* pathway was invoked for the formation of the *meta* substituted derivative, whereas the *para* analogues arose from a radical ion pair via electron transfer from the amine to the triplet excited state [159]. The last mechanism was also involved for the selective *para* substitution observed with piperidine [160]. Further evidence on the mechanism was obtained by laser kinetic spectroscopy and by measurement of transient charges in electrical conductivity [161–162].

A dual pathway was also confirmed by steady state and time-resolved experiments for the irradiation of 1-methoxy-4-nitronaphthalene in the presence of amines. In each case, photosubstitution was demonstrated to arise from the triplet excited state,

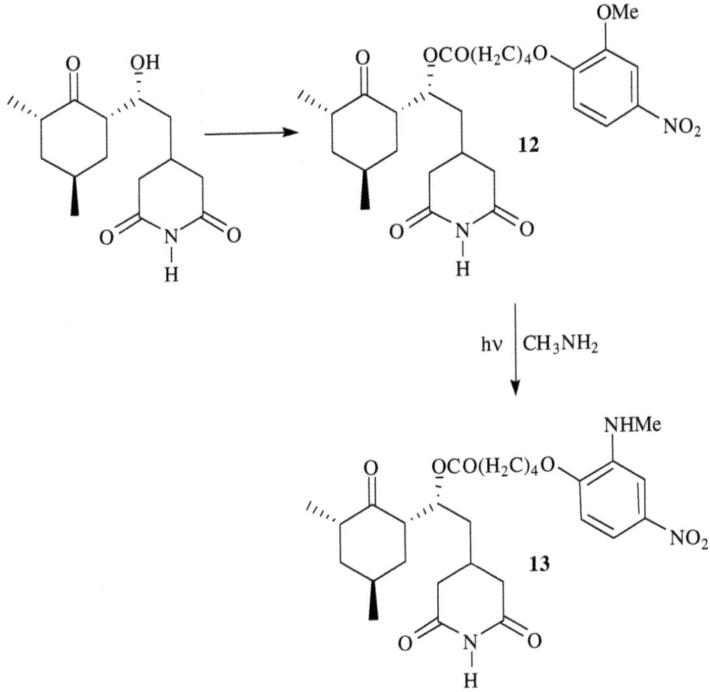

SCHEME 4.28

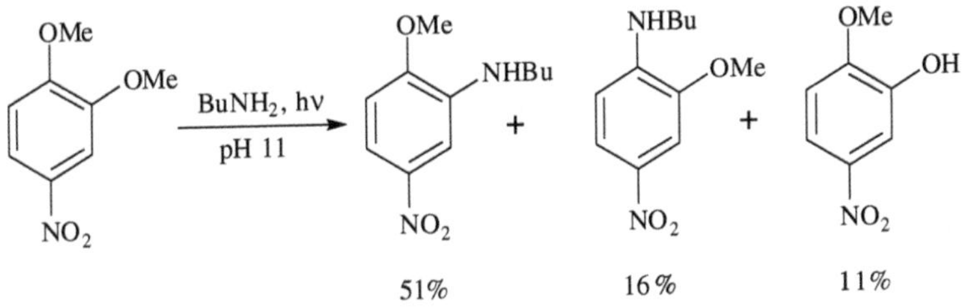

51%   16%   11%

SCHEME 4.29

but primary amines caused the replacement of the nitro group by a S_N2Ar^* process, whereas secondary amines displaced the methoxy group by an electron-transfer process [163].

Other candidates as possible photoaffinity probes were studied. As an example, when using N-butyl-3,4-dimethoxy-6-nitrobenzamide an insignificant decrease of the overall reactivity occurred with respect to 4-nitroveratrole. However, the degree of the selectivity in the substitution of one of the methoxy groups was unsatisfactory

[164]. Adding a nitro group, as in the case of 4,5-dinitroveratrole, photosubstitution occurred as well, but only when using amines with a relatively high ionization potential. On the contrary, photoreduction of the nitro group took place with easily oxidized amines such as dimethyl- or triethylamine [165].

More recently fluorinated aromatic nitroethers were considered as well. As an example, in the photoreaction of 2-fluoro-4-nitroanisole (14) there were no significant changes either on the selectivity of the position substituted or on the mechanism involved [166–169]. Nevertheless, compound 14 showed a higher and broader spectrum of photoreactivity toward the nucleophiles present in the different amino acid residues of RNAase A [169]. Imidazole used as model nucleophile in similar experiments was able to replace both the nitro and the fluoro group, with a preference for the former one, leading to the formation of an aryl-nitrogen bond (see Scheme 4.30) [169]. Amino derivative 15 was thought to be formed by photodegradation of the imidazolyl photoproduct (16).

The competition of reduction of the nitro group with photosubstitution is one of the main drawbacks in the photolysis of a biological photoprobe. This deleterious side reaction can be in part overcome by increasing the number of fluorine atom in the aromatic ring, as evidenced in the photoreactions of 2,6-difluoro-4-nitroanisole [170]. Investigations were also carried out on pentafluoronitrobenzene, though in this case the reaction was for the largest part thermal [171].

Substrate inactivated toward photosubstitution, viz. monosubstituted arenes such as fluoro- and chlorobenzene, anisole, benzonitrile, and related compounds,

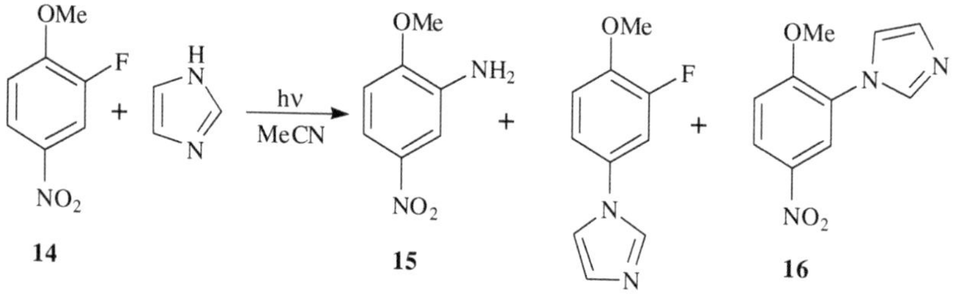

SCHEME 4.30

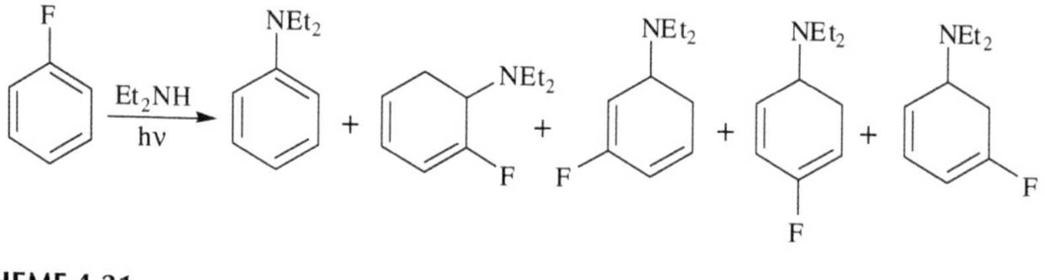

SCHEME 4.31

gave a complex mixture of photoproducts when irradiated in the presence of amines including small amounts of photosubstitution and addition products [172–174] as illustrated in the photoreaction of fluorobenzene in the presence of diethylamine (see Scheme 4.31). A complex mixture was formed analogously in the photolysis of 1-fluoro-3,5-dimethylbenzene [175].

The photolysis of 4-chlorobiphenyl and diisopropylamine in MeCN gave likewise a poor yield of the corresponding aniline [176]. Replacing the nitro group with a phosphate induced an improvement of the yield of photosubstitution. This was observed in the case of 4-nitrophenyl phosphate, where 1-arylpyridinium salts were formed upon photolysis in an aqueous solution of pyridine (see Scheme 4.32). Under the same conditions p-dinitrobenzene failed to react, however [177].

Despite the drawbacks often encountered in photosubstitution reactions, the synthesis of 1-methoxy-4-methylamino-9,10-anthraquinone was successfully carried out starting from the corresponding dimethoxy analogue in more than 90% yield [178].

Only few examples are present in the literature on aryl-N bond formation through the $S_{RN}1$ mechanism. 2-Bromomesitylene has been found to undergo bromine substitution by amide anion affording 2-aminomesitylene in 70% yield, with only a small amount of dehalogenated derivative (see Scheme 4.33). The success of the reaction in this case was probably due to the precluded formation of benzyne intermediates from this reagent [179].

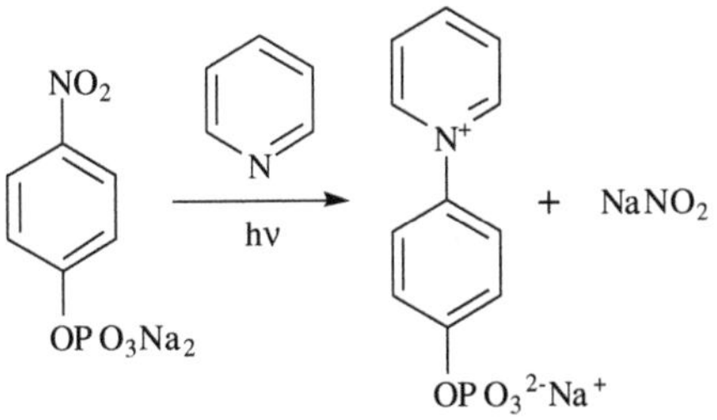

SCHEME 4.32

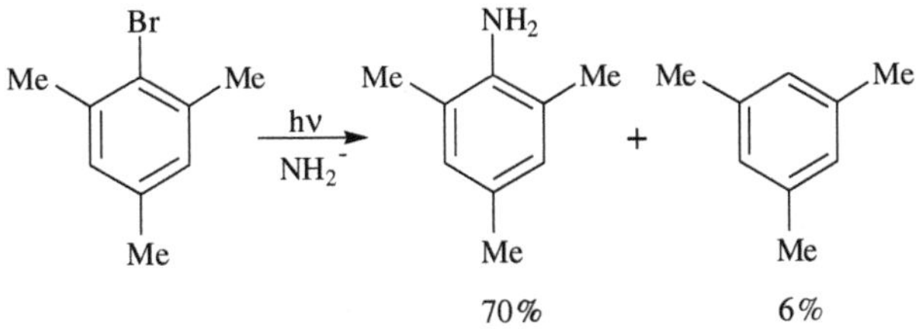

SCHEME 4.33

4.4.2 FORMATION OF NITRO COMPOUNDS

Nitroaromatics can be obtained, though only in a low to moderate yield, in the photolysis of haloaromatics bearing electron-donor substituents [180–183]. Haloanilines underwent conversion to nitroanilines upon irradiation in a 4% sodium nitrite aqueous solution. Better yields were obtained with chloroderivatives especially when the halogen was present in the *para* position with respect to the amino group. The same held for halophenols (*p*-nitrophenol was isolated in 45% yield), whereas a negligible amount of photoreplacement of the halo group took place in the case of *p*-chloroanisole (see Scheme 4.34). Noteworthy, carrying out the reaction in aqueous media avoided further reduction of the nitrite ion to ammonia observed in an alcoholic environment [180–183].

4.4.3 FORMATION OF A C-P BOND

Photolysis of 1,2-dichlorobenzene in trimethylphosphite at 60°C for 5 days gave 1,2-bis(dimethoxyphosphoryl)benzene (**17**) in 50% yield along with the monosubstitution product in 20% yield [184] (see Scheme 4.35). The same reaction, when applied to the case of 1-iodo-2-bromobenzene, led to disubstitution affording product **17** in 87% yield [185].

A more versatile method for the introduction of the phosphorus-linked group is the $S_{RN}1$ reaction. In fact, various mono- and dihalobenzenes underwent photoinduced substitution under $S_{RN}1$ conditions in the presence of the diethylphosphite anion [186–190] (see Scheme 4.36).

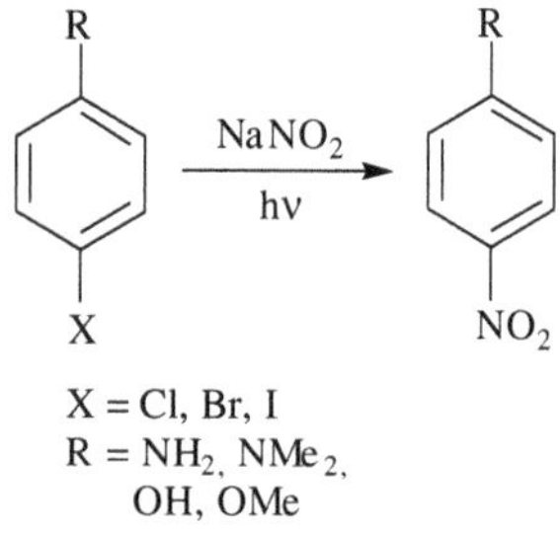

$$X = Cl, Br, I$$
$$R = NH_2, NMe_2,$$
$$OH, OMe$$

SCHEME 4.34

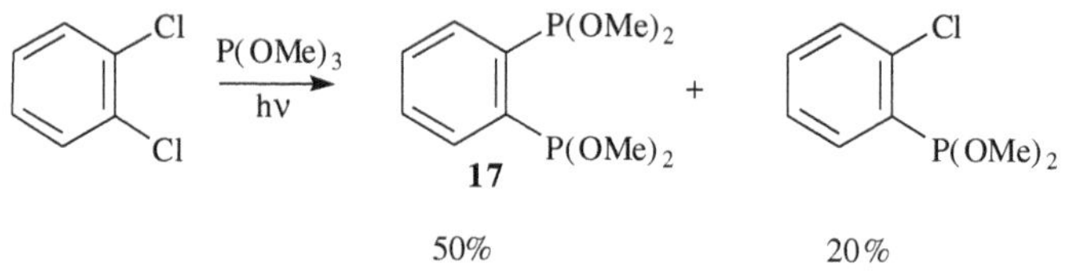

SCHEME 4.35

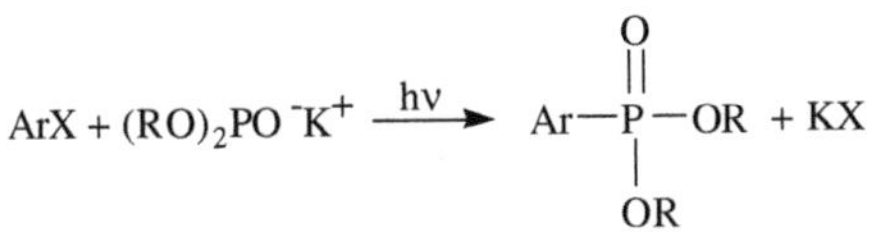

$$\text{ArX} + (\text{RO})_2\text{PO}^-\text{K}^+ \xrightarrow{h\nu} \text{Ar} - \overset{\overset{\displaystyle O}{\|}}{\underset{\underset{\displaystyle OR}{|}}{P}} - OR + KX$$

SCHEME 4.36

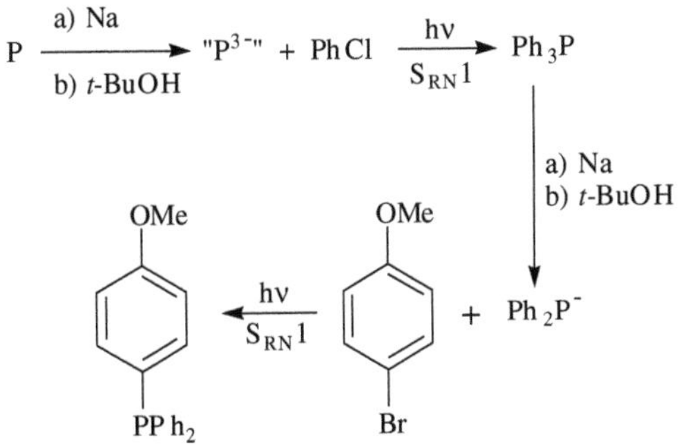

SCHEME 4.37

Iodoaromatics (e.g., iodoanisoles and iodotoluenes) reacted smoothly under such conditions and gave the desired photoproducts in a high yield. The possibility of extending the reaction to aryldihalides and the formation of mono or bisadducts depended on the halogen and on the nucleophile used. The weaker was the aryl-halogen bond, the higher was the yield of photosubstitution [186–190]. Several phosphoanion nucleophiles can be employed. As an example, PhP(OBu)O⁻, (EtO)₂PS⁻, and (Me₂N)₂PO⁻ reacted photochemically both with bromo- and with iodobenzene to give respectively Ph₂P(OBu)O, PhPS(EtO)₂, and PhPO(NMe₂)₂ [190]. In some cases dark reactions occurred as well, but with a different course [189]. The diphenyl- and dibenzylphosphinite anions were used successfully for the synthesis of aryldiphenyl- and aryldibenzylphosphine oxides under photostimulated conditions and by using aryl halides as the reagents [191]. Another route to symmetrical and unsymmetrical triarylphosphine started directly from elemental phosphorous through a "P³⁻" species generated with sodium in liquid ammonia [192]. t-BuOH was added after each reduction step in order to neutralize any amide ion eventually present that could interfere with the $S_{RN}1$ step (see Scheme 4.37).

The synthesis of 5-membered heterocycles containing phosphorus was achieved through the intermediate formation of an arylphosphonate through a $S_{RN}1$ step [193–195]. Thus, bromoaniline (**18**) was transformed into the benzazaphosphole derivative **19** as illustrated in Scheme 4.38 [194].

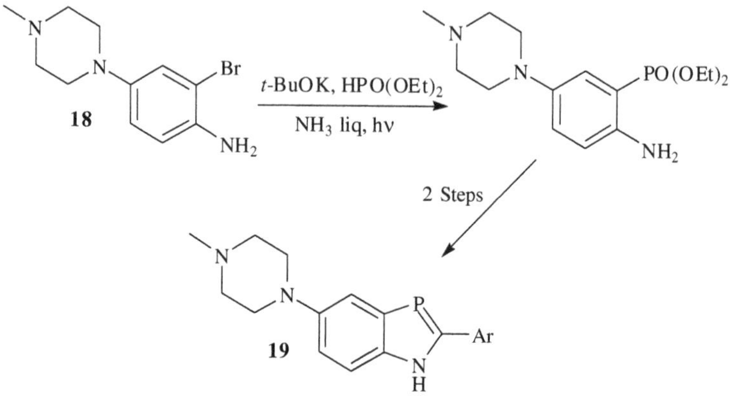

SCHEME 4.38

4.4.4 Formation of a C-As and C-Sb Bond

Potassium diphenylarsenide and diphenylstibide reacted with haloaromatics under photostimulation following the $S_{RN}1$ mechanism, though the process was rather unselective since respectively four arsines and four stilbines were isolated [196,197].

4.5 FORMATION OF A C-VI GROUP ELEMENTS BOND, EXCEPT OXYGEN

4.5.1 Formation of a C-S Bond

Most of the examples belonging to this category concern the application of $S_{RN}1$ reactions to aromatic halides [31]. Unfortunately, these reactions are not always clean. This is due to the fragmentation of the S-C alkyl bond at the radical anion stage. This competes with electron transfer (yielding the desired aryl ether) and forms a benzenethiolate anion as a side product (see Scheme 4.39).

In any case the reaction of the ethanethiolate anion with bromobenzenes bearing an electron-withdrawing substituent gave photosubstitution in 70 to 95% yield [198] and satisfactory results were also achieved with various other sulfur nucleophiles [199]. Furthermore, dianions from polymethylenedisulfides gave interesting bridged compounds when reacting with halonaphthalenes as illustrated in Scheme 4.40 [200].

Moreover, thianthrene derivatives have been synthesized starting from *o*-diiodobenzene through the photoreaction with arenedithiolate anions [201]. Arene and heteroarenethiolate anions have been further used as suitable nucleophiles for $S_{RN}1$ reactions. Starting from aryl iodides, several arylphenyl sulfides were obtained, in some cases in excellent yields [202,203]. Very recently, the anion of thiourea

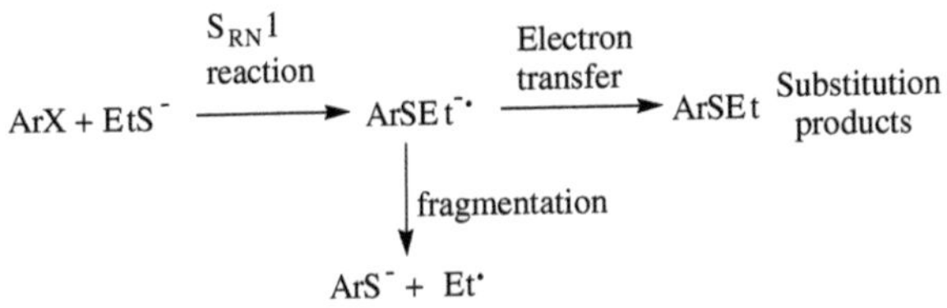

SCHEME 4.39

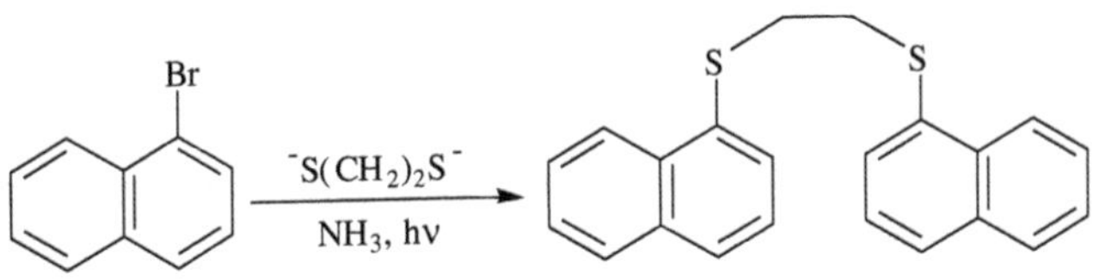

SCHEME 4.40

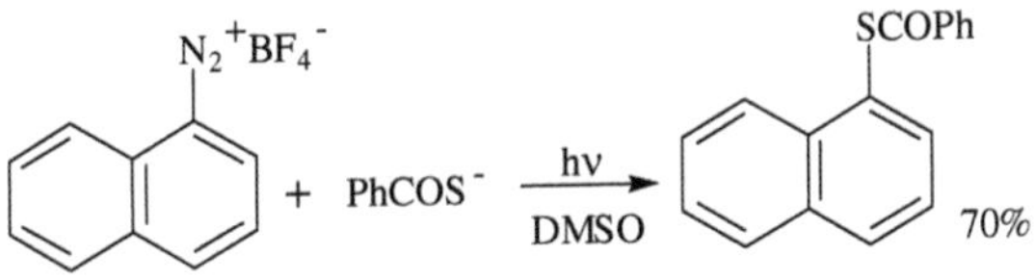

SCHEME 4.41

in DMSO was used to generate photochemically arene thiolate anions from aryl halides. The anions were subsequently employed in several in situ derivatization reactions; as a consequence, this method can be considered an easy entry for various diaryl and arylalkyl sulfides [204]. Arenediazonum salts can be suitable precursors for the synthesis of aryl thioesters when photolyzed in the presence of thioacetate and thiobenzoate anions (see Scheme 4.41) [205,206].

Irradiation of 4-chloro-1-hydroxynaphthalene in aqueous solution of sodium sulfite caused the substitution of the chlorine atom by sulfite. The mechanism was rationalized as involving the sulfite radical anion in a chain $S_{RN}1$ process [207]. The reaction led to sulfonic acids in the photolysis of various para substituted anilines; a halogen or a SO_2X (X = CH_3, NH_2, CF_3) group was replaced, thus leading to the corresponding sulfonic acid [208]. As for the p-sulfonyl precursors, experiments with labelled sulfite confirmed that the formation of the products involved cleavage of the aryl-S bond.

The synthesis of 1,3-oxathiole-2-thiones could be easily achieved by the photoinduced reaction of zwitterionic iodonium compounds such as **20** (see Scheme 4.42) with carbon disulfide. The reaction is thought to occur through the intermediacy of a trivalent iodine compound [209].

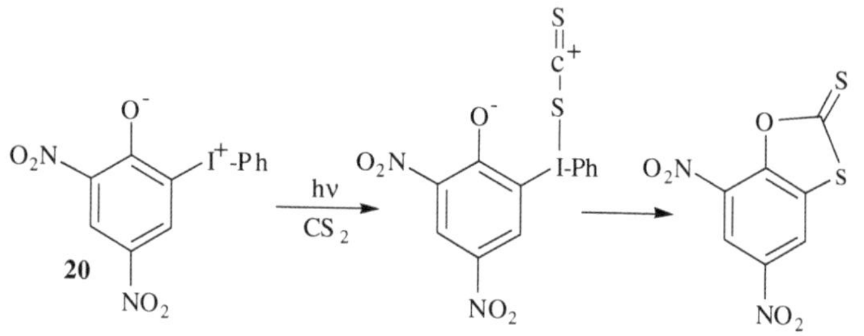

SCHEME 4.42

At present, it appears that the best way to form aryl-sulfur bonds is the recently developed metal-catalyzed route from aryl halides [210,211]; this method appears to be more versatile than the photochemical methods. The latter ones are in practice limited to the $S_{RN}1$ reaction, a procedure that is seriously limited by the concurrent fragmentation of the radical anion intermediate as discussed above.

4.5.2 FORMATION OF A C-SE OR C-TE BOND

The $S_{RN}1$ route was the best way also for forming aryl-selenium and aryl-tellurium bonds [31]. Thus, the selenide anion, prepared through the reaction of selenium with sodium in liquid ammonia, gave diphenyl selenide and diphenyldiselenide under irradiation in the presence of iodobenzene [212,213] or of p-haloanisoles. The aryl selenide ion can also be trapped by alkyl halides to form aryl alkyl selenides. The telluride ion gave somewhat lower yield of arylated derivatives under the same conditions [212]. The reaction gave better yields using potassium tellurocyanate in DMSO. This method could be a valid alternative to the classical synthesis via the usual arylmagnesium–lithium and –mercury precursors that has limited application for the synthesis of aryltellurium compounds carrying nitrogen-containing moieties [214].

4.6 FORMATION OF A C-HALOGEN BOND

Aryl halides, as shown previously, have been generally used as precursors in photoinduced nucleophilic substitution. However, in particular cases, a halogen can be the entering rather than (or as well as) the leaving group from the aromatic ring. The typical situation is the synthesis of aromatic fluorides starting from aryl diazonium salts. The fluoro-dediazoniation through thermal decomposition of these salts, the Balz–Schiemann reaction, has been known for 80 years [215]. However, this is not generally applicable due to the nonreproducibility of the yields and the difficult application to the preparation of aromatic fluorides bearing polar substituents [216]. Shifting to photoinduced decomposition of arenediazonium tetrafluoroborates can improve the yields of fluorination, though the reaction may be rather unselective. As an example, the photodecomposition of some of these salts when carried out in alcoholic solvents such as TFE or EtOH gave rise to a mixture in which the corresponding aryl fluoride was in no case the main photoproduct, due to the competition of the solvent as the nucleophile or hydrogen donor [217]. As a consequence, either the formation of the corresponding aryl ethers or that of hydrodediazoniated photoproducts were observed depending on the solvent and the substrate used (see Scheme 4.43) (see also Section 4.3.1). Fluoroarenes were supposed to be formed by fluoride abstraction by the photogenerated aryl cation in part by the solvent and in part by the counterion of the salt [217].

Better results were obtained by using n moles of HF combined with pyridine as the reaction medium; yields were high in these cases and quite independent from the nature of the substituents present on the aromatic ring, as depicted in Scheme 4.44 [216,218–220].

In some cases, fluorination was photoaccelerated in the reaction of aryl diazonium salts in the presence of etherated boron trifluoride [221]. Photodecomposition of diazonium salts in the presence of other halide anions (bromide or chloride) was in practice not explored [222,223]. Although aryl iodides can be obtained thermally from diazonium and iodonium precursors, some photoinduced syntheses were also reported [224,225]. The reaction seemed to involve aryl radicals [225] in the former case and aryl cations in the latter one [224].

Formation of aryl chlorides from aryl bromides was described by Wubbels and coworkers [226,227] using 3- and 4-bromonitrobenzenes under irradiation in aqueous

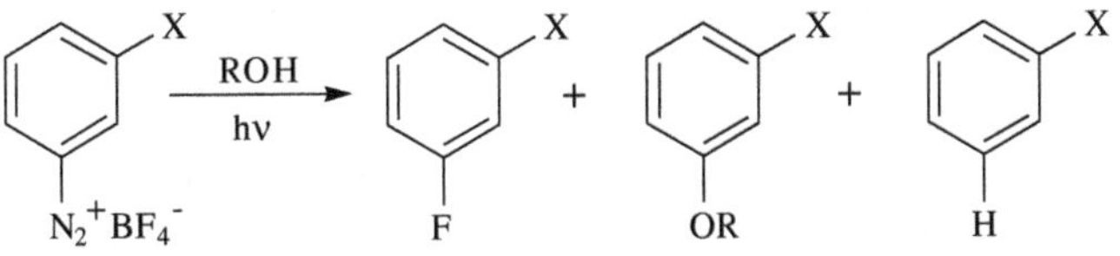

	F	OR	H
X = H; TFE	43	57	
X = F; TFE	33	67	< 0.1
X = NO₂; TFE	43	57	
X = NO₂; MeOH	< 0.1	15	85

SCHEME 4.43

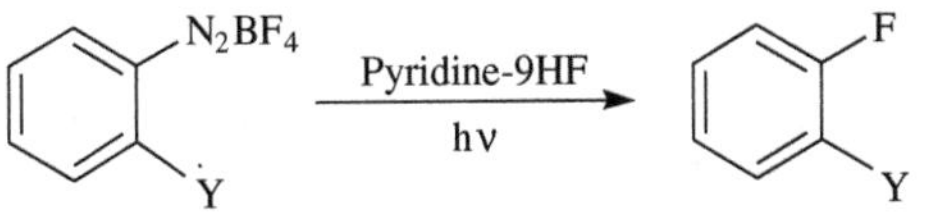

Y = H, Cl, F, OH, OMe   Yields > 80%

SCHEME 4.44

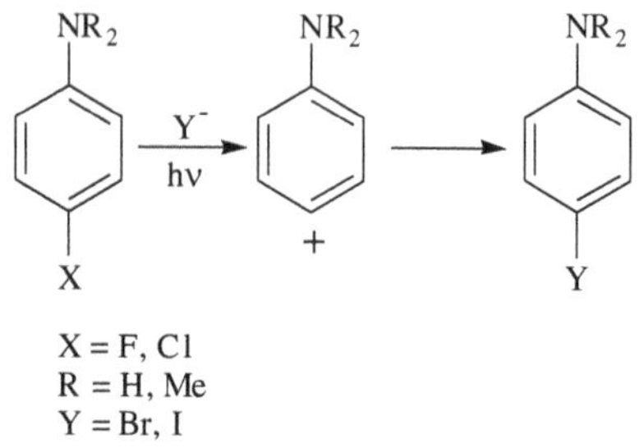

X = F, Cl
R = H, Me
Y = Br, I

SCHEME 4.45

solution. In both cases photosubstitution was quite inefficient ($\Phi = 0.011$), though clean, and was catalyzed by the hydronium ion. It was proposed that a $S_N 2Ar^{3*}$ mechanism was operating [226,227]. On the other hand, a $S_{RN}1$ mechanism was involved in the formation of iodo (and bromo) anilines from the corresponding chloro derivatives, a reaction occurring via the phenyl cation (see Scheme 4.45) [16].

4.7 FORMATION OF A C-O BOND

As mentioned in the introduction, the first reported aromatic photo S_N reaction has been the photohydrolysis of nitro aryl phosphates and sulfates [1]. In the almost 50 years intervened since then, many more examples of photoreactions involving the formation of an aryl-oxygen bond have been reported. Indeed, most photo S_N reactions are carried out in water, alcohols, or mixed solvents containing them and accordingly phenols and phenyl ethers are often formed as a by product when other nucleophiles are used.

Photohydrolysis reactions have been extensively studied on chloroaromatics since such procedures are involved in the transformation of many pesticides and herbicides in the environment. In fact, their large use in agriculture causes water pollution and more information on the photochemical behavior of these halo compounds is desirable. In this context, it is important to establish whether the photoproducts formed have themselves an effect on the environment and are more or less persistent than the starting pesticide or whether they undergo deep-seated degradation and mineralization. It is therefore important to understand the mechanism of their transformation and the effect of conditions on them.

Chloro-, bromo- and fluorobenzenes were hydrolyzed to phenol upon photolysis in water at different pHs; chlorobenzene reacted markedly faster than the other halo derivatives [228]. Photosubstitution likewise occurred to some extent with various *meta*-substituted halobenzenes [229]. Interesting was the case of monochlorobiphenyls in water; in all cases variable amounts of hydroxybiphenyls were formed, but through a different mechanism according to the position of the chlorine on the aromatic ring [230,231].

Heterolytic fragmentation occurred with 2-chlorobiphenyl (2-CBP) and in part with 3-chlorobiphenyl (3-CBP) but with the latter substrate competing isomerization (to 2-CBP and 4-CBP) and homolytic fragmentation were also observed. In the case of 4-CBP, it was supposed that the photofragmentation was homolytic but subsequent electron transfer between aryl and chlorine radicals generated the aryl cation [230,231]. Halogenated anions were not the only electrofugal groups. Various nitrophenols and nitroanisoles underwent photosubstitution in water and, according to the structure, either the nitro or the methoxy group was replaced. 3-Nitrophenolate could be prepared from the irradiation of 3-nitroanisole in alkaline aqueous solution [145] but both the presence of methanol in solution [232] and the complexation by β-cyclodextrin [233] led to inhibition of the photohydroxylation path. SDS micelles influenced strongly the photochemical behavior of n-alkyl 3-nitrophenyl ethers in a way that depended on the chain length [234,235]. Although photosubstitution of the n-decyl ether was practically suppressed by SDS, the reaction of the analogous methyl and ethyl ethers was not affected and this was rationalized on the basis of a different hydrogen bond basicity of the triplet states of such ethers. This caused a change of the micellar solubility and thus a change of the microenvironment of the reactivity [236]. The replacement of one of the groups in o- and p-nitroanisole by hydroxide anion occurred in a nonregioselective fashion. However, the quantum yield of formation of 4-methoxyphenol was temperature independent, while that of 4-nitrophenol dramatically changed in the region -20 to 196°C [237,238]. The methoxy group was replaced exclusively upon irradiation in water of both 3,5-dinitroanisole [239] and dimethoxynitrobiphenyls [240]. As for the last compounds, regioselective substitution was observed in the photohydrolysis of 3,4-dimethoxy-3'-nitrobiphenyl, which yielded 4-hydroxy-3-methoxy-3'-nitrobiphenyl, whereas in the case of 3,4-dimethoxy-4'-nitrobiphenyl there was no selectivity and both the 3- and the 4-hydroxy derivative were obtained [240]. Photosubstitution of the nitro group by the hydroxyl group occurred upon irradiation of both 2- and 3-nitrophenols in water; however, several other photoproducts were identified [241,242]. Both reactions were envisaged as involving heterolytic fragmentation as the primary step.

Several studies were devoted to the mechanism of photodegradation of halophenols and -anisoles since especially the former substrates are being largely employed as antimicrobials. For monochlorophenols, C-Cl bond cleavage took place but the position of the chlorine on the ring strongly influenced the course of the reaction [243]. 3-Chlorophenol gave resorcinol with a quantum yield virtually independent on the pH of the aqueous solution and a S_N2 type reaction was suggested [243]; the same mechanism was proposed for 4-bromophenol [244]. 2-Chlorophenol gave different products according to the basicity of the medium. In fact, when the compound reacted in the undissociated form, pyrocatechol was the main product, whereas the anionic form yielded cyclopentadienyl carboxylic acid that spontaneously dimerized during the isolation [244]. Such transformation could be sensitized by hydroquinone. Besides the other products, 2,5,2'-trihydroxybiphenyl was detected [245].

Oligomeric biphenyls were also formed in the photolysis of 4-chlorophenol [243]. Laser flash photolysis studies on the mechanism of photodegradation of this compound evidenced the intermediacy of 4-oxocyclohexa-2,5-dienylidene carbene, in turn formed by deprotonation of an aryl cation arising by initial heterolytic fragmentation of the C-Cl bond [246]. This carbene was thought to be the key intermediate of the reaction leading to hydroquinone and 5-chloro-2,4'-dihydroxybiphenyl as main photoproducts. Oxygenation of the carbene took place in aerated solution and led to p-benzoquinone as the end product, thus limiting the extent of photosubstitution occurring [246] (see Scheme 4.46). Analogous considerations have been applied for the photolysis of the other three 4-halophenols [244,247]. Polyhalogenated phenols such as 2,4,5-trichlorophenols when irradiated in water gave a mixture of products including polychlorobenzodioxins, i.e., compounds considerably more toxic than the starting material [248].

The photochemistry of both 4-chloro- and 4-fluoroanisoles in water-acetonitrile mixtures dramatically changed increasing the amount of water. In neat acetonitrile a tight ion pair was formed from an excited and a ground state molecule but only the presence of water allowed the dissociation of this intermediate into free radical ions. The photohydrolysis end products were thought to arise from the radical cation [249]. A clean photohydrolysis was also observed in the irradiation of fluoromethoxybenzenes, where water addition onto the aryl cation intermediate was invoked [250].

Alcoholic solvents can be an alternative medium for photosubstitution reactions. As an example chlorobenzene yielded anisole (with a quantum yield of 0.049) when irradiated in MeOH [251] and phenetole in aqueous ethanolic solutions, though in the latter case phenol was the main photoproduct [252]. Analogously, monochlorotoluenes underwent photosubstitution in MeOH forming methylanisoles [253,254] and the quantum yield of decomposition ranged from 0.1 to 0.4 with the smallest values applying to the *para* and *ortho* isomers. Biacetyl sensitized experiments confirmed

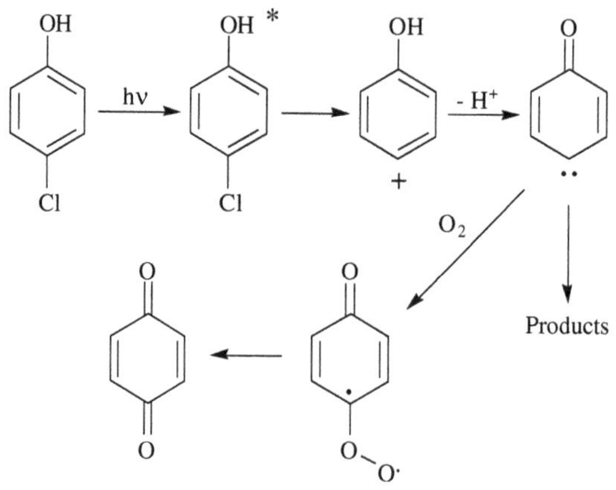

SCHEME 4.46

the involvement of the triplet state in the reaction. *o*-Chlorobiphenyl was the most photoreactive substrates in the monochlorobiphenyl series due to the relief of steric strain between the chlorine atom and the hydrogen on the adjacent aromatic ring [255]. Methoxybiphenyl were obtained in MeOH, whether via a radical or via a cation intermediate. Photolysis of nitro[^{14}C]methoxybenzenes or [^{14}C]methoxy-labeled-4-nitroveratroles in the presence of methoxyde anion underwent substitution of the labeled methoxyl group by the same unlabeled group [145,256]. Although the consumption of *o*-nitroanisole was negligible, with the *para* isomer photosubstitution to give dimethoxybenzene occurred along with photoreduction of the nitro to nitroso group (see Scheme 4.47). The reduction was found also for the *meta* isomer. As shown previously, nitroveratroles underwent substitution of both of the methoxy group [145,256].

3-Haloanilines were also investigated. From the photolysis in methanol 3-anisidine was isolated along with variable amounts of the photoreduction derivative aniline [257]. Substitution by the methoxy group took place after heterolytic cleavage of the C-X bond from the excited singlet state and addition of MeOH; with 3-bromoaniline, homolytic cleavage of the C-Br bond played a role. Competition between photoreduction and photosubstitution was also important in the photochemistry of *p*-methoxyanisoles [258,259] with a predominance of the former path; a photoinduced electron transfer between the precursor and the solvent may account for the formation of the photosubstitution products [259].

An alternative way for the synthesis of phenols and arylethers is the photosubstitution of the diazo group. This reaction was first reported at the end of the

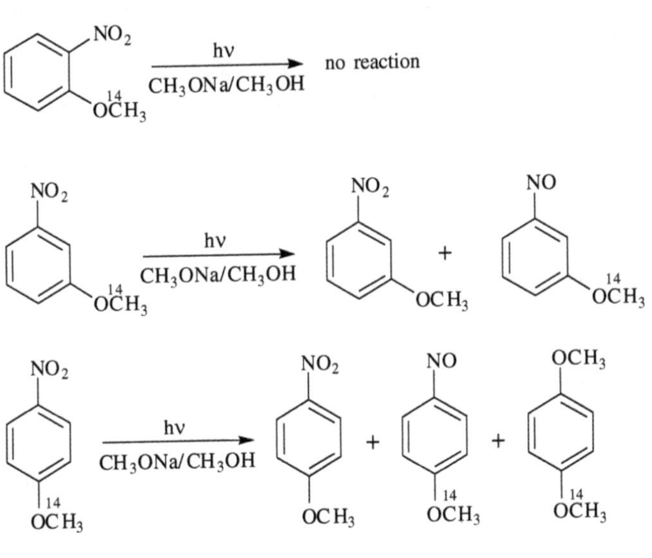

SCHEME 4.47

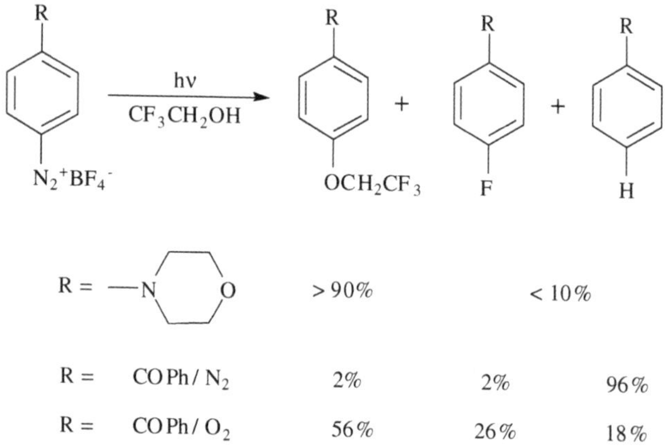

R = —N(morpholine) > 90% < 10%

| R = | COPh/ N_2 | 2% | 2% | 96% |
| R = | COPh/ O_2 | 56% | 26% | 18% |

SCHEME 4.48

nineteenth century [260] but it has found up to now little application [222,261–264]. More recently this process was reinvestigated and found to be a side process in the photo-Schiemann reaction (217, see Section 4.6).

An important difference here is that the spin of the phenyl cation formed determines the chemistry. The singlet state smoothly adds to alcohols or to nucleophilic solvents, whereas the triplet is reduced by hydrogen donating solvents. Thus, formation of ethers could be taken as a measure of the fraction of the singlet phenyl cation formed. This, in turn, depended on the singlet or triplet path followed in the fragmentation of the diazonium salts (see also Section 4.3.1) [23,265,266]. The presence of electron donating rather than electron withdrawing substituents on the aromatic ring seemed to increase the chemical yield of the ether [23,265,266]. As for the latter compounds, the presence of oxygen limited dramatically the photoreduction path increasing the ether yield (see Scheme 4.48) [265]. Since 2,2,2-trifluoroethanol was often employed as the solvent, this can be considered an interesting route for the synthesis of the corresponding aryl trifluoroethylethers [266].

4.8 SYNTHETIC SIGNIFICANCE OF PHOTONUCLEOPHILIC SUBSTITUTIONS

In most cases, the substituted aromatic derivatives described in the previous sections can be obtained from the corresponding aryl halides in the presence of organometallic species (especially palladium and copper) in good yields in a highly versatile fashion [267–270]. Nevertheless, the toxicity and the price of the metal employed as well as the nontrivial synthetic procedures of the ligands required in most cases for the reaction leaves room for alternative procedures. Photonucleophilic substitutions have the great advantage of using a mild nonpolluting reagent (the light) as the initiator, low reaction temperatures, and metal free conditions. Moreover, the large choice of possible leaving groups (halogen, methoxy, nitro, diazo, cyano, iodonium, sulfonyl) on the aromatic ring widens the field of applicability of this reaction.

Concerning the efficiency and selectivity of the reaction, some rules can be summarized which have general validity at least for a rough prediction of the trend:

- Monosubstituted aromatics compounds undergo photosubstitution reactions inefficiently, except for the case of diazonium and iodonium groups, which are in any case the best leaving groups.
- In polysubstituted aromatics, halides are excellent leaving groups and are displaced exclusively.
- In anilines the amino group does not undergo substitution.
- When a nitro and an alkoxy group are both present, either/or can be substituted depending on the position, the presence of further substituents, and the entering nucleophile.
- Nucleophilic solvents (water and alcohols) compete in the reaction with any other nucleophile used.

Moreover, photo S_N reactions offer a range of exclusive procedures for the introduction of carbon-linked substituent.

As an example, a revealing comparison between thermal and photochemical methods can be made for the synthesis of 4-cyanoanisole. This compound can be easily obtained by both methods in high yields starting from the corresponding 4-haloanisoles. Palladium catalyzed cyanation was successful when applied to iodo- and bromoanisoles using $CuCN/Et_4NCN$ [271], NaCN [272], acetone cyanohydrine [273], trimethylsilylcyanide [274], and $Zn(CN)_2$ (under microwave irradiation) [275] as the cyanating agents. Recently, 4-chloroanisole was likewise used as the precursor; however, in this case the reaction required a high temperature (150°C) to occur [276].

These metal mediated reactions occurred on average in a very high yield (from 62 to 98%). However, the overall process is quite expensive because of the cost both of the palladium catalyst (though used in < 4 mol % amount) and the ligand necessary for the reaction that requires in most cases a nontrivial synthesis. Furthermore, the quite inexpensive aryl chloride could be used only under drastic reaction conditions in comparison with the more expensive bromide and iodide. Further drawbacks can be indicated for the above thermal cyanations. Thus, the reactions were more efficient when the aryl ring bore electron-withdrawing rather than electron-donating substituents and further the relatively high amount of catalyst used generated a nonnegligible amount of metal waste.

On the other hand, the photochemical synthesis of 4-cyanoanisole was carried out under very mild conditions in a nonpolluting environment (in aqueous-alcoholic medium) using UV light as the initiator rather than a toxic catalyst. Moreover, the reaction was more efficient when starting from 4-chloro- and 4-fluoroanisole, where in both cases the photosubstitution yields were higher than 90% [114]. Noteworthy, even strong aryl-fluoride bonds could be easily broken by photochemical means. A drawback is that these photochemical reactions were carried out in diluted solutions of the halide (5×10^{-4} M) and required a high excess of the cyanide ion (ca. 30 to 50 times with respect to the haloanisole). The use of such a low reagent concentration was due to the mainly mechanistic target of the investigation, as is mostly the case with photochemical studies. It is certainly possible to obtain satisfactory results when using much larger starting concentrations, arriving, e.g., at a concentration in the range 0.02 to 0.1 M, which is comparable to those used in thermal catalytic procedures. Pursuing research in this direction is certainly worthwhile.

REFERENCES

1. E Havinga, RO de Jongh, W Dorst, *Rec Trav Chim Pays-Bas* 75: 378–383, 1956.
2. E Havinga, JLMA Schlatmann, *Tetrahedron* 16: 146–152, 1961.
3. ND Epiotis, S Shaik, *J Am Chem Soc* 100: 29–33, 1978.
4. K Mutai, R Nakagaki, H Tukada, *Bull Chem Soc Jpn* 58: 2066–2070, 1985.
5. K Mutai, R Nakagaki, *Bull Chem Soc Jpn* 58: 3663–3664, 1985
6. E Havinga, ME Kronenberg, *Pure Appl Chem* 16: 137–152, 1968.
7. J Cornelisse, E Havinga, *Chem Rev* 75: 353–388, 1975.
8. E Havinga, J Cornelisse, *Pure Appl Chem* 47: 1–10, 1976.
9. J Cornelisse, GP de Gunst, E Havinga, *Adv Phys Org Chem* 11: 225–266, 1975.
10. J Cornelisse, G Lodder, E Havinga, *Rev Chem Interm* 2: 231–265, 1979.
11. J Cornelisse, In WM Horspool, PS Song, Eds., *Handbook of Organic Photochemistry and Photobiology,* Boca Raton, FL: CRC Press, 1995, pp. 250–265.
12. C Karapire, S Icli, In WM Horspool, F Lenci, Eds., *Handbook of Organic Photochemistry and Photobiology, 2nd Edition,* Boca Raton, FL: CRC Press, 2004, pp. 37/1–37/14.
13. D Bryce-Smith, A Gilbert, Eds., *Specialistic Periodical Reports, Photochemistry, Vol. 1–34,* Cambridge: The Royal Society of Chemistry, 1970–2002.
14. M Freccero, M Fagnoni, A Albini, *J Am Chem Soc* 125: 13182–13190, 2003.
15. M Fagnoni, M Mella, A Albini, *Org Lett* 1: 1299–1301, 1999.
16. B Guizzardi, M Mella, M Fagnoni, M Freccero, A Albini, *J Org Chem* 66: 6353–6363, 2001.
17. M Mella, P Coppo, B Guizzardi, M Fagnoni, M Freccero, A Albini, *J Org Chem* 66: 6344–6352, 2001.
18. P Coppo, M Fagnoni, A Albini, *Tetrahedron Lett* 42: 4271–4273, 2001.
19. B Guizzardi, M Mella, M Fagnoni, A Albini, *J Org Chem* 68: 1067–1074, 2003.
20. B Guizzardi, M Mella, M Fagnoni, A Albini, *Chem Eur J* 9: 1549–1555, 2003.
21. A Fraboni, M Fagnoni, A Albini, *J Org Chem* 68: 4886–4893, 2003.
22. B Guizzardi, M Mella, M Fagnoni, A Albini, *Tetrahedron* 56: 9383–9390, 2000.
23. S Milanesi, M Fagnoni, A Albini, *Chem Commun* 216–217, 2003.
24. S Spyroudis, P Tarantili, *Tetrahedron* 50: 11541–11552, 1994.
25. SP Spyroudis, *J Org Chem* 51: 3453–3456, 1986.
26. A Annunziata, C Galli, M Marinelli, T Pau, *Eur J Org Chem* 1323–1329, 2001.

27. C Amatore, C Combellas, J Pinson, MA Oturan, S Robveille, JM Savéant, A Thiébault, *J Am Chem Soc* 107: 4846–4853, 1985.
28. B Branchi, C Galli, P Gentili, *Eur J Org Chem* 2844–2854, 2002.
29. JK Kim, JF Bunnett, *J Am Chem Soc* 92: 7463–7464, 1970.
30. RA Rossi, RH de Rossi, *Aromatic Substitution by the $S_{RN}1$ mechanism,* Washington: American Chemical Society, 1983.
31. RA Rossi, AB Pierini, AN Santiago, *Org React* 54: 1–271, 1999.
32. RA Rossi, AB Pierini, AB Peñéñory, *Chem Rev* 103: 71–167, 2003.
33. R Beugelmans, in W Horspool, PS Song, Eds., *CRC Handbook of Photochemistry and Photobiolology,* Boca Raton, FL: CRC Press, 1995, pp. 1200–1217.
34. RA Rossi, AB Peñéñory In: W Horspool, F Lenci, Eds., *CRC Handbook of Photochemistry and Photobiolology, 2nd Ed.* Boca Raton, FL: CRC Press, 2004, pp 47/1–47/18.
35. LM Tolbert, S Siddiqui, *J Org Chem* 49: 1744–1751, 1984.
36. RA Rossi, JF Bunnettt, *J Org Chem* 38: 3020–3025, 1973.
37. LM Tolbert, DP Martone, *J Org Chem* 48: 1185–1190, 1983.
38. RA Rossi, RH de Rossi, AF Lopez, *J Am Chem Soc* 98: 1252–1257, 1976.
39. JW Wong, KJ Natalie, GC Nwokogu, JS Pisipati, PT Flaherty, TD Greenwood, JF Wolfe, *J Org Chem* 62: 6152–6159, 1997.
40. GA Lotz, SM Palacios, RA Rossi, *Tetrahedron Lett* 35: 7711–7714, 1989.
41. BQ Wu, FW Zeng, MJ Ge, XZ Cheng, GS Wu, Sci China 34: 777–786, 1991; *Chem Abst* 116: 58463h, 1992.
42. R Beugelmans, J Chastanet, *Tetrahedron* 49: 7883–7890, 1993.
43. R Beugelmans, M Bois-Choussy, *Tetrahedron Lett* 29: 1289–1292, 1988.
44. C Dell'Erba, M Novi, G Petrillo, C Tavani, *Tetrahedron* 49: 235–242, 1993.
45. MT Baumgartner, AB Pierini, RA Rossi, *J Org Chem* 64: 6487–6489, 1999.
46. SM Barolo, AE Luckach, RA Rossi, *J Org Chem* 68: 2807–2811, 2003.
47. CG Ferrayoli, SM Palacios, RA Alonso, *J Chem Soc Perkin Trans 1* 1635–1638, 1995.
48. SE Vaillard, A Postigo, RA Rossi, J *Org Chem* 67: 8500–8506, 2002.
49. QY Chen, ZT Li, *J Chem Soc Perkin Trans 1* 1705–1710, 1993.
50. QY Chen, ZT Li, *J Org Chem* 58: 2599–2604, 1993.
51. M Freccero, M Mella, A Albini, *Tetrahedron* 50: 2115–2130, 1994.
52. P Maslak, *Top Curr Chem* 168: 1–46, 1993.
53. R Popielarts, DR Arnold, *J Am Chem Soc* 112, 3068–3082, 1990.
54. A Albini, M Mella, M Freccero, *Tetrahedron* 50: 575–607, 1994.
55. M Freccero, AC Pratt, A Albini, C Long, *J Am Chem Soc* 120: 284–297, 1998.
56. AN Frolov, *Russ J Org Chem* 34: 139–161, 1998.
57. M Mella, M Fagnoni, M Freccero, E Fasani, A Albini, *Chem Soc Rev* 27: 81–89, 1998.
58. R Borg, D Arnold, T Cameron, *Can J Chem* 62: 1785–1802, 1984.
59. M Vanossi, M Mella, A Albini, *J Am Chem Soc* 116: 10070–10075, 1994.
60. N d'Alessandro, E Fasani, M Mella, A Albini, *J Chem Soc Perkin Trans 2* 1977–1980, 1991.
61. A Sulpizio, A Albini, N d'Alessandro, E Fasani, S Pietra, *J Am Chem Soc* 111: 5773–5777, 1989.
62. L Bardi, E Fasani, A Albini, *J Chem Soc Perkin Trans 1* 545–549, 1994.
63. A Albini, E Fasani, M Freccero, *Adv El Tr Chem* 5: 103–140, 1996.
64. K Tsujimoto, N Nakao, M Ohashi, *J Chem Soc Chem Commun* 366–367, 1992.

65. E Fasani, D Peverali, A, Albini, *Tetrahedron Lett* 35: 9275–9278, 1994.
66. E Fasani, M Mella, A Albini, *J Chem Soc Perkin Trans 2* 449–452, 1995.
67. M Mella, M Freccero, A Albini, *J Org Chem* 59: 1047–1052, 1994.
68. K Mizuno, T Nishiyama, N Takahashi, H Inoue, *Tetrahedron Lett* 37: 2975–2978, 1996.
69. S Kyushin, Y Matsuda, K Matsushita, Y Nadaira, M Ohashi, *Tetrahedron Lett* 31: 6395–6398, 1990.
70. M Mella, E Fasani, A Albini, *J Org Chem* 57: 6210–6216, 1992.
71. M Mella, N d'Alessandro, M Freccero, A Albini, *J Chem Soc Perkin Trans 2* 515–519, 1993.
72. J Lan, G Schuster, *J Am Chem Soc* 107: 6710–6711, 1985.
73. K Mizuno, M Ikeda, Y Otsuji, *Tetrahedron Lett* 26: 461–464, 1985.
74. K Mizuno, K Nakanishi, Y Otsuji, *Chem Lett* 1833–1836, 1988.
75. K Mizuno, T Nishiyama, K Terasaka, M Yasuda, K Shima, Y Otsuji, *Tetrahedron* 48: 9673–9686, 1992.
76. K Mizuno, K Terasaka, M Ikeda, Y Otsuji, *Tetrahedron Lett* 26: 5819–5822, 1985.
77. K Nakanishi, K Mizuno, Y Otsuji, *Bull Chem Soc Jpn* 66: 2371–2379, 1991.
78. M Fagnoni, M Mella, A Albini,, *Tetrahedron* 50: 6401–6410, 1994.
79. M Mella, M Freccero, A Albini, *Tetrahedron* 52: 5533–5548, 1996.
80. M Mella, M Freccero, T Soldi, E Fasani, A Albini, *J Org Chem* 61: 1413–1422, 1996.
81. M Fagnoni, M Vanossi, M Mella, A Albini, *Tetrahedron* 52: 1785–1796, 1996.
82. R Torriani, M Mella, E Fasani, A Albini, *Tetrahedron* 53: 2573–2580, 1997.
83. E Fasani, N d'Alessandro, A Albini, PS Mariano, *J Org Chem* 59: 829–835, 1994.
84. T Nishiyama, K Mizuno, Y Otsuji, H Inoue, *Tetrahedron* 51: 6695–6706, 1995.
85. J Lan, G Schuster,*Tetrahedron Lett* 27: 4261–4264, 1986.
86. T Tamai, K Mizuno, I Hashida, Y Otsuji, *Bull Chem Soc Jpn* 66: 3747–3754, 1993.
87. M Fagnoni, M Mella, A Albini, *J Am Chem Soc* 117: 7877–7881, 1995.
88. M Mella, M Fagnoni, A Albini, *J Org Chem* 59: 5614–5622, 1994.
89. D Mangion, DR Arnold, *Acc Chem Res* 35: 291–304, 2002.
90. D Mangion, DR Arnold, in WM Horspool, F Lenci, Eds., *Handbook of Organic Photochemistry and Photobiology, Second Edition,* Boca Raton, FL: CRC, 2004, pp 40/1–40/18.
91. DR Arnold, MS Snow, *Can J Chem* 66: 3012–3026, 1988.
92. RA Neunteufel, DR Arnold, *J Am Chem Soc* 95: 4080–4081, 1973.
93. Y Shigemitsu, DR Arnold, *J Chem Soc, Chem Commun* 407–408, 1975.
94. RM Borg, D Franke, A Vella, *Can J Chem* 81: 723–726, 2003.
95. DR Arnold, MSW Chan, KA McManus, *Can J Chem* 74: 2143–2166, 1996.
96. MSW Chan, DR Arnold, *Can J Chem* 75: 1810–1819, 1997.
97. DR Arnold, KA McManus, MSW Chan, *Can J Chem* 75: 1055–1075, 1997.
98. HJP de Lijser, DR Arnold, *J Org Chem* 62: 8432–8438, 1997.
99. KA McManus, DR Arnold, *Can J Chem* 72: 2291–2304, 1994.
100. DR Arnold, KA McManus, X Du, *Can J Chem* 72: 415–429, 1994.
101. DA Connor, DR Arnold, PK Bakshi, TS Cameron, *Can J Chem* 73, 762–771, 1995.
102. KA McManus, DR Arnold, *Can J Chem* 73: 2158–2169, 1995.
103. H Weng, C Scarlata, HD Roth, *J Am Chem Soc* 118: 10947–10953, 1996.
104. D Mangion, DR Arnold, *J Chem Soc Perkin Trans 2* 48–60, 2001.
105. DR Arnold, X Du, *Can J Chem* 72: 403–414, 1994.

106. DR Arnold, X Du, HJP de Lijser, *Can J Chem* 73: 522–530, 1995.
107. M Sundermeier, A Zapf, M Beller, *Eur J Inorg Chem* 3513–3526, 2003, and references cited therein.
108. GP Ellis, TM Romney-Alexander, *Chem Rev* 87: 779–794, 1987.
109. JA Barltrop, NJ Bunce, A Thomson, *J Chem Soc C* 1142–1145, 1967.
110. P Kuzmič, M, Souček, *Collect Czech Chem Commun* 51: 358–367, 1986.
111. RL Letsinger, RR Hautala, *Tetrahedron Lett* 48: 4205–4208, 1969.
112. JAJ Vink, PL Verheijdt, J Cornelisse, E Havinga, *Tetrahedron* 28: 5081–5087, 1972.
113. P Kuzmič, M Souček, *Collect Czech Chem Commun* 52: 980–988, 1987.
114. J Den Heijer, OB Shadid, J Cornelisse, E Havinga, *Tetrahedron* 33: 779–786, 1977.
115. JH Liu, RG Weiss, *J Photochem* 30: 303–314, 1985.
116. JH Liu, RG Weiss, *J Org Chem* 50: 3655–3657, 1985.
117. M Cervera, J Marquet, *Can J Chem* 76: 966–969, 1998.
118. AN Frolov, OV Kul'bitskaya, AV El'tsov, *Russ J Org Chem* 8: 436–437, 1972.
119. AV El'tsov, OV Kul'bitskaya, EV Smirnov, AN Frolov, *Russ J Org Chem* 9: 2542–2545, 1973.
120. OV Kul'bitskaya, AN Frolov, AV El'tsov, *Russ J Org Chem* 9: 2331–2334, 1973.
121. R Beugelmans, H Ginsburg, A Lecas, MT Le Goff, J Pusset, G Roussi, *J Chem Soc Chem Commun* 885–886, 1977.
122. N Suzuki, Y Ayaguchi, Y Izawa, *Bull Chem Soc Jpn* 55: 3349–3350, 1982.
123. A Kostantinov, CA Kingsmill, G Ferguson, NJ Bunce, *J Am Chem Soc* 120: 5464–5468, 1998.
124. A Kostantinov, AN Johnston, NJ Bunce, *Can J Chem* 77: 1366–1373, 1999.
125. M Novi, G Petrillo, C Dell'Erba, *Tetrahedron Lett* 28: 1345–1348, 1987.
126. G Petrillo, M Novi, G Garbarino, C Dell'Erba, *Tetrahedron* 43: 4625–4634, 1987.
127. M Novi, G Garbarino, G Petrillo, C Dell'Erba, *Tetrahedron* 46: 2205–2212, 1990.
128. C Dell'Erba, A Houmam, M Novi, G Petrillo, J Pinson, *J Org Chem* 58: 2670–2677, 1993.
129. J J Brunet, C Sidot, P Caubère, *J Organomet Chem* 204: 229–241, 1981.
130. J J Brunet, C Sidot, P Caubère, *J Org Chem* 48: 1166–1171, 1983.
131a. J J Brunet, C Sidot, P Caubère, *J Org Chem* 44: 2199–2202, 1979.
131b. T Kishimura, K Kudo, S Mori, N Sugita, *Chem Lett* 299–302, 1986.
132. S Kyushin, Y Ehara, Y Nakadaira, M Ohashi, *J Chem Soc Chem Commun* 279–280, 1989.
133. CC Yammal, JC Podestà, RA Rossi, *J Org Chem* 57: 5720–5725, 1992.
134. CC Yammal, JC Podestà, RA Rossi, *J Organomet Chem* 509: 1–8, 1996.
135. EF Corsico, RA Rossi, *Synlett* 227–229, 2000.
136. A Postigo, SE Vaillard, RA Rossi, *J Phys Org Chem* 15: 889–893, 2002.
137. MT Lockhart, AB Chopa, RA Rossi, *J Organomet Chem* 582: 229–234, 1999.
138. AB Chopa, MT Lockhart, G Silbestri, *Organometallics* 19: 2249–2250, 2000.
139. AB Chopa, MT Lockhart, VB Dorn, *Organometallics* 21: 1425–1429, 2002.
140. AB Chopa, MT Lockhart, G Silbestri, *Organometallics* 20: 3358–3360, 2001.
141. K Matsui, N Maeno, S Suzuki, *Tetrahedron Lett* 11: 1467–1469, 1970.
142. GG Wubbels, DM Celander, *J Am Chem Soc* 103: 7669–7670, 1981.
143. GG Wubbels, AM Halverson, JD Oxman, VH De Bruyn, *J Org Chem* 50: 4499–4504, 1985.
144. GG Wubbels, AM Halverson, JD Oxman, *J Am Chem Soc* 102: 4848–4849, 1980.

145. HCHA van Riel, G Lodder, E Havinga, *J Am Chem Soc* 103: 7257–7262, 1981.
146. K Mutai, S Kanno, K Kobayashi, *Tetrahedron Lett* 19: 1273–1276, 1978.
147. K Mutai, K Kobayashi, *Bull Chem Soc Jpn* 54: 462–465, 1981.
148. K Mutai, K Yokoyama, S Kanno, K Kobayashi, *Bull Chem Soc Jpn* 55: 1112–1115, 1982.
149. K Yokoyama, J Nakamura, K Mutai, S Nagakura, *Bull Chem Soc Jpn* 55: 317–318, 1982.
150. R, Nakagaki, K Mutai, *Bull Chem Soc Jpn* 69: 261–274, 1996.
151. GG Wubbels, BR Sevetson, H Sanders, *J Am Chem Soc* 111: 1018–1022, 1989.
152. K Mutai, H Tukada, R Nakagaki, *J Org Chem* 56: 4896–4903, 1991.
153. J Cervello, M Figueredo, J Marquet, M Moreno-Mañas, J Bertran, JM Lluch, *Tetrahedron Lett* 25: 4147–4150, 1984.
154. A Cantos, J Marquet, M Moreno-Mañas, A Gonzalez-Lafont, JM Lluch, J Bertran, *J Org Chem* 55: 3303–3310, 1990.
155. A Cantos, J Marquet, M Moreno-Mañas, *Tetrahedron Lett* 30: 2423–2426, 1989.
156. A Castello, J Marquet, M Moreno-Mañas, X Sirera, *Tetrahedron Lett* 26: 2489–2492, 1985.
157. A Castello, J Cervello, J Marquet, M Moreno-Mañas, X Sirera, *Tetrahedron* 42: 4073–4082, 1986.
158. P Kuzmič, L Pavličkova, J Velek, M Souček, *Collect Czech Chem Commun* 51: 1665–1670, 1986.
159. A Cantos, J Marquet, M Moreno-Mañas, *Tetrahedron Lett* 28: 4191–4194, 1987.
160. A Cantos, J Marquet, M Moreno-Mañas, A Castello, *Tetrahedron* 44: 2607–2618, 1988.
161. AMJ van Eijk, AH Huizer, CAGO Varma, J Marquet, *J Am Chem Soc* 111: 88–95, 1989.
162. J Marquet, A Cantos, M Moreno-Mañas, E Cayon, I Gallardo, *Tetrahedron* 48: 1333–1342, 1992.
163. NJ Bunce, SR Cater, JC Scaiano, LJ Johnston, *J Org Chem* 52: 4214–4223, 1987.
164. P Kuzmič, L Pavličkova, M Souček, *Collect Czech Chem Commun* 52: 1781–1785, 1987.
165. J Marquet, M Moreno-Mañas, A Vallribera, A Virgili, J Bertran, A Gonzalez-Lafont, JM Lluch, *Tetrahedron* 43: 351–360, 1987.
166. R Pleixats, M Figueredo, J Marquet, M Moreno-Mañas, A Cantos, *Tetrahedron* 45: 7817–7826, 1989.
167. M Figueredo, J Marquet, M Moreno-Mañas, R Pleixats, *Tetrahedron Lett* 30: 2427–2428, 1989.
168. R Pleixats, J Marquet, *Tetrahedron* 46: 1343–1352, 1990.
169. J Marquet, L Rafecas, A Cantos, M Moreno-Mañas, M Cervera, F Casado, MV Nogues, CM Cuchillo, *Tetrahedron* 49: 1297–1306, 1993.
170. F Casado, M Cervera, J Marquet, M Moreno-Mañas, *Tetrahedron* 51: 6557–6564, 1995.
171. J Marquet, Z Jiang, I Gallardo, A Battle, E Cayon, *Tetrahedron Lett* 34: 2801–2804, 1993.
172. A Gilbert, S Krestonosich, DL Westover, *J Chem Soc Perkin Trans 1* 295–302, 1981.
173. D Bryce-Smith, A Gilbert, S Krestonosich, *J Chem Soc Chem Commun* 405–406, 1976.
174. A Gilbert, S Krestonosich, *J Chem Soc Perkin Trans 1* 1393–1399, 1980.
175. NJ Bunce, SR Cater, *J Chem Soc Perkin Trans 2* 169–173, 1986.
176. Y Tanaka, K Tsujimoto, M Ohashi, *Bull Chem Soc Jpn* 60: 788–790, 1987.
177. RL Letsinger, O Bertrand Ramsay, JH McCain, *J Am Chem Soc* 87: 2945–2950, 1965.

178. G Green-Buckley, J Griffiths, *J Photochem* 27: 119–121, 1984.
179. RA Alonso, A Bardon, RA Rossi, *J Org Chem* 49: 3584–3587, 1984.
180. AV El'tsov, AN Frolov, OV Kul'bitskaya, *Russ J Org Chem* 6: 1955, 1970.
181. AN Frolov, OV Kul'bitskaya, AV El'tsov, *Russ J Org Chem* 9: 2335–2345, 1973.
182. OV Kul'bitskaya, AN Frolov, AV El'tsov, *Russ J Org Chem* 15: 389–390, 1979.
183. AN Frolov, OV Kul'bitskaya, AV El'tsov, *Russ J Org Chem* 15: 1915–1925, 1979.
184. EP Kyba, S-T Liu, RL Harris *Organometallics* 2: 1877–1879, 1983.
185. R Obrycki, CE Griffin *Tetrahedron Lett* 7: 5049–5052, 1966.
186. JF Bunnett, X Creary, *J Org Chem* 39: 3612–3614, 1974.
187. JF Bunnett, RP Traber, *J Org Chem* 43: 1867–1872, 1978.
188. JF Bunnett, SJ Shafer, *J Org Chem* 43: 1877–1883, 1978.
189. RR Bard, JF Bunnett, RP Traber, *J Org Chem* 44: 4918–4924, 1979.
190. JE Swartz, JF Bunnett, *J Org Chem* 44: 4673–4677, 1979.
191. ERN Bornancini, RA Rossi, *J Org Chem* 55: 2332–2336, 1990.
192. ERN Bornancini, RA Alonso, RA Rossi, *J Organomet Chem* 270: 177–183, 1984.
193. R Beugelmans, M Chbani, *Bull Soc Chim Fr* 132: 290–305, 1995.
194. R Beugelmans, M Chbani, *Bull Soc Chim Fr* 132: 306–313, 1995.
195. R Beugelmans, M Chbani, *Bull Soc Chim Fr* 132: 729–733, 1995.
196. RA Rossi, RA Alonso, SM Palacios, *J Org Chem* 46: 2498–2502, 1981.
197. RA Alonso, RA Rossi, *J Org Chem* 47: 77–80, 1982.
198. R Beugelmans, M Bois-Choussy, B Boudet, *Tetrahedron* 39: 4153–4161, 1983.
199. RA Rossi, SM Palacios, *J Org Chem* 46: 5300–5304, 1981.
200. R Beugelmans, H Ginsburg, *Tetrahedron Lett* 28: 413–414, 1987.
201. AB Pierini, MT Baumgartner, RA Rossi, *J Org Chem* 52: 1089–1092, 1987.
202. JF Bunnett, X Creary, *J Org Chem* 39: 3173–3174, 1974.
203. DW Hobbs, WC Still, *Tetrahedron Lett* 28: 2805–2808, 1987.
204. JE Arguello, LC Smith, AB Peñéñory, *Org Lett* 5: 4133–4136, 2003.
205. G Petrillo, M Novi, G Garbarino, M Filiberti, *Tetrahedron Lett* 29: 4185–4188, 1988.
206. G Petrillo, M Novi, G Garbarino, M Filiberti, *Tetrahedron* 45: 7411–7420, 1989.
207. VL Ivanov, SY Lyashkevich, H Lemmetyinen, *J Photochem Photobiol A* 109: 21–24, 1997.
208. AN Frolov, EV Smirnov, AV El'tsov, *Russ J Org Chem* 10: 1702–1706, 1974.
209. M Papadopoulou, S Spyroudis, A Varvoglis, *J Org Chem* 50: 1509–1511, 1985.
210. FY Kwong, SL Buchwald, *Org Lett* 4: 3517–3520, 2002 and references cited therein.
211. T Kondo, T Mitsudo, *Chem Rev* 100, 3205–3220, 2000.
212. RA Rossi, AB Peñéñory, *J Org Chem* 46: 4580–4582, 1981.
213. AB Peñéñory, RA Rossi, *J Phys Org Chem* 3: 266–272, 1990.
214. T Junk, FR Fronczek, *Tetrahedron Lett* 37: 4361–4362, 1996.
215. G Balz, G Schiemann, *Chem Ber* 60: 1186–1190, 1927.
216. M Sawaguchi, T Fukuhara, N Yoneda, *J Fluorine Chem* 97: 127–133, 1999.
217. PSJ Canning, K McCrudden, H Maskill, B Sexton, *Chem Commun* 1971–1972, 1998.
218. N Yoneda, T Fukuhara, *Tetrahedron* 52: 23–36, 1996.
219. N Yoneda, T Fukuhara, T Kikuchi, A Suzuki, *Synth Comm* 19: 865–871, 1989.
220. T Fukuhara, M Sekiguchi, N Yoneda, *Chem Lett* 1011–1012, 1994.
221. K Shinama, S Aki, T Furuta, J Minamikawa, *Synth Comm* 23: 1577–1582, 1993.
222. ES Lewis, RE Holliday, LD Hartung, *J Am Chem Soc* 91: 430–433, 1969.
223. O Süs, *Liebigs Ann Chem* 557: 237–242, 1947.

224. RJ Devoe, MRV Sahyun, N Serpone, DK Sharma, *Can J Chem* 65: 2342–2349, 1987.
225. D Kosynkin, TM Bockman, JK Kochi, *J Am Chem Soc* 119: 4846–4855, 1997.
226. GG Wubbels, DP Susens, EB Coughlin, *J Am Chem Soc* 110: 2538–2543, 1988.
227. GG Wubbels, EJ Snyder, EB Coughlin, *J Am Chem Soc* 110: 2543–2548, 1988.
228. A Tissot, P Boule, J Lemaire, *Chemosphere* 12: 859–872, 1983.
229. MHL Stegeman, WJGM Peijnenburg, H Verboom, *Chemosphere* 26: 837–849, 1993.
230. T Moore, RM Pagni, *J Org Chem* 52: 770–773, 1987.
231. J Orvis, J Weiss, RM Pagni, *J Org Chem* 56: 1851–1857, 1991.
232. CH Evans, G Arnadottir, JC Scaiano, *J Org Chem* 62: 8777–8783, 1997.
233. CH Evans, T Gunnlaugsson, *J Photochem Photobiol A* 78: 57–62, 1994.
234. AC Tedesco, RMZG Naal, JC Carreiro, JBS Bonilha, *Tetrahedron* 50: 3071–3080, 1994.
235. AC Tedesco, LC Nogueira, JBS Bonilha, EO Alonso, FH Quina, *Quim Nova* 16: 275–279, 1993.
236. AC Tedesco, LC Nogueira, JC Carreiro, AB da Costa, JBS Bonilha, *Langmuir* 16: 134–140, 2000.
237. M Sawaura, T Mukai, *Bull Chem Soc Jpn* 54: 3213–3214, 1981.
238. P Klan, R Ružička, D Heger, J Literak, P Kulhanek, A Loupy, *Photochem Photobiol Sci* 1: 1012–1016, 2002.
239. CAGO Varma, JJ Tamminga, J Cornelisse, *J Chem Soc Faraday Trans* 2 265–284, 1982.
240. J Urban, P Kuzmič, D Šaman, M Souček, *Collect Czech Chem Commun* 52: 2482–2491, 1987.
241. A Alif, P Boule, J Lemaire, *J Photochem Photobiol A* 50: 331–342, 1990.
242. A Alif, JF Pilichowski, P Boule, *J Photochem Photobiol A* 59: 209–219, 1991.
243. P Boule, C Guyon, J Lemaire, *Chemosphere* 11: 1179–1188, 1982.
244. E Lipczynska-Kochany, *Chemosphere* 24: 911–918, 1992.
245. K Oudjani, P Boule, *New J Chem* 17: 567–571, 1993.
246. G Grabner, C Richard, G Koehler, *J Am Chem Soc* 116: 11470–11480, 1994.
247. AP Durand, RG Brown, D Worrall, F Wilkinson, *J Chem Soc Perkin Trans* 2 365–370, 1998.
248. YI Skurlatov, LS Ernestova, EV Vichutinskaya, DP Samsonov, IV Semenova, IY Rod'ko, VO Shvidky, RI Pervunina, TJ Kemp, *J Photochem Photobiol A* 107: 207–213, 1997.
249. H Lemmetyinen, J Konijnenberg, J Cornelisse, CAGO Varma, *J Photochem* 30: 315–338, 1985.
250. G Zhang, P Wan, *J Chem Soc Chem Commun* 19–20, 1994.
251. S Nagaoka, T Takemura, H Baba, *Bull Chem Soc Jpn* 58: 2082–2087, 1985.
252. V Avila, HE Gsponer, CM Previtali, *J Photochem* 27: 163–170, 1984.
253. T Ichimura, M Iwai, Y Mori, *J Photochem* 39: 129–134, 1987.
254. T Ichimura, M Iwai, Y Mori, *J Phys Chem* 92: 4047–4052, 1988.
255. A Mamantov, *Chemosphere* 14: 901–904, 1985.
256. PHM van Zeil, LMJ van Eijk, CAGO Varma, *J Photochem* 29: 415–433, 1985.
257. K Othmen, P Boule, C Richard, *New J Chem* 23: 857–861, 1999.
258. JP Soumillion, BD Wolf, *J Chem Soc Chem Commun* 436–437, 1981.
259. JR Siegman, JJ Houser, *J Org Chem* 47: 2773–2779, 1982.
260. M Andressen, *Chem Zentr* 66: 550, 1895.

261. L Hörner, H Stohr, *Chem Ber* 85: 993–999, 1952.
262. VG Adam, D Voigt, K Schreiber, *J Prakt Chem* 315: 739–756, 1973.
263. R Knorr, T Phung Hoang, J Mehlstäubl, M Hintermeyer-Hilpert, HL Lüdemann, E Lang, G Sextl, W Rattay, P Böhrer, *Chem Ber* 126: 217–224, 1993.
264. M Nomayo, A Wokaun, *Ber Bunsen-Ges Phys Chem* 99: 1495–1503, 1995.
265. SM Gasper, C Devadoss, GB Schuster, *J Am Chem Soc* 117: 5206–5211, 1995.
266. PSJ Canning, H Maskill, K McCrudden, B Sexton *Bull Chem Soc Jpn* 75: 789–800, 2002.
267. JF Hartwig, *Acc Chem Res* 31: 852–860, 1998.
268. JP Wolfe, S Wagaw, JF Marcoux, SL Buchwald, *Acc Chem Res* 31: 805–818, 1998.
269. G Mann, JF Hartwig, *J Org Chem* 62: 5413–5418, 1997.
270. K Kunz, U Scholz, D Ganzer, *Synlett* 2003, 2428–2439.
271. T Sakamoto, K Ohsawa, *J Chem Soc Perkin Trans 1* 2323–2326, 1999.
272. T Okano, M Iwahara, J Kiji, *Synlett* 243–244, 1998.
273. M Sundermeier, A Zapf, M Beller, *Angew Chem Int Ed* 42: 1661–1664, 2003.
274. M Sundermeier, S Mutyala, A Zapf, A Spannenberg, M Beller, *J Organomet Chem* 684: 50–55, 2003.
275. M Alterman, A Halberg, *J Org Chem* 65: 7984–7989, 2000.
276. F Jin, N Confalone, *Tetrahedron Lett* 41: 3271–3273, 2000.

5 Mechanistic and Synthetic Aspects of SET-Promoted Photocyclization Reactions of Silicon Substituted Phthalimides

Ung Chan Yoon and Patrick S. Mariano

CONTENTS

5.1 INTRODUCTION

5.1.1 SET Photochemistry

Owing to the large energetic driving force provided by the high energies of electronic excited states, single electron transfer (SET) is the key mechanistic event in a wide variety of photochemical processes [1]. Photoinduced SET between neutral donors and acceptors results in the generation of radical ions, highly reactive intermediates which participate in facile and selective reactions as part of mechanistic routes leading to product formation. Consequently, the unique feature of SET-photochemical processes arises from the fact that the chemistry is governed principally by the chemical properties of these radical cation and radical anion intermediates.

One of the most common reaction pathways open to radical ions is α-heterolytic fragmentation. In processes of this type, either an electrofugal or nucleofugal group is expelled from a position adjacent to the charged radical center and this process produces a stabilized neutral radical intermediate (Scheme 5.1). Owing to their pivotal role in SET photochemistry, α-heterolytic fragmentation reactions have been subjected to a wide variety of studies aimed at elucidating the rates of the processes and determining how the rates depend on the structure of the ion radicals and the reaction medium.

SCHEME 5.1

5.1.2 Cation Radical Desilylation Reactions

Research carried out in the authors' laboratories in this area of photochemistry has focused on SET-promoted photochemical processes which are promoted by α-heterolytic fragmentation reactions of radical cation intermediates that are generated from α-trialkylsilyl substituted electron donors. At the outset of this work, the authors recognized that silicon containing organic compounds participate in numerous chemical reactions [2] in which they serve as equivalents of allylic and enolate anions. For example, activation of allylsilanes by using potent silophiles leads to formation of desilylated allyl anions, which participate in reactions with electrophilic reagents to yield products resulting from nucleophilic addition and substitution. In addition, when partnered with strong electrophiles, allylsilanes and enolsilanes undergo addition reactions proceeding via the intermediacy of β-silicon stabilized carbenium ions which readily transfer their silyl groups to even weakly silophilic substances. The latter observations led the authors to suggest that cation radicals derived from α-trialkylsilyl donors would likewise undergo facile desilylation to generate stabilized, carbon centered radicals (Scheme 5.2) [3–6].

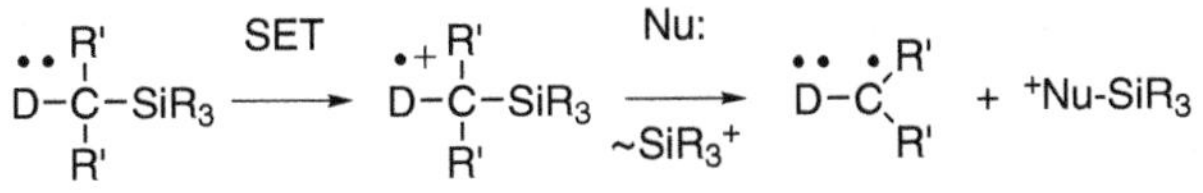

$$D-\underset{\underset{R'}{|}}{\overset{\overset{\bullet\bullet}{|}}{C}}-SiR_3 \xrightarrow{SET} D-\underset{\underset{R'}{|}}{\overset{\overset{\bullet+}{|}}{C}}-SiR_3 \xrightarrow[\sim SiR_3^+]{Nu:} \overset{\bullet\bullet}{D}-\overset{\bullet}{\underset{R'}{\overset{R'}{C}}} + {}^+Nu\text{-}SiR_3$$

SCHEME 5.2

Photochemical and electrochemical investigations [7,8] in the ensuing years confirmed this proposal. The results of these efforts show that organosilanes, which possess silyl-substitution at sites adjacent to electron donor centers, undergo ready oxidation to generate silicon stabilized cation radical intermediates. Stabilization by trialkylsilyl groups in these species is a consequence of hyperconjugative interactions resulting from overlap of the relatively (compared to C-C bonded analogs) high energy doubly occupied σ_{C-Si} orbitals with half-filled heteroatom atomic orbitals or alkene π-orbitals. Electron delocalization, resulting from this interaction, weakens the σ_{C-Si} bond and makes the silicon center more electropositive. As a result, these intermediates undergo fast, silophile promoted desilylation to produce neutral, carbon-centered free radical intermediates. When viewed from the perspective of SET photochemistry, α-silyl substituted electron donors are chemical equivalents of carbon centered free radical intermediates. This conceptual connection has enabled the design of new excited state reactions, in which C-C bond formation occurs by radical coupling and addition pathways.

5.1.3 α-SILYLAMINIUM RADICALS

SET oxidation of α-silylamines is thermodynamically more favorable than that of its nonsilicon containing counterparts owing to the thermodynamic stability of the resulting α-silylaminium radicals. Owing to their kinetic instability, these cation radicals undergo facile trialkylsilyl group transfer reactions with silophiles to form carbon centered α-amino radicals. The dynamics of aminium radical desilylation has been probed by using time resolved laser flash photolysis methods [9,10]. This work has shown that the rate of desilylation of the anilinium radical **2** (R=Me, Scheme 5.3), formed by SET from the corresponding aniline **1** to the singlet excited state of DCB, is slow when weak silophiles like MeCN are present. In this case, bimolecular α-deprotonation of the α-silylanilinium radical by even weak bases (e.g., acetate) occurs more rapidly. However, in the presence of more silophilic species, such as MeOH or water, desilylation of **2** (R=Me) occurs more rapidly. The second order

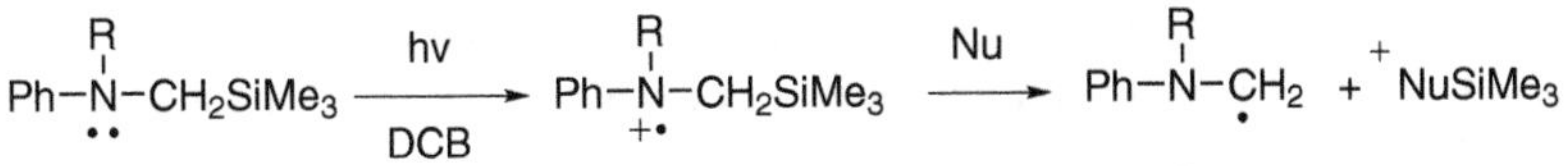

$$\underset{\mathbf{1}}{\overset{\overset{R}{\overset{|}{N}}}{Ph-\underset{\bullet\bullet}{N}-CH_2SiMe_3}} \xrightarrow[DCB]{hv} \underset{\mathbf{2}}{\overset{\overset{R}{\overset{|}{N}}}{Ph-\underset{+\bullet}{N}-CH_2SiMe_3}} \xrightarrow{Nu} \underset{\mathbf{3}}{\overset{\overset{R}{\overset{|}{N}}}{Ph-N-\overset{\bullet}{C}H_2}} + {}^+NuSiMe_3$$

SCHEME 5.3

rate constants for the desilylation processes are dependent on the silophilicity of the silophile (e.g., 8.9×10^5 M^{-1} s^{-1} for MeOH, 1.3×10^6 M^{-1} s^{-1} for H_2O, and 3.2×10^9 $M^{-1} s^{-1}$ for fluoride ion in MeCN).

Electron withdrawing substituents on nitrogen significantly enhance the rates of desilylation of α-silylaminium radicals. This is seen in comparisons of the second order rate constants for methanol promoted desilylation of anilinium radicals **2** having N-methyl (R=Me, 7×10^5 M^{-1} s^{-1}), N-ethoxycarbonyl (R= CO_2Et, 2×10^7 $M^{-1} s^{-1}$), and N-acetyl (R=COMe, 6×10^7 $M^{-1} s^{-1}$) substituents. Since an N-electron withdrawing group pronouncedly increases the oxidation potential of the silylaniline **1** while having little effect on the stability of the resulting α-amino radical **3**, the rate acceleration is attributable to destabilization of the amine cation radical **2**. The ability to control the nature of fragmentation reactions of silylamine cation radicals by the choice of the solvent/silophile serves as a key component in the design of new and highly efficient SET-promoted photocyclization reactions.

5.2 SET-PHOTOCHEMISTRY OF SILICON CONTAINING PHTHALIMIDES

5.2.1 PHOTOADDITION AND PHOTOCYCLIZATION REACTIONS OF PHTHALIMIDES

From the beginning, SET photochemical investigations have focused on a wide variety of electron acceptors which are used either as SET-photosensitizers or reactants. Included in this group are members of the cyanoarene, conjugated ketone, and iminium and related N-heteroaromatic salt families [1,2–5]. Early pioneering studies by Kanaoka [11–14] and Coyle [15,16] provided the first examples of SET-photochemical reactions in which phthalimides serve as electron acceptors. Owing to their large positive excited state reduction potentials (E*(–) ca. 1.6 to 2.1 V) [17], phthalimides participate as electron acceptors in photoinduced SET processes with a variety of donors having ground state oxidation potentials lower than ca. 2 V.

Investigations of phthalimide photochemistry in the authors' laboratories have concentrated on SET-induced excited-state reactions with α-trialkylsilyl substituted ethers, thioethers, and amines. These efforts have uncovered several interesting photochemical reactions. For example, in an early effort simple, α-silyl-substituted ethers, thioethers, and amines were observed to undergo efficient photoaddition reactions with phthalimide and its N-methyl derivative (Scheme 5.4) [20]. In these processes, thermodynamically/kinetically driven SET from the α-silyl donors **13** to excited phthalimide leads to formation of ion radical pairs **16** (Scheme 5.5). Solvent (MeOH) promoted desilylation of the cation radicals **16** and protonation of the phthalimide anion radicals **15** then provides radical pairs **17**, the direct precursor of adduct **14**.

In an extension of this effort, the authors and their coworkers showed that photocyclization reactions of phthalimides containing N-tethered α-silyl donors

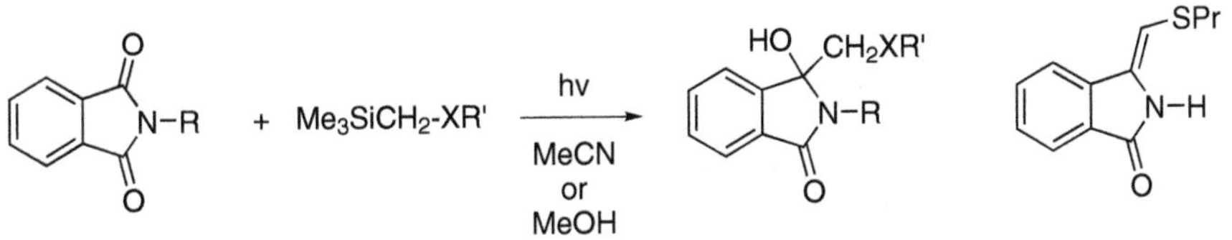

4 (R = H)
5 (R = Me)

6 (XR' = OEt)
7 (XR' = SPr)
8 (XR' = NEt₂)

9 XR' = OEt
10 (XR' = SPr)
11 (XR' = NEt₂)

12

Reactants	Products (Yield)
4 + 6	**9** (R = H, 40%)
4 + 7	**10** (12%), **12** (36%)
5 + 7	**11** (R = Me, 78%)
5 + 8	**12** (R = Me, 41%)

SCHEME 5.4

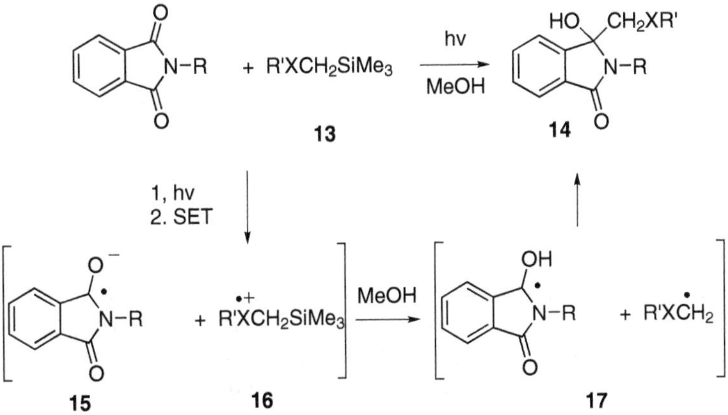

SCHEME 5.5

proceed with high chemical efficiencies to generate a variety of novel heterocyclic products. For example, irradiation of MeOH solutions of the phthalimide linked silyl ethers **18** leads to formation of the amidol containing, oxygen heterocycles **19** in excellent yields (Scheme 5.6) [21]. Analogous photocyclization reactions are observed with the phthalimido thioethers **20** [22]. Although dehydration of the primary photoproducts (**21** → **22**) occurs with varying ease in these systems, the overall yields of the reactions remain high even when the size of the ring that is formed in the ultimate diradical cyclization steps is large.

Studies with α-silylamine and α-silylamide linked phthalimides have given insight into the design of preparatively useful SET-photochemical reactions [23,24].

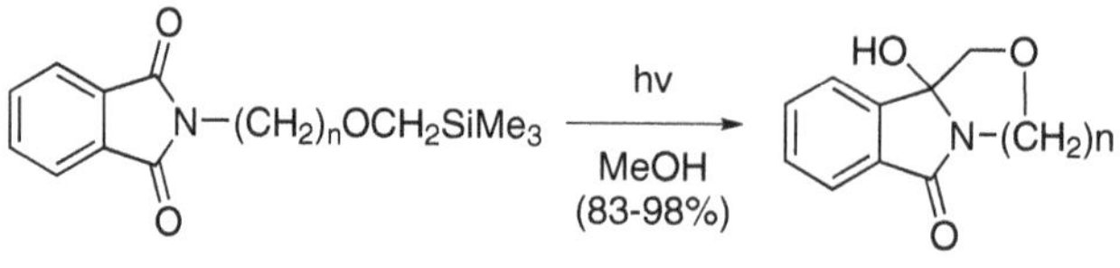

18 (n = 2-4)   **19** $\left\{\begin{array}{l} n = 2, 98\% \\ n = 3, 99\% \\ n = 4, 83\% \end{array}\right\}$

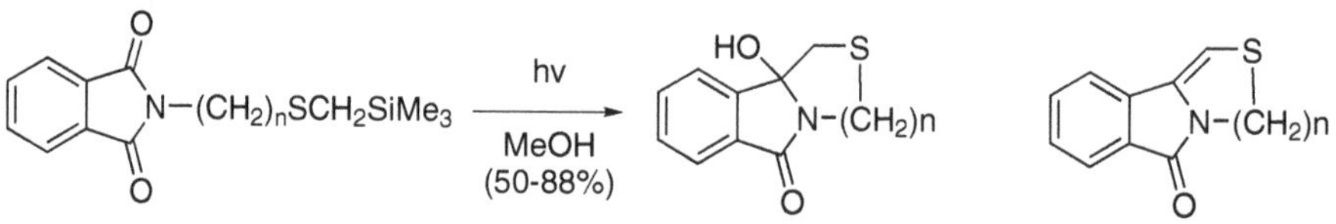

20 (n = 2-6)   **21**   **22**

n	21	22
2	50%	---
3	45%	19%
4	75%	12%
5	88%	---
6	---	62%

SCHEME 5.6

For example, irradiation of a MeOH solution of the silylamino-phthalimides **23** was found to promote unselective formation of a mixture of products including the fused diazines **29** (R=Me) and amidol **30** (R=Me) and azetidinol **31** (Scheme 5.7). On the other hand, the tricyclic acetamide or sulfonamide **29** (R=Ac or Ms) is formed in high yield when the silylacetoamido- and silylsulfonamido-phthalimides **24–28** are irradiated in MeOH (Scheme 5.7). These results demonstrated that the preparative utility of SET-promoted photocyclization reactions of nitrogen containing electron donors can be enhanced by using substrates that contain α-silylacetamide or α-silylsulfonamide rather than α-silylamine donor sites. Additional studies showed that the quantum yields of these photocyclization reactions are also dependent on the nitrogen-substituent with the acetamide or sulfonamide substrates **24** ($\phi = 0.22$) and **28** ($\phi = 0.12$) reacting with a greater efficiency than the amine analog **23** ($\phi = 0.05$).

An analysis of the factors that govern the chemical and quantum efficiencies of all excited-state reactions, which proceed via pathways where initial SET is followed by secondary reactions of charged radical intermediates, provides a solid rationale for these observations. SET from the α-silylsulfonamide donor in **28** ($\Delta G_{SET} = -0.1$ V) to the singlet excited state of the phthalimide chromophore, for example, should be less

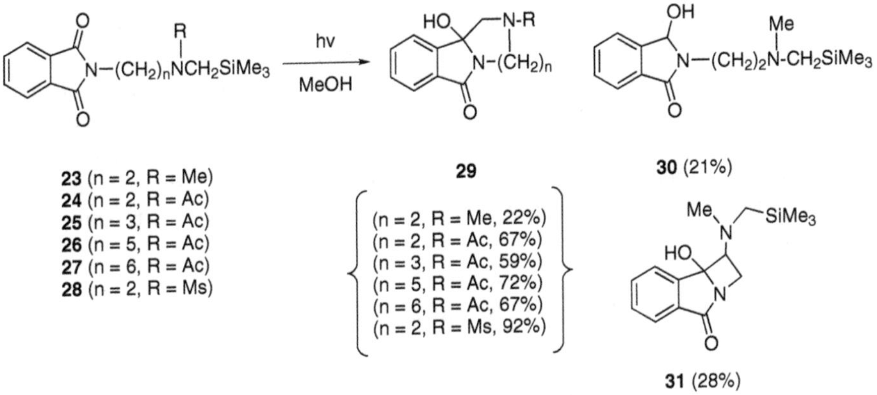

SCHEME 5.7

exothermic than that from the α-silylamine donor in **23** ($\Delta G_{SET} = -1.5$ V) [24]. As a result, it should occur more slowly than from the α-silylamine donor to the excited phthalimide in **23**. As a result, SET in the excited state of **28** is less competitive with other decay pathways. Thus, if the rates of excited state electron transfer were solely responsible for determining the quantum efficiencies of SET-promoted photoreactions, reaction of **28** would have been less efficient than that for **23**.

The intermediate zwitterionic biradicals **32** (Scheme 5.8) formed by excited-state SET in the respective silylamino- and silylsulfonamido-phthalimides **23** and **28** can react by several different routes. Highly exothermic back-SET to produce ground state reactants should take place with rates which are essentially independent of the nitrogen substituent. Competitive with this decay mode is methanol-induced

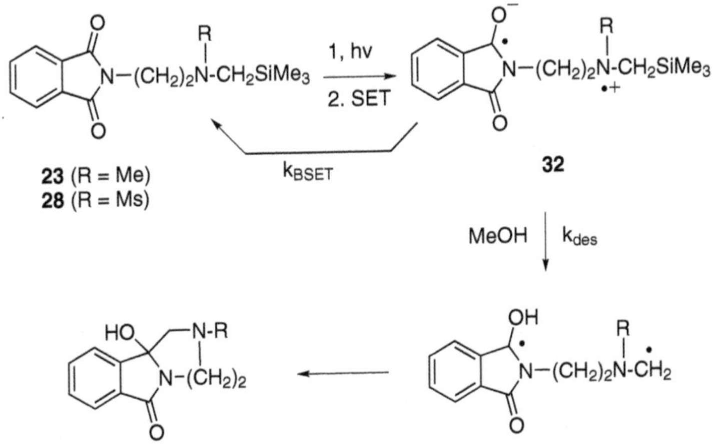

SCHEME 5.8

α-desilylation, which the authors' earlier studies [9,10] demonstrated occur with rates that are highly dependent on the nitrogen substituent. Specifically, methanol induced desilylation of silylamine cation radicals is two orders of magnitude slower than the analogous reaction of α-silylamide cation radicals. Thus, the larger rate constant for desilylation of zwitterionic biradicals derived by photoinduced intramolecular SET in α-silylsulfonamido-phthalimide **28** translates into a larger rate constant ratio, $k_{des}/(k_{des} + k_{BSET})$ and, consequently, a larger quantum efficiency for SET-promoted photocyclization.

5.2.2 Photomacrocyclization Reactions of Silicon Substituted Phthalimides

As a result of their high chemical yields, SET-promoted photocyclization reactions of linked phthalimide acceptors and α-silyl-substituted polyheteroatom containing donors can be incorporated into novel strategies for the preparation of interesting cyclic substances. This proposal is based on an analysis of the mechanistic pathways followed in the potentially competitive photoreactions of substrates in this family. As depicted in Scheme 5.9, irradiation of polydonor-linked phthalimides **33**, which

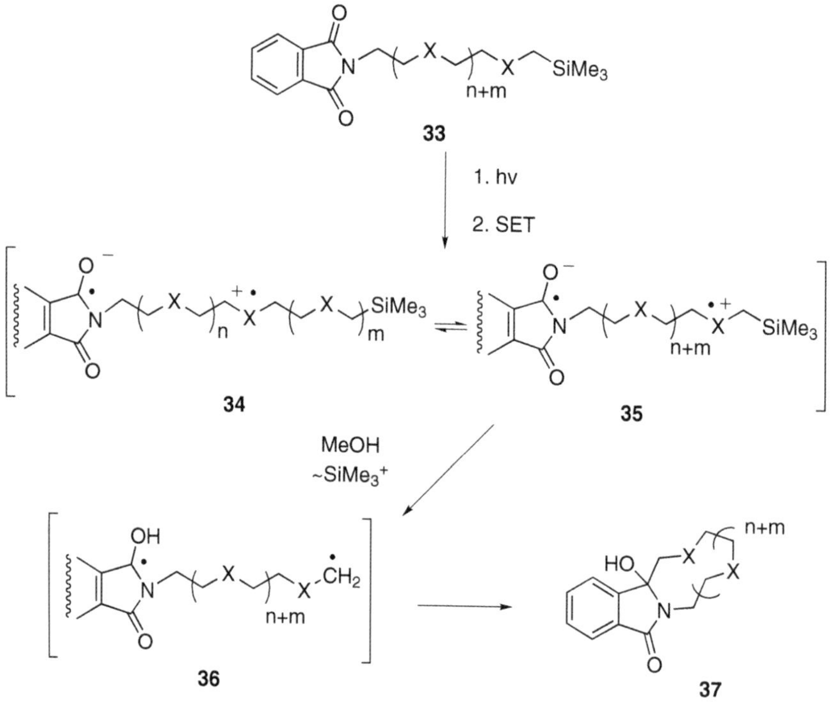

SCHEME 5.9

possesses trimethylsilyl groups at terminal donor sites, is expected to result in regioselective production of macrocyclic products **37**. The key excited state SET steps in these pathways should generate mixtures of possibly rapidly interconverting zwitterionic biradicals, **34** and **35**. The relative populations of these biradicals could be governed either by the rates of the intramolecular SET from each donor site or by the relative energies of the cation radicals, as judged by the oxidation potential at each heteroatom center. Importantly, the rates of secondary α-deprotonation and α-desilylation reactions that take place at centers adjacent to each of the cation radical sites would not depend on the relative populations of the zwitterionic biradicals. Rather, the relative rates of these processes which produce the 1,n-biradical precursors of cyclization products, should be determined by the rates of the competing fragmentation reactions. Since desilylation of the terminal cation radical **35** should be the most rapid process, selective generation of the diradical precursors **36** of the macrocyclic products **37** should be observed.

5.2.2.1 Photomacrocyclization Reactions of Polydonor-Linked Phthalimides

A three-pronged approach has been used to (1) probe the scope and limitations of the SET-promoted photomacrocyclization processes, (2) gain information about the factors that govern chemical yields and quantum efficiencies, and (3) investigate applications to the preparation of potentially important materials.

Exploratory studies with a variety of phthalimides, containing trimethylsilyl-terminated poly-sulfonamide, poly-ether and poly-thioether chains, showed that these substances undergo modestly to highly efficient photocyclization reactions to produce the corresponding macrocyclic products. Examples of these processes are shown in Scheme 5.10 through Scheme 5.15 [25,26].

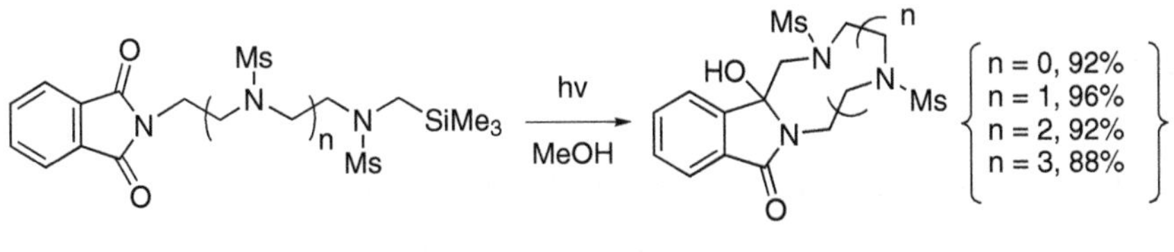

SCHEME 5.10

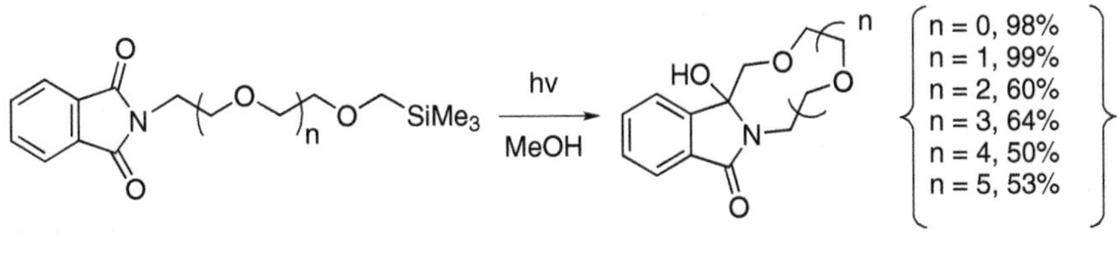

SCHEME 5.11

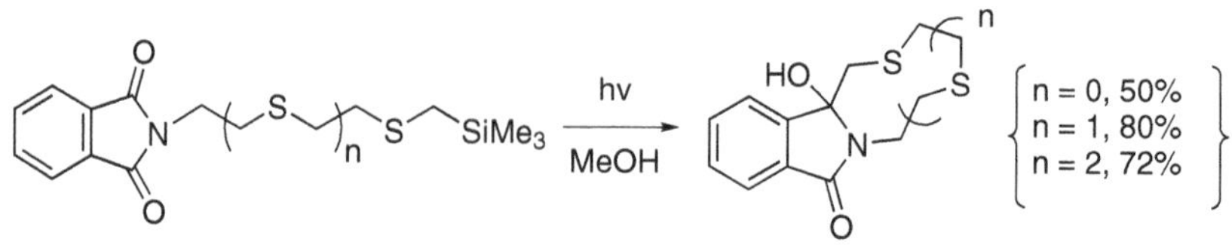

SCHEME 5.12

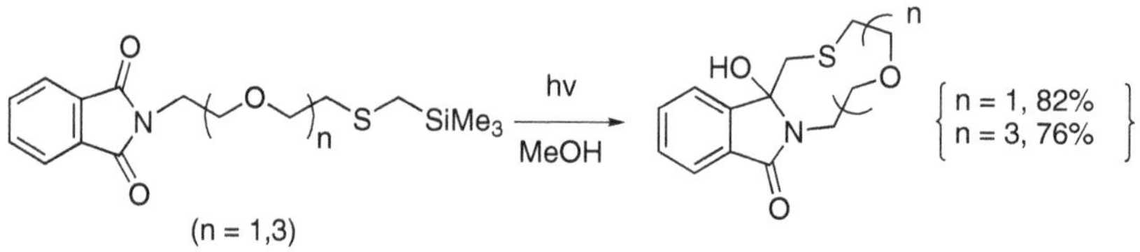

(n = 1,3)

SCHEME 5.13

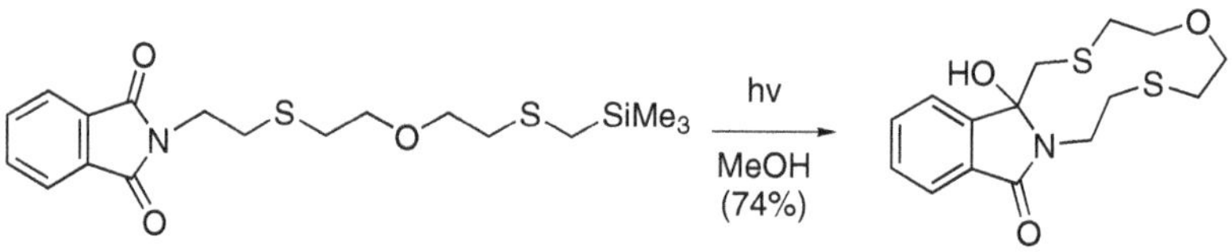

SCHEME 5.14

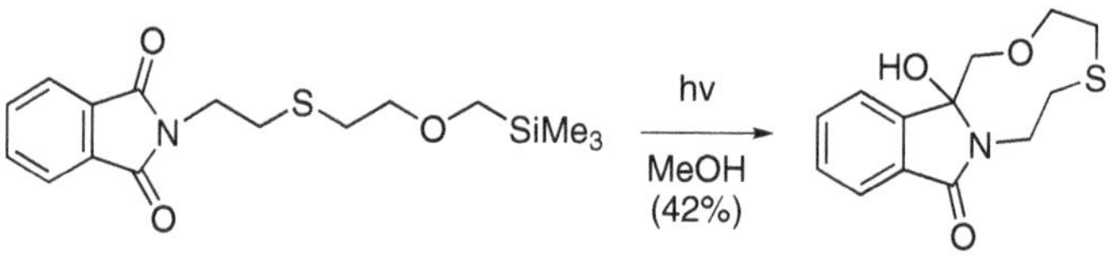

SCHEME 5.15

5.2.2.2 Factors Governing the Chemical and Quantum Yields of Photomacrocyclization Reactions of Silicon Substituted Phthalimides

As a consequence of their structural outcomes and high yields, the photomacrocyclization reactions shown above can serve as the foundation of practical synthetic routes for the preparation of polyfunctionalized macrocycles. However, before these methods can be practically implemented, a detailed understanding of the factors that govern chemical yields and quantum efficiencies of these photochemical reactions must be developed. As mentioned above, several factors can contribute in controlling the chemical and quantum yields in photoreactions of the polydonor-substituted phthalimides **38** (Scheme 5.16). For example, photoproduct distributions (yields) in these processes should be governed by the relative rates of intramolecular SET from the respective donor sites to the excited phthalimide chromophore and the rates of cation radical fragmentation reactions of the intermediate zwitterionic biradicals **39** and **40**. This would be the case when interconversion of the initially formed zwitterionic biradicals is slower than the fragmentation reactions. In another limiting situation where zwitterionic biradical interconversion is faster than fragmentation, an equilibrium mixture of the zwitterionic biradicals will be generated in which the mole fraction of each is governed by the redox potential at each donor site. The chemical efficiencies for formation of products **41** and **42** in these cases will be dependent on the relative rates of the competing fragmentation reactions, if the energy barriers for

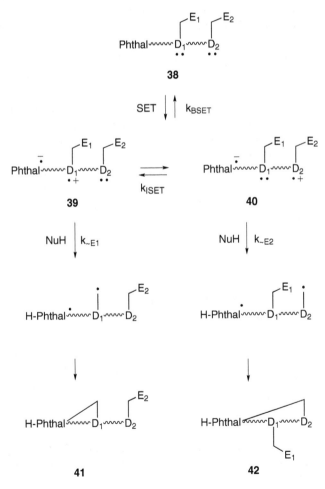

SCHEME 5.16

fragmentation are higher than those for interconversion of the zwitterionic biradical. More complex situations exist in cases where the rates of zwitterionic biradical interconversion are either in the same range as or somewhat less than the rates of cation radical fragmentation.

This mechanistic analysis demonstrates that the chemical efficiencies of SET-promoted reactions of polydonor-linked phthalimides could depend on the number, types, location, and reactivity of ion radical centers formed by either initial excited state SET or intrasite-SET. Clearly, knowledge about how these factors govern product yields, regiochemical selectivities, and quantum efficiencies is crucial to the design of synthetically useful photochemical reactions of these substrates. Recent studies in the authors' laboratories [27] with substrates **43** (Scheme 5.17) containing light absorbing phthalimide and naphthalimide acceptor moieties and a variety of N-linked bis-donor sites have provided information about these issues. A standard donor group ($NMsCH_2SiMe_3$) with known oxidation potential and rate of methanol induced cation radical fragmentation was incorporated at the terminal position in each of these substances. Finally, the oxidation potentials and fragmentation rates at the other donor sites in these substances were varied by using different heteroatoms and/or substituents.

Fluorescence spectroscopic studies with donor linked 2,3-naphthalimides have shown that the rates of intramolecular SET from donors to the naphthalimide singlet excited state vary over three orders of magnitude range and are directly dependent on the electron donating ability of the donor. For example, first-order rate constants for SET from tertiary amine centers in naphthalimides **43** (D = NMe) is 4×10^9 s^{-1}. In contrast, SET from the α-silylsulfonamide donor site in **43** (D = NMs) is much slower (1×10^8 s^{-1}). Thus, the rates of intramolecular SET from the donor sites in the linked naphthalimides parallel those predicted by using simple free energy calculations. Accordingly, SET from the tertiary amine centers ($E_{1/2}(+)$ = ca. 0.6 V) to the singlet excited 2,3-naphthalimide chromophore ($E_{1/2}^{S1}(-)$ = $ca.$ 1.8 V) [28] is highly exothermic in contrast to SET from more weakly donating sulfonamide ($E_{1/2}(+)$ = ca. 2 V) [24,29] and ether ($E_{1/2}(+)$ = ca. 2 V) [30] donors. As a consequence, the relative populations of zwitterionic biradicals **44** and **45** (Scheme 5.17), formed initially by intramolecular excited state electron transfer, should be biased in favor of those that contain lower energy cation radical sites.

Several interesting trends are noted in the results of preparative photochemical reactions with these substrates. Firstly, α-silylsulfonamido-phthalimide and -naphthalimide that contain other heteroatom donor sites (e.g., O, NMs) with higher or nearly equal oxidation potentials as compared to the terminal sulfonamide, undergo chemically efficient/highly regioselective photocyclization reactions via sequential SET-desilylation pathways. Examples are found in photocyclization reactions of phthalimides **50–53** (Scheme 5.18) and naphthalimides **54–57** (Scheme 5.19). Secondly, α-silylsulfonamide substrates, which possess an internal, strongly electron donating tertiary amine site, undergo low yielding unselective photoreactions. In these cases, exemplified by the amino-sulfonamides **58** and **59** (Scheme 5.20), the

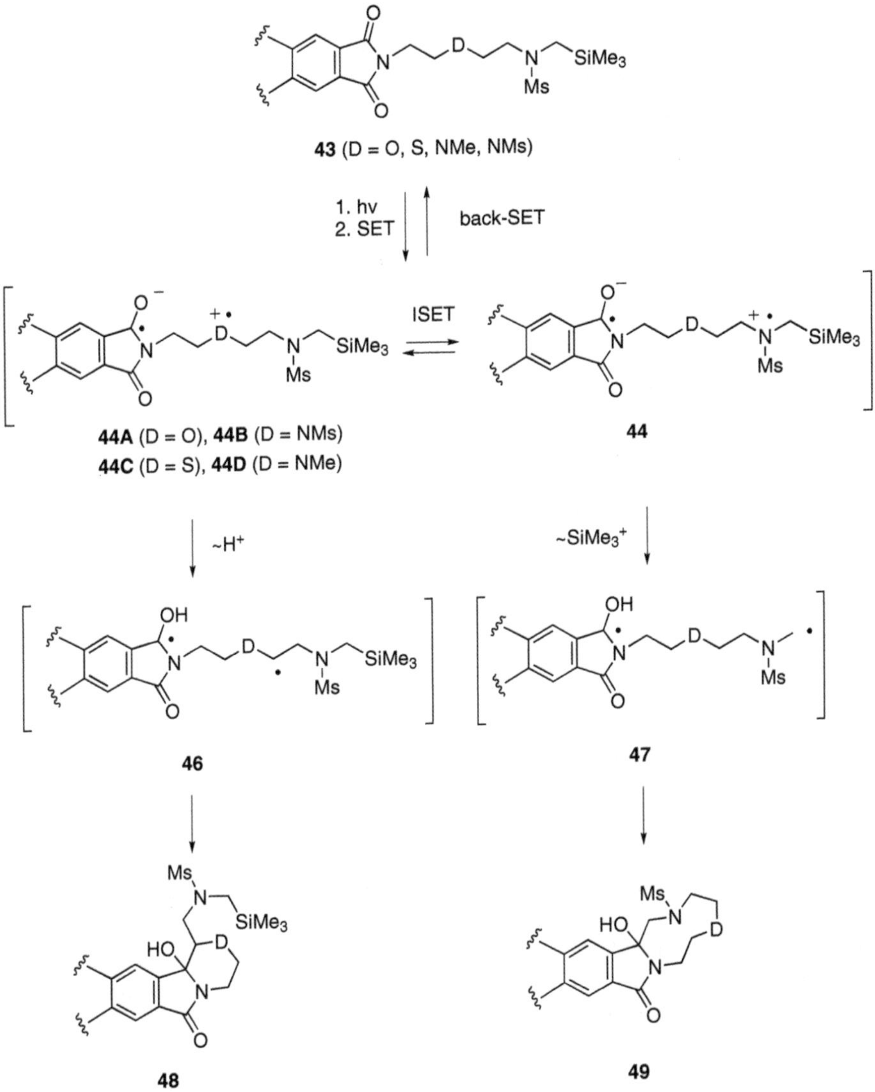

43 (D = O, S, NMe, NMs)

1. hv
2. SET   back-SET

ISET

44A (D = O), **44B** (D = NMs)
44C (D = S), **44D** (D = NMe)

44

~H+   ~SiMe3+

46   **47**

48   **49**

SCHEME 5.17

major pathway involves α-proton transfer from the internal tertiary aminium radical centers followed by biradical cyclization or disproportionation. Finally, the thioether containing phthalimide-sulfonamide **60** (Scheme 5.21) represents an intermediary case where highly efficient photocyclization takes place via a sequential SET-desilylation route despite the presence of a sulfur donor site that has a significantly lower oxidation potential than the terminal α-silylsulfonamide.

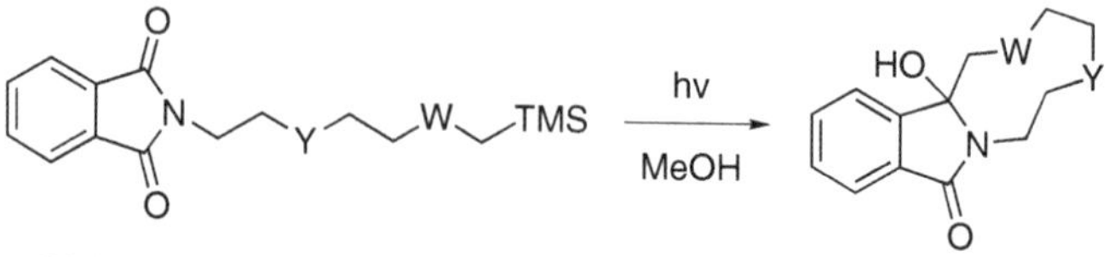

50 (Y = W = NMs) (96%)
51 (Y = O, W = NMs) (91%)
52 (Y = O, W = NAc) (93%)
53 (Y = OCH$_2$CH$_2$O, W = NMs) (89%)

SCHEME 5.18

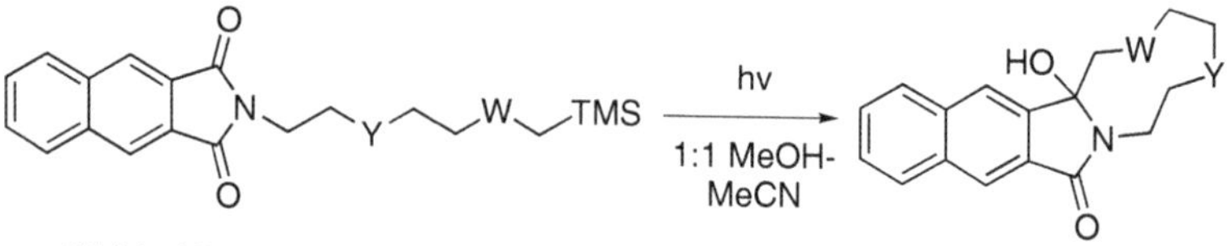

54 (Y = W = NMs) (100%)
55 (Y = O, W = NMs) (88%)

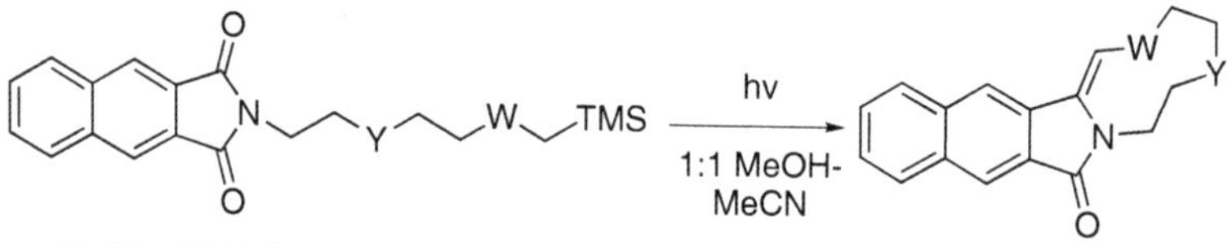

56 (Y = OCH$_2$CH$_2$O, W = NMs) (70%)
57 (Y = O, W = NAc) (80%)

SCHEME 5.19

These observations suggest that at least two factors contribute in governing the chemical efficiencies/regioselectivities of photocyclization reactions of the bis-donor linked, α-silylsulfonamide terminated phthalimides and naphthalimides. The first contribution comes from the relative intrinsic rates of the α-heterolytic fragmentation processes (~H$^+$, ~SiMe$_3^+$) which can occur at each cation radical site. A second factor is the relative energies of the zwitterionic biradicals which not only control the initial and final populations but also govern the energy barriers for the competing α-heterolytic fragmentation processes. Thus, fast α-desilylations at α-silylsulfonamide sites of zwitterionic biradicals **45** (Scheme 5.17) are the dominant reaction pathways followed when the competitively formed, zwitterionic biradicals **44**, which can only undergo slow α-deprotonation, are of near equal or higher energy than **45**. However, when the internal zwitterionic biradicals **44** have significantly lower energies than their terminal counterparts **45**, pathways involving α-deprotonation at the internal cation radical site can predominate.

Important information about how the energies and fragmentation rates of zwitterionic biradical intermediates govern the nature and efficiencies of SET-promoted

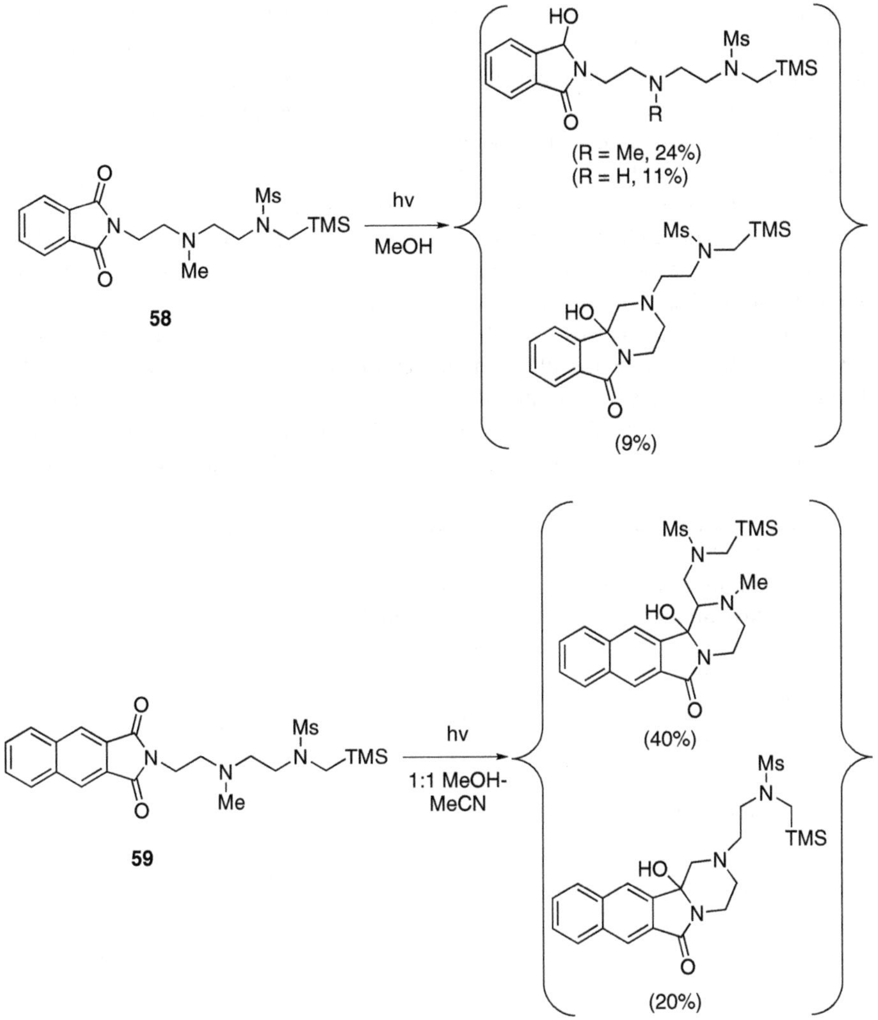

SCHEME 5.20

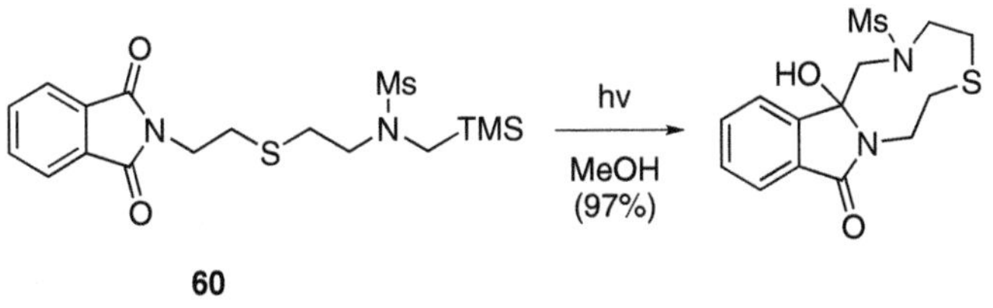

SCHEME 5.21

photoreactions of acceptor-polydonor substrates, has been gained from reaction quantum yield measurements. Based on the mechanistic sequence shown in Scheme 5.17, it is expected that the quantum efficiencies for reactions of the bis-donor substrates will depend on the relative rates of the fragmentation reactions which compete with back-SET for deactivation of the zwitterionic biradical intermediates. Specifically, the fraction of the biradical intermediates, which react to form products, is dependent on the combined rates of α-deprotonation and α-desilylation as compared to that for back-SET. Also, depending on whether interconversion of the zwitterionic biradicals is faster or slower than their decay, the equilibrium constants for interconversion of these intermediates will also contribute in governing reaction quantum yields.

Data accumulated earlier in studies of photocyclization reactions of mono-donor substituted aminoethyl-phthalimide derivatives provide a framework for the development of a plausible interpretation of the quantum yields for reactions of the bis-donor substituted phthalimides (Table 5.1). Earlier, we observed that the quantum efficiencies for photocyclization reactions of substituted aminoethyl-phthalimides

TABLE 5.1
Quantum Yields for Reactions of the
α-Sillylsulfonamido-Terminated Phthalimides

50 (D = NMs))
51 (D = O)
58 (D = NMe₃)
60 (D = S)

61

62

Compounds	Quantum Yield (Φ)[a]
50	0.12
51	0.14
58	0.04
60	0.02
61	0.12
62	0.003

[a] Measured in CH_3OH.

vary in a regular manner with the nature of the α-electrofugal group (E=H, TMS) and nitrogen substituent (R=Me, Ac, or Ms) [10,24]. Accordingly, the quantum yields for photocyclization reactions of the α-silylsulfonamido terminated bis-donor-phthalimides should be in the range of those for the mono-silylsulfonamido-phthalimide 61 ($\phi = 0.12$, Table 5.1) if the radical cation sites in zwitterionic biradicals derived from the former substances reside exclusively at the terminal sulfonamide moieties. Interestingly, this is the case for photoreactions of the bis-sulfonamide 50 ($\phi = 0.12$) and ether-sulfonamide 51 ($\phi = 0.14$), which have photoreaction quantum yields that are nearly equal to that of the mono-α-silylsulfonamidoethyl-phthalimide 61 ($\phi = 0.12$). In contrast, photoreaction of the tertiary amine containing α-silyl-sulfonamide 58 ($\phi = 0.04$) has a quantum efficiency that matches those of tertiary-aminoethylphthalimides that undergo photoreactions via a sequential SET-proton transfer pathway (e.g., 62, $\phi = 0.003$). Moreover, although proceeding with a high chemical yield, photoreaction of the thioether-silylsulfonamide 60 has a much lower quantum efficiency ($\phi = 0.02$) than those of the model mono-silyl-sulfonamide (e.g., 61, $\phi = 0.12$).

The combined results of this effort provide a view of the factors that control the chemical and quantum efficiencies of photochemical reactions of polydonor-substituted phthlimides. This is best seen by using the energy profiles shown in Figure 5.1. Assignments of the relative energies of the zwitterionic biradicals, 44A–D and 45 (Scheme 5.17), which serve as intermediates in photoreactions of the bis-donor linked phthalimides probed in this effort, are made on the basis of the known oxidation potentials of model ethers (ca. + 2.5 V) [30], methanesulfonamides (ca. + 2.5 V) [29], α-silyl-methanesulfonamides (ca. + 2.0 V) [24], thioethers (ca. + 1.4 V) [31], and tertiary amines (ca. + 0.6 V) [32]. In addition, the relative heights of the intrinsic energy barriers for competing α-heterolytic fragmentation reactions (~H$^+$, ~SiMe$_3$$^+$) of the zwitterionic biradicals are estimated by using kinetic data obtained from studies of α-deprotonation and α-desilylation reactions of tertiary amine and amide derived cation radicals [9,10]. Accordingly, the intrinsic barriers for MeOH induced desilylation are lower than those for deprotonation of amine, thioether, ether, and sulfonamide cation radicals. The kinetic and thermodynamic estimates presented in Figure 5.1 lead to the experimentally verified conclusion that reactions of the (potentially) rapidly interconverting mixture of zwitterionic biradicals, 44A (D = O)–45 and 44B (D = NMs)–45, formed from SET in the corresponding ether-sulfonamides 43 (D = O) and bis-sulfonamide 43 (D = NMs), will be dominated by α-desilylation. Moreover, in cases where the lowest energy zwitterionic biradicals 45 (D = O or NMs) which have α-silyl substituents are also the most reactive, the energy barriers (thus rates) for desilylation (ΔE_{45-E}) will be nearly the same as that for desilylation of the zwitterionic biradicals derived from the mono-silylsulfonamide 61. This prediction is consistent with the observation that the quantum yields for photocyclization reactions of ether-sulfonamide 51 ($\phi = 0.14$) and bis-sulfonamide 50 ($\phi = 0.12$) are in the same range as that for photocyclization of the corresponding mono-donor model 61.

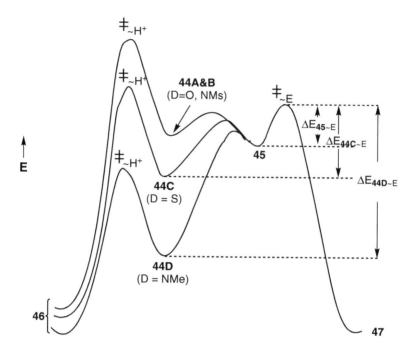

FIGURE 5.1 Energy profiles for competing α-deprotonation and α-desilylation reactions of zwitterionic biradicals 44 and 45 serving as intermediates in SET-promoted photochemical reactions of bis-donor substituted phthalimides. See Scheme 5.17 for the structures of the zwitterionic biradicals 44A–D and 45 and biradicals 46–47.

The situation is different in the case of the zwitterionic biradicals **44C** (D = S)–**45** arising from the thioether-sulfonamide substituted phthalimide **60**. The observation that **60** undergoes chemically efficient photocyclization by a sequential SET-desilylation pathway suggests that the energy of the transition state for desilylation ($\neq_{\sim E}$) of **45** (D = S) is lower than the transition state for α-deprotonation ($\neq_{\sim H}$) at the internal thioether cation radical center in **44C** (D = S). The unique/distinguishing feature of photoreaction of thioether-sulfonamide **60** is that the energy of the apparently reactive zwitterionic biradical **45** (D = S) is significantly higher (ca. 14 kcal/mol) than that of its apparently unreactive partner **44C** (D = S). Although this difference does not influence the chemical yield/regioselectivity of the process, it has a profound effect on the quantum efficiency. Specifically, the energy barrier for desilylation of the zwitterionic biradicals **44C** (D = S)–**45** (ΔE_{44C-E}) is now much greater than that (ΔE_{45-E}) for desilylation of either **44A** or **44B** (D = O or NMs)–**45** and, as a result, desilylation is less competitive with back-SET.

Photoreaction of the tertiary amine containing α-silylsulfonamide **58** represents the third scenario for photoreactions of polydonor-substituted phthalimides, which proceed via the intermediacy of two or more potentially interconverting zwitterionic biradicals. In this case, the energy of the tertiary aminium radical containing zwitterionic biradical **44D** (D = NMe) is much lower (ca. 44 kcal/mol) than that of its α-silylsulfonamide cation radical partner **45** (D = NMe). As a result, the energy barrier for desilylation of **44D** (D = NMe)–**45** is insurmountably high. Consequently, photoreaction of this substrate takes place by the lower energy pathway involving slow deprotonation of **44D** (D = NMe). Also, this process has the same low quantum efficiency ($\emptyset = 0.04$) as do photoreactions of tertiary-aminoethyl-phthalimides which take place by sequential SET-deprotonation routes.

This investigation has provided some general guidelines for predicting the chemical yields/regioselectivities and quantum efficiencies of SET-promoted photochemical reactions of acceptor-polydonor systems. For example, when the lowest energy zwitterionic biradical, derived by intramolecular and/or intrasite SET in a substrate of this type, also contains the most reactive cation radical site, a highly chemically efficient/regioselective photoreaction will ensue. Also, the quantum efficiency of the process will be in the range of that for photoreaction of a simple acceptor-monodonor model. On the other hand, photoreactions of substrates in which the lowest energy zwitterionic biradical is also the least reactive will have low quantum efficiencies. In addition, these processes will have either high or low chemical yields/regioselectivities, depending on whether the relative energies of the transition states for competing cation radical fragmentation processes are large or small.

5.2.2.3 Application of SET-Promoted Photocyclization Reactions of Phthalimide to the Preparation of Cyclic Peptide Mimetics

As stated above, the unique features of SET-promoted photocyclization reactions of polydonor-linked phthalimides make these processes compatible with strategies for the efficient construction of highly functionalized macrocyclic targets. This is exemplified by the plan for synthesis of cyclic peptide mimics. The substrates for the key photomacrocyclization processes (**65** or **66**, Scheme 5.22) in the approach would be prepared by linking a carboxylic acid containing phthalimide to the N-terminal amino group of an intact peptide (**63** or **64**). The synthetic plan used for preparation of the phthalimide-linked polypeptides would be guided by the structural requirements of the target cyclic peptides. Specifically, a precursor to a selectively reactive α-trimethylsilylamide cation radical site would be incorporated at a preselected location within or at the end of the peptide chain. Irradiation of the substrate is then expected to initiate a reaction cascade involving near neighbor SET, intra-chain SET, and desilylation at the reactive amide cation radical position. Cyclization of the biradical, formed in this fashion, then produces the target cyclopeptide mimic (**67** or **68**).

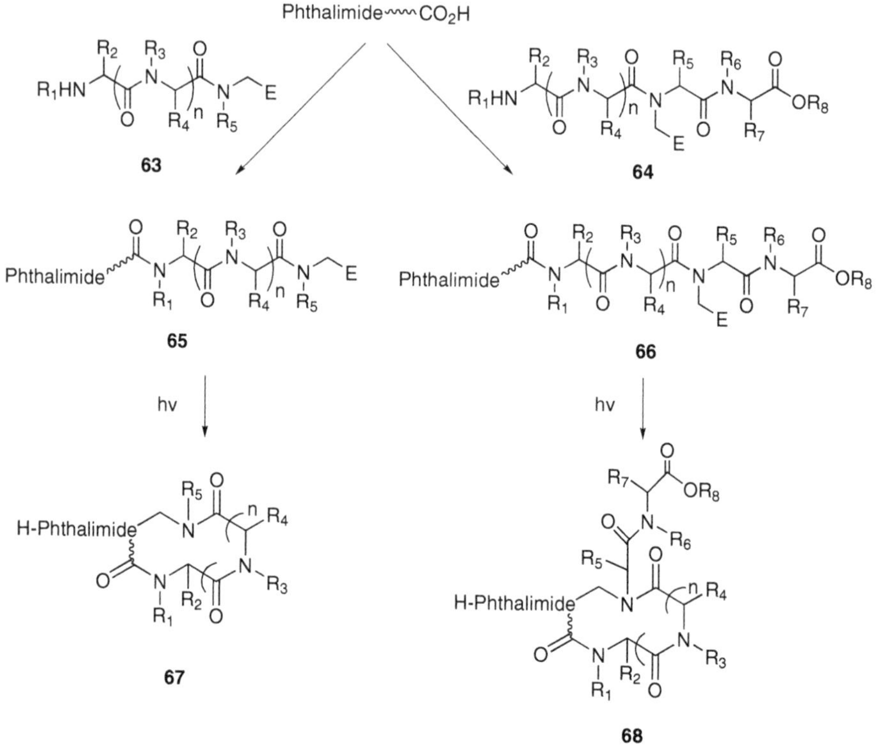

SCHEME 5.22

Studies with a series of TMS-terminated phthalimidoamides and phthalimidopeptides **69–76** were carried out in the authors' laboratories in order to test the scope and potential limitations of this methodology for the preparation of cyclopeptide mimetics [33]. As seen by viewing the results summarized in Scheme 5.23, irradiation of the TMS-terminated phthalimidopeptides leads to modestly efficient production of the cyclopeptide analogs **77–84**.

In addition, photocyclization reactions of the TMS-terminated glycine-((S)-alanine)$_n$ peptides **85–87** were studied to show that the SET-photocyclization based methodology can be used to synthesize stereoregular cyclic peptide analogs (Scheme 5.24). An important feature of photoreactions of these substrates resides in the fact that a new chiral center is created at the bicyclic amidol carbon in the products **88–90**. Earlier studies with a wide variety of related phthalimide derived photoproducts have demonstrated that configurational inversion at amidol centers of this type, via either reversible N-acyliminium ion or amido-ketone forming pathways, is slow under neutral conditions [34]. As a consequence, stereochemical preferences in photocyclization reactions of alaninyl peptides **85–87** would need to be the result of kinetic factors governing the rates of cyclization of the ultimate

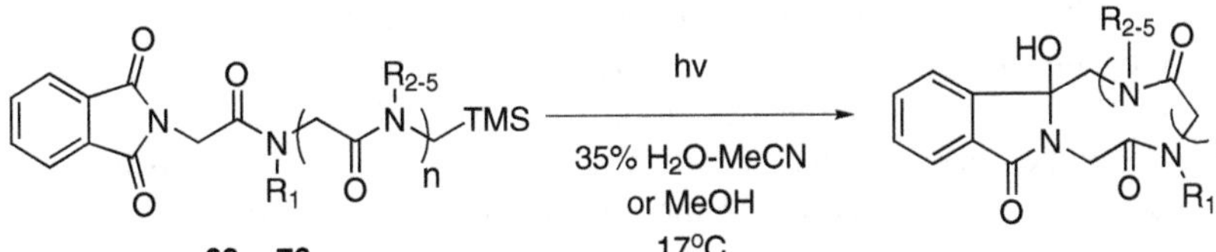

SCHEME 5.23

Substrate	Product	n	R_1	R_2	R_3	R_4	R_5	Yield
69	77	0	H	-	-	-	-	18%
70	78	0	Bn	-	-	-	-	60%
71	79	1	Bn	Bn	-	-	-	40%
72	80	1	Me	Bn	-	-	-	40%
73	81	2	Bn	Bn	Bn	-	-	55%
74	82	2	Me	Me	Bn	-	-	15%
75	83	3	Bn	Bn	Bn	Bn	-	70%
76	84	4	Bn	Bn	Bn	Bn	Bn	61%

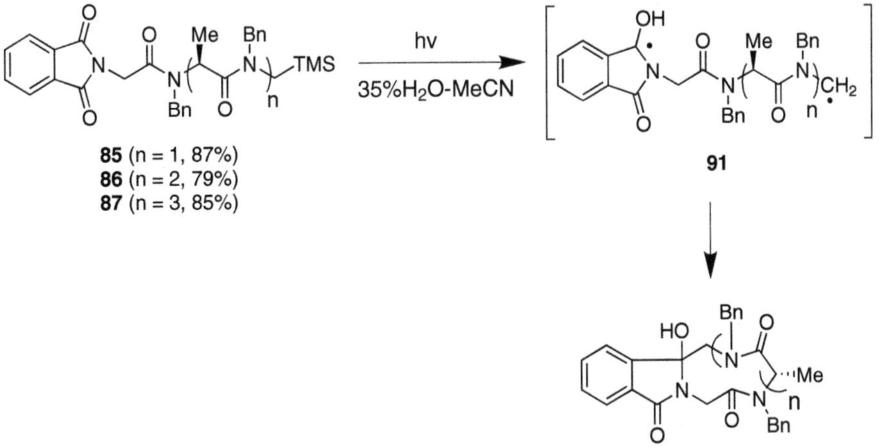

85 (n = 1, 87%)
86 (n = 2, 79%)
87 (n = 3, 85%)

91

88 (n = 1, β-OH, 55%)
89 (n = 2, α-OH, 74%)
90 (n = 3, α-OH, 38%)

SCHEME 5.24

biradical intermediates **91**. In light of these considerations, it was interesting to find that irradiation of the peptides **85–87** in 35% H_2O-MeCN leads to modestly efficient formation of the cyclic peptides **88–90**, each as a single diastereomer.

The results of the investigation described above demonstrate the feasibility of the photochemical based strategy for preparation of cyclic peptide analogs. The key step in routes, which follow this design, involves SET-photoinduced cyclization of N-acceptor linked-peptides that contain C-terminal α-amido-trimethylsilyl groups. Photomacrocyclization reactions of the phthalimide linked-peptides take place by a sequence of events (Scheme 5.25) involving (1) intramolecular SET from near neighbor amide donor sites to the excited phthalimide chromophore, (2) amide cation radical migration to the α-amido-silane centers, (3) desilylation to form 1,ω-biradical intermediates, and (4) biradical cyclization.

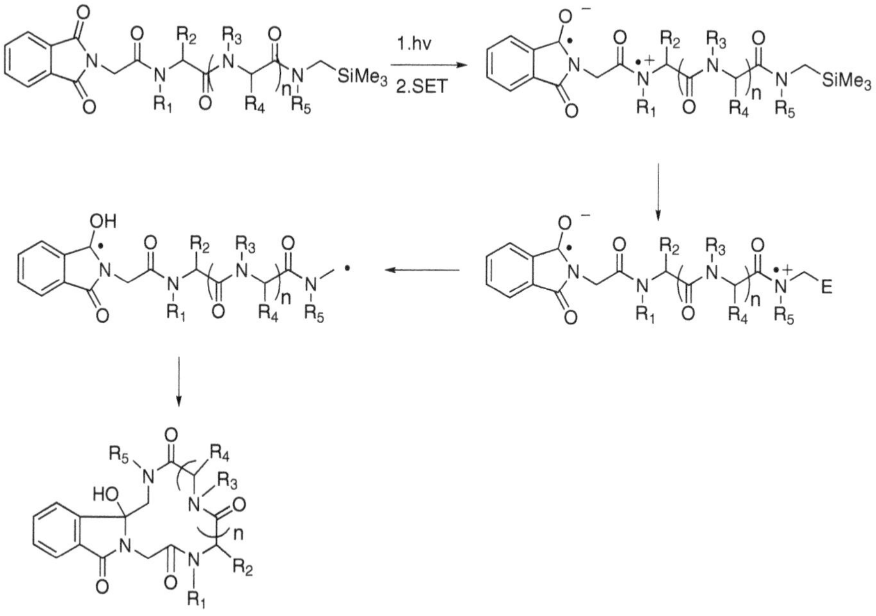

SCHEME 5.25

The modestly high yields observed for photocyclization reactions of substrates, which have a phthalimide acceptor group suggest that SET from amide donor sites in the peptide chains to the excited phthalimide chromophore occurs more rapidly than other typical excited state reaction modes (e.g., H-atom abstraction) [35]. In addition, the efficiencies of these processes are not significantly affected by the length of the polypeptide chain separating the centers at which bonding occurs. This result indicates that, following the initial SET event, migration of the radical cation center (hole migration) to the position in the peptide chain where the reactive alkylsilyl electrofugal group is located takes place at a rate which is competitive with both back electron transfer (leading to the ground state reactant) and proton loss from benzylic sites in intervening cation radicals [36–39]. Furthermore, the apparent chain length independence of the efficiencies of these processes suggests that the

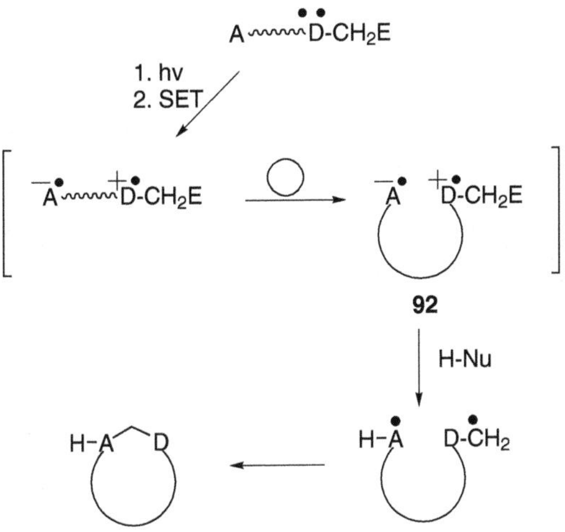

SCHEME 5.26 Schematic for conformational control of SET-promoted photocyclization reactions of linked acceptor-donor systems.

rates of the biradical cyclization reactions that serve as ultimate mechanistic steps in the reaction sequences are not significantly influenced by entropy [15,40–41]. It is tempting to propose a universal explanation for this phenomenon, which invokes the intermediacy of conformationally preorganized biradicals as precursors of the cyclic products in intramolecular SET-promoted cyclization reactions. A unique feature of cyclization reactions of linked donor-acceptor systems, promoted in this manner, is that the final neutral biradical intermediates arise by fragmentation reactions of zwitterionic biradicals. The electronic nature of the zwitterionic biradicals (**92** in Scheme 5.26) could cause them to exist in folded conformations that minimize the distance between the oppositely charged centers. Thus, if the rates of ion radical fragmentation and biradical coupling are in the range of those for complete conformational randomization, the cyclization processes would not be as entropically disfavored as conventional non-SET promoted cyclization reactions.

5.2.2.4 *Application SET-Promoted Photocyclization Reactions of Phthalimide to the Preparation of Novel Crown Ethers*

Another potentially interesting application of SET-promoted photocyclization reactions of TMS-terminated, polydonor-linked phthalimides is found in the synthesis of novel metal cation complexing agents related to the well-known crown ethers. For example, by using this process it should be possible to construct lariate type crown ethers **94**, bis-crowns **96** and **98**, and polymeric crowns **100** and bis-crowns **102** from the corresponding phthalimide derivatives (Scheme 5.27).

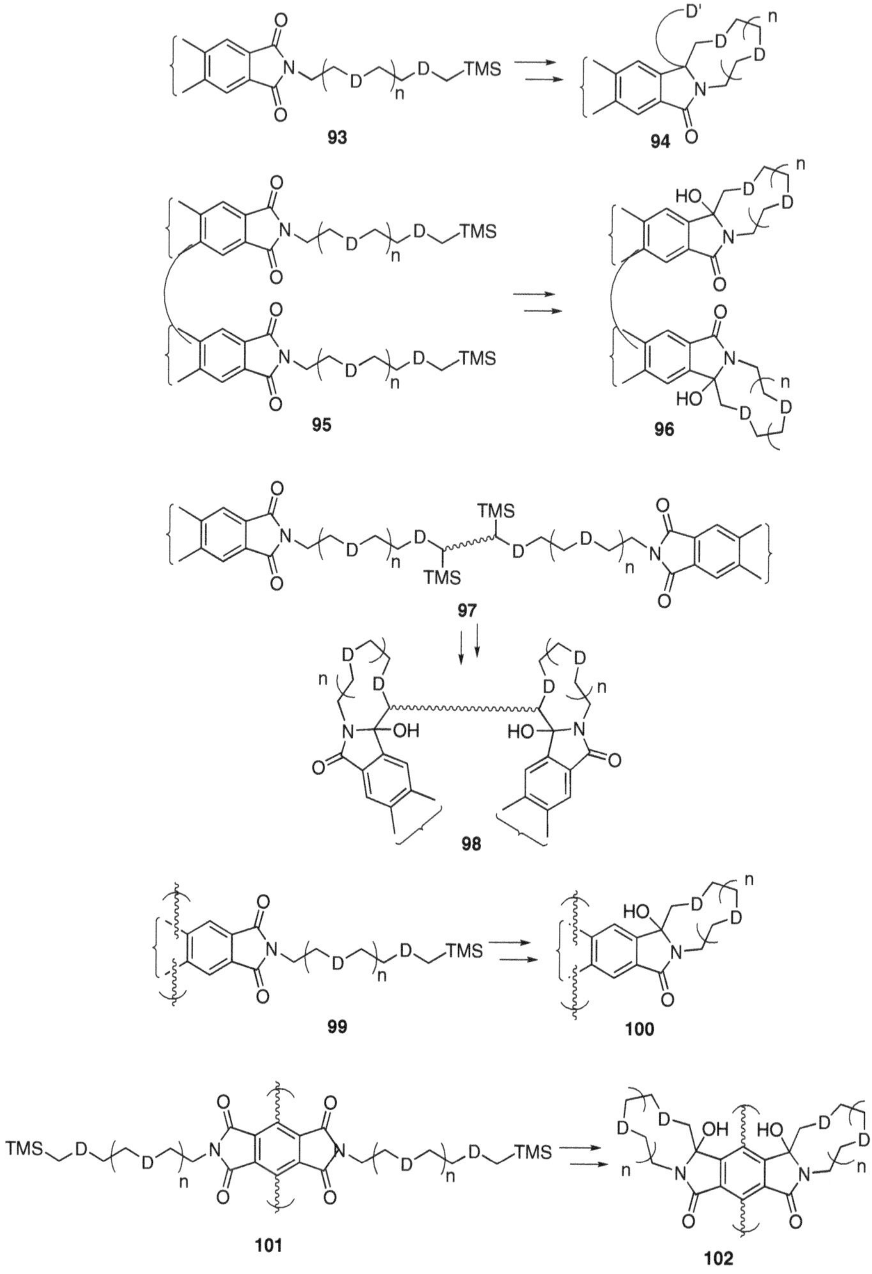

SCHEME 5.27

Although only at preliminary stages, studies in the authors' laboratories [42] have already demonstrated that photocyclization reactions of polydonor-linked phthalimides can be used to generate a number of different crown ethers and their aza- and thia-analogs. An example of this is found in the route used for preparation of the possibly metal cation selective, fluorescence sensing, lariat-type crown ethers **104** shown in Scheme 5.28. In this effort, the authors and their coworkers have shown that irradiation of the polyether-linked naphthalimides **105–107** results in highly efficient photocyclization reactions that produce the polyether ring containing cyclic amidols **108–110** (Scheme 5.29). Introduction of an *N*,*N*-dimethylaminopropyl

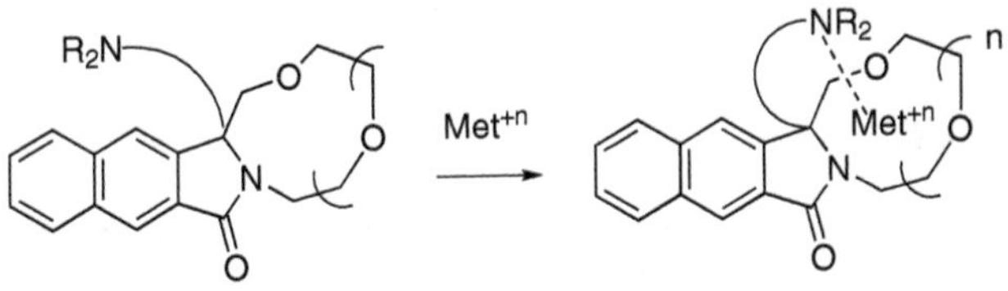
R2N
Met+n
NR2
Met+n
n

103 (non-fluorescent)

104 (fluorescent)

SCHEME 5.28

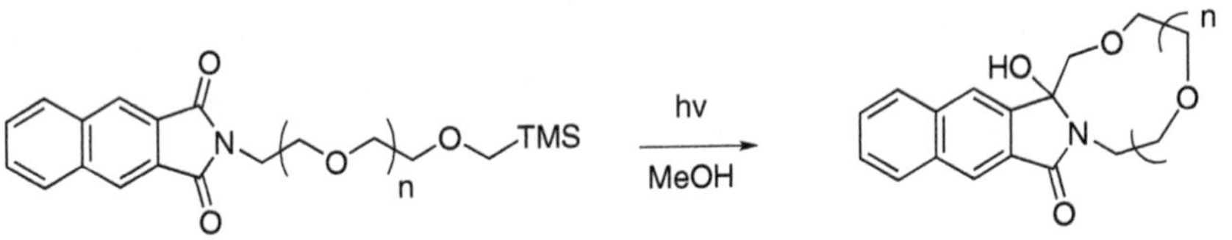
TMS
n
hv
MeOH
HO
n

105 (n = 3)
106 (n = 4)
107 (n = 5)

108 (76%)
109 (71%)
110 (74%)

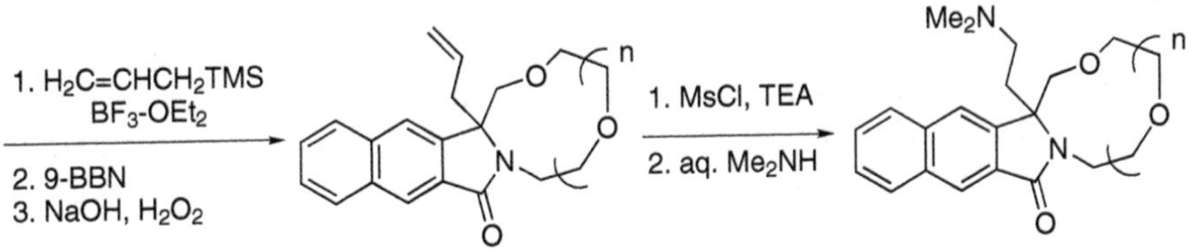
1. H2C=CHCH2TMS
BF3-OEt2
2. 9-BBN
3. NaOH, H2O2
n
1. MsCl, TEA
2. aq. Me2NH
Me2N
n

111 (60%)
112 (71%)
113 (65%)

114 (60%)
115 (69%)
116 (63%)

SCHEME 5.29

side chain into these substances is initiated by Lewis acid catalyzed allylation with allyltrimethylsilane. This is followed by hydroboration-oxidation to form the hydroxypropyl intermediates **111–113**, which are then smoothly converted to the target aza-lariat crown ether analogs **114–116**. Owing to efficient SET quenching of the singlet excited naphthalene chromophore by the conformationally mobile tertiary amine groups, **114–116** have very low fluorescence quantum efficiencies. As expected, protonation of the amine groups in these substances results in ca. 20-fold enhancements in their fluorescence quantum yields. Thus, it is anticipated that metal cation complexes of the lariat crowns **114–116** will utilize the side chain amine group for additional binding to the metal cation. If so, these substances could represent a new class of metal cation selective fluorescence sensors.

5.3 SUMMARY

The studies described above, carried out in the authors' laboratories over the past decade, have contributed to an understanding of the factors that govern the nature and efficiencies of SET-promoted photochemical reactions. In these efforts, phthalimides have been used as excited state acceptors. The well-characterized photophysical properties of phthalimides guided this choice. However, the conclusions drawn from this work about the factors that govern chemical and quantum efficiencies should be generally applicable to a wide variety of photochemical processes promoted by SET from donors to excited states of acceptors. In addition, the synthetic strategies that have been developed for preparation of new phthalimide based materials should be generally applicable to a host of different targets.

ACKNOWLEDGMENTS

The authors express their deep appreciation to their enthusiastic, hard-working, and productive coworkers whose studies in the area of amine electron transfer photochemistry established the basis for the results presented in this review. Also, the generous financial support given to the PSM research group by the National Institutes of Health and the National Science Foundation and to the UCY research group by the Korea Science and Engineering Foundation (Basic Research Program 2001-1-12300-006-3 and International Cooperative Research Program) is acknowledged.

NOTES

1. P.S. Mariano, J. Stavinoha. In: W.M. Horspool, Ed. *Synthetic Organic Photochemistry.* New York: Plenum, 1984, pp 259–284.
2. E. Colvin. *Silicon in Organic Synthesis.* London: Butterworth, 1981.
3. K. Ohga, P.S. Mariano. *J Am Chem Soc* 104, 617–619, 1982.

4. K. Ohga, U.C. Yoon, P.S. Mariano. *J Org Chem* 49, 213–219, 1984.
5. U.C. Yoon, P.S. Mariano. *Acc Chem Res* 25, 233–240, 1992.
6. W. Xu, X.M. Zhang, P.S. Mariano. *J Am Chem Soc* 113, 8863–8878, 1991.
7. J. Yoshida, T. Maekawa, T. Murata, S. Matsunaya, S. Isoe. *J Am Chem Soc* 112, 1962–1970, 1990.
8. J. Yoshida, S. Matsunaga, T. Murata, S. Isoe. *Tetrahedron* 47, 615–624, 1991.
9. X.M. Zhang, S.R. Yeh, S. Hong, M. Freccero, A. Albini, D.E. Falvey, P.S. Mariano. *J Am Chem Soc* 116, 4211–4220, 1994.
10. Z. Su, D.E. Falvey, U.C. Yoon, S.W. Oh, P.S. Mariano. *J Am Chem Soc* 120, 10676–10686, 1998.
11. Y. Kanaoka. *Acc Chem Res* 11, 407–413, 1978.
12. Y. Kanaoka, Y. Migita, K. Koyama, Y. Sato, H. Nakai, T. Mizoguchi. *Tetrahedron Lett* 14, 1193–1196, 1973.
13. Y. Kanaoka, K. Koyama, J.L. Flippen, I.L. Karle, B. Witkop. *J Am Chem Soc* 96, 4719–4721, 1974.
14. M. Machida, H. Takechi, Y. Shishido, Y. Kanaoka. *Synthesis* 12, 1078–1080, 1982.
15. For a recent comprehensive review, see J.D. Coyle. In W.M. Horspool, Ed. *Synthetic Organic Photochemistry*. New York: Plenum, 1984, pp 259–284.
16. J.D. Coyle, G.L. Newport. *Synthesis* 5, 381–382, 1979.
17. The excited state reduction potentials are calculated by use of the relationships $E^{S1}(-) = E^{S1} + E_{1/2}(-)$ and $E^{T1}(-) = E^{T1} + E_{1/2}(-)$, where $E_{1/2}(-)$ is the ground state reduction potential (-1.4 V vs. SCE, Ref. 18) of N-methylphthalimide, and E^{S1} and E^{T1} are the excited singlet (79.5 kcal/mol, Ref. 19) and triplet (68.5 kcal/mol) state energies of N-methylphthalimide.
18. D.W. Leedy, D.L. Muck. *J Am Chem Soc* 93, 4264–4270, 1971.
19 J.D. Coyle, G.L. Newport, A. Harriman. *J Chem Soc Perkin II* 2, 133–137, 1978.
20. U.C. Yoon, H.J. Kim, P.S. Mariano. *Heterocycles* 29, 1041–1064, 1989.
21. U.C. Yoon, J.H. Oh, S.J. Lee, D.U. Kim, J.G. Lee, K-T. Kang, P.S. Mariano. *Bull Korean Chem Soc* 13, 166–172, 1992.
22. U.C. Yoon, S.J. Lee, K.J. Lee, S.J. Cho, C.W. Lee, P.S. Mariano. *Bull Korean Chem Soc* 15, 154–161, 1994.
23. U.C. Yoon, S.J. Cho, J.H. Oh, K-T. Kang, J.G. Lee, P.S. Mariano. *Bull Korean Chem Soc* 12, 241–243, 1991.
24. U.C. Yoon, J.W. Kim, J.Y. Ryu, S.J. Cho, S.W. Oh, P.S. Mariano. *J Photochem Photobiol A* 106, 145–154, 1997.
25. U.C. Yoon, S.W. Oh, C.W. Lee. *Heterocycles* 41, 2665–2682, 1995.
26. U.C. Yoon, S.W. Oh, J.H. Lee, J.H. Park, K.-T. Kang, P.S. Mariano. *J Org Chem* 66, 939–943, 2001.
27. U.C. Yoon, H.C. Kwon, T.G. Hyung, K.H. Choi, S.W. Oh, S. Yang, Z. Zhao, P.S. Mariano. *J Am Chem Soc* 126, 1110–1124, 2004.
28. C. Somich, P.H. Mazzocchi, M. Edwards, T. Morgan, H.L. Ammon. *J Org Chem* 55, 2624–2630, 1990.
29. T. Shono, Y. Matsumura, K. Tsubata, K. Uchida, T. Kanazawa, K. Tsuda. *J Org Chem* 49, 3711–3716, 1984.
30. Estimated based on oxidation potentials of alcohols given in H. Lund. *Acta Chem Scand* 11, 491, 1957.
31. P.T. Cottrell, C.K. Mann. *J Electrochem Soc* 116, 1499–1503, 1969.

32. C.K. Mann. *Anal Chem* 36, 2424–2426, 1964.
33. U.C. Yoon, Y.X. Jin, S.W. Oh, C.H. Park, J.H. Park, C.F. Campana, X. Cai, E.N. Duesler, P.S. Mariano. *J Amer Chem Soc* 125, 10664–10671, 2003.
34. U.C. Yoon, D.U. Kim, C.W. Lee, Y.S. Choi, Y.-J. Lee, H.L. Ammon, P.S. Mariano. *J Am Chem Soc* 117, 2698–2710, 1995.
35. For a recent comprehensive review, see J.D. Coyle. In: W.M. Horspool, Ed. *Synthetic Organic Photochemistry*. New York: Plenum, 1984, pp 259–284.
36. S.S. Isied, I. Moreira, M.Y. Ogawa, B.A. Vassilian, J. Sun, *J Photochem Photobiol A: Chem* 82, 203–210, 1994.
37. M.R. Defelippis, M. Faraggi, M.H. Klapper. *J Am Chem Soc* 112, 5640–5642, 1990.
38. H.B. Gray, J.R. Winkler. *Ann Rev Biochem* 65, 537–561, 1996.
39. M.C.R. Symons. *Free Radical Biology and Medicine* 22, 1271–1276, 1997.
40. J.D. Coyle, G.L. Newport. *Synthesis* 381–382, 1979.
41. Y. Sato, H. Nakai, H. Ogiwara, T. Mizoguchi, Y. Migita, Y. Kanaoka. *Tetrahedron Lett* 4565–4568, 1973.
42. Unpublished results of U.C. Yoon, Z. Zhou, P.S. Mariano.

6 Photoamination with Ammonia and Amines

Masahide Yasuda, Tsutomu Shiragami,
Jin Matsumoto, Toshiaki Yamashita,
and Kensuke Shima

CONTENTS

6.1 INTRODUCTION AND BACKGROUND

Ammonia is the simplest nitrogen-containing compound and an essential nitrogen source in biological and industrial synthesis. In biological synthetic pathways, ammonia is playing a crucial role as a biologically available nitrogen source to produce amino acids, nucleotides, and so on in the nitrogen cycle [1]. Ammonia has been industrially produced from hydrogen and nitrogen since 1908 with the invention of the Haber–Bosch process operating at temperatures of 400 to 500°C and pressures of 1000 atm in the presence of an iron catalyst. Therefore, ammonia is an easily available industrially important raw material to synthesize nitrogen-containing molecules.

Amino compounds are potentially important substrates because of their significant chemical and biological activities. Usually the preparations of amino compounds are carried out by means of indirect methods such as reductions of nitro, azo, and azide or substitution of halogen, hydroxy, and alkoxy groups [2]. Direct amination of a carbon–carbon double bond, however, is limited to the Friedel–Crafts reaction with activated amination reagents or nucleophilic addition of amide anion to highly activated substrates, showing no common amination method in thermal reactions. Therefore, more convenient direct aminations using ammonia have been required.

Ammonia and amines are photochemically reactive species acting as a base, a reductant, and a nucleophile. Therefore, the photochemical reactions may make direct amination with ammonia and amines possible to provide more convenient and powerful synthetic method among many synthetic methodologies.

We review here photoamination based on a survey of the literature up to 2003.

6.1.1 Amines as Electron Donors

It is well known that amines are typically good electron donors in the photochemical reaction, since their oxidation potentials are relatively low and their chemical reactivity is potentially high. During the 1970s, the donor–acceptor photoreactions using amines as electron donor have been extensively investigated. In 1973, Bryce-Smith et al. developed the *meta*-photocyclization of 1-dimethylamino-3-phenylpropane, which gave the azacyclic compounds via a C-C bond formation (Scheme 6.1) [3].

The photoreactions of anthracene [4] and stilbene [5] with amines have been reported by Yang in 1973 and Lewis in 1977, respectively. In 1974, Hixson found the photoreaction of 1,2-diphenylcyclopropane with primary and secondary amines, which gave the addition products [6]. The electron-transfer quenching of the excited singlet of substrates with amines results in the formation of the exciplex or radical ion pair. These reactive species undergo proton transfer from α-C-H or N-H of amines

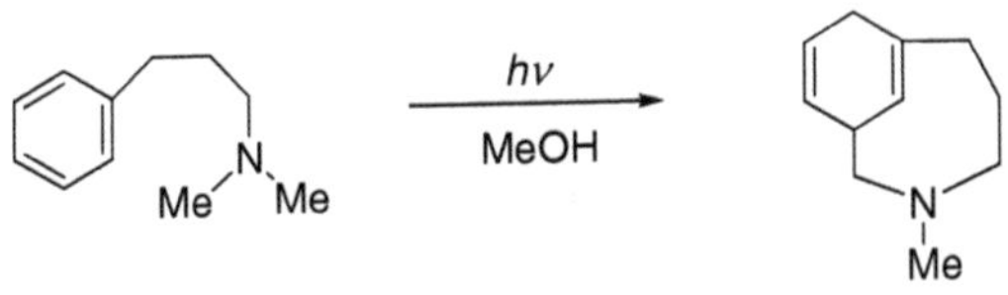

SCHEME 6.1

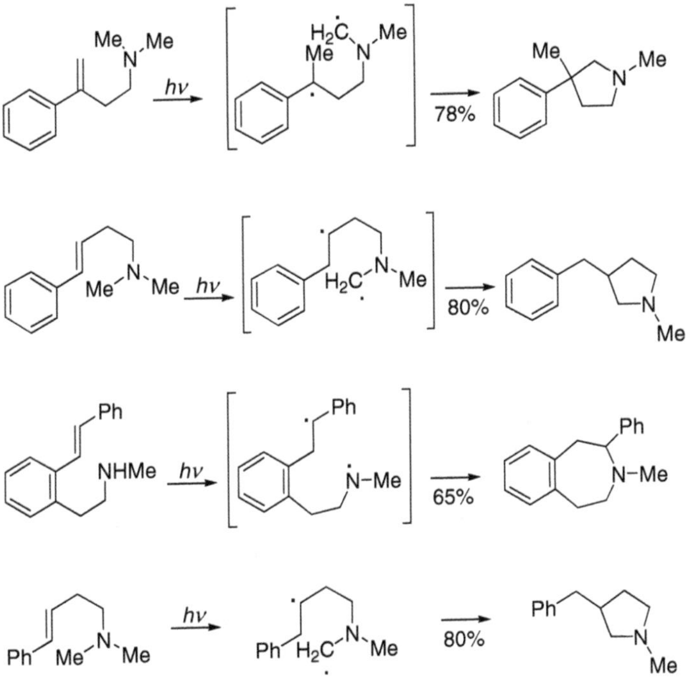

SCHEME 6.2

to substrates, resulting in the radical pair, which can combine to give the aminated products.

It is synthetically noteworthy that the intramolecular donor–acceptor photoreaction is applied to synthesis of the azacyclic compounds. Lewis and coworkers have applied to synthesize the pyrrolidine and piperidine derivatives (Scheme 6.2) [7]. Sugimoto and coworkers have investigated the intramolecular photoreaction of 9-(ω-aminoalkyl)phenanthrenes which gave the cyclized products depending on the alkyl chain length under irradiation (Scheme 6.3) [8]. The donor–acceptor photochemical reaction with α-silylamines acting as electron donor was intended to undergo the fission of the C-Si bond from the cation radical of α-silylamines, giving the efficient α-aminoalkyl radical by Mariano and coworkers [9].

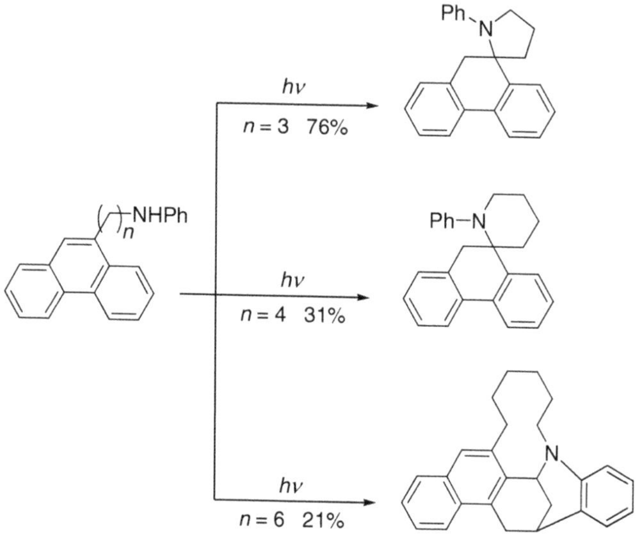

SCHEME 6.3

In the case of photoreaction of amines with substrates having strong electron-accepting ability, the amines work as a reductant. For example, the photoreaction of 9-cyanophenanthrene with Et_2NH in MeCN gave 9-cyano-9,10-dihydrophenanthrene [10]. The photoreaction of anthracene with N,N-dimethyl-p-toluidine in MeCN gave mainly 9,10-dihydroanthracene [11]. The proton transfer and subsequent hydrogen transfer from N-H and/or C-H in photogenerated ion pairs of the substrate and the amines resulted in the reduction products. Thus, donor–acceptor photochemistry is one of the direct introduction methods of the amino group. Lewis has reviewed the donor–acceptor photochemistry of the amines [12].

6.1.2 Amines as Nucleophiles

Another type of photoamination is achieved by using amines as nucleophiles. In general, photoactivation of the substrates can induce the generation of reactive species toward the nucleophiles, resulting in the nucleophilic addition reaction. This reaction has been named *photochemical polar addition* (PPA). A variety of PPA reactions have been investigated since 1960s. As mentioned below, there are three Categories in the classification of PPA. Among a large number of PPA reactions, PPA via electron transfer (Category III) is the most popular reaction in organic photochemistry. The photoreaction of Category III was found by Arnold in 1973 for the photoaddition of MeOH to 1,1-diphenylethene in the presence of methyl p-cyanobenzoate [13]. The nucleophilic addition of MeOH to the cation radical of 1,1-diphenylethene generated by the photoinduced electron transfer to methyl

TABLE 6.1
Comparison of Amines with Alcohols

	Amines	Alcohols
Nucleophilicity	Strong	Moderately strong
Basicity	Strong	Weak
Oxidation potentials	Low	High
Diversity	High	Moderately high
First report on the PPA of Category III	1985	1973

p-cyanobenzoate occurred. Photoamination of Category III reaction using ammonia and amines as nucleophile has been firstly reported by us in 1985 (Table 6.1) [14]. Here we review photoamination via PPA reaction using amines as nucleophiles.

6.2 PHOTOCHEMICAL POLAR ADDITION

6.2.1 CLASSIFICATION

The PPA can be classified into three categories (Categories I–III) from the standpoints of methodology to generate the reactive species toward nucleophiles (Scheme 6.4) [15].

Category I involves the photochemical unimolecular process to generate the reactive species toward nucleophiles. The photoexcitation of the substrates causes the intramolecular polarization or the heterolytic bond-fission of carbon–halogen and carbon–hetero atom bonds to generate the cationic species, which enhances their affinity toward the nucleophiles, especially strong nucleophiles. Category II involves the participation of the addition of proton to the excited states of the substrates. The protonated cationic species allow the nucleophilic addition. In phenolic molecules, for example, the photoexcitation causes the intramolecular proton transfer and/ or tautomerization to generate the zwitter ion intermediates, which undergo the nucleophilic addition. In Category III, the generation of the cationic species of the substrates takes place by the photoinduced electron transfer from the substrates to the electron-accepting molecules. The photoexcitation of either electron-donating substrates or electron acceptors causes the electron-transfer to generate the cation radical of the substrates and the anion radical of electron acceptor. The transformation from the substrates to the final products occurs by the successive process involving the nucleophilic addition to the cation radical of substrates, the reduction by the anion radical of electron acceptor, and the protonation and/or other processes.

Thus, the PPA reactions have been extensively investigated using a variety of nucleophiles such as alcohols, water, cyanide ion, and carboxylic acid. Among these PPA reactions, the present review deals with the PPA reaction with ammonia and the amines.

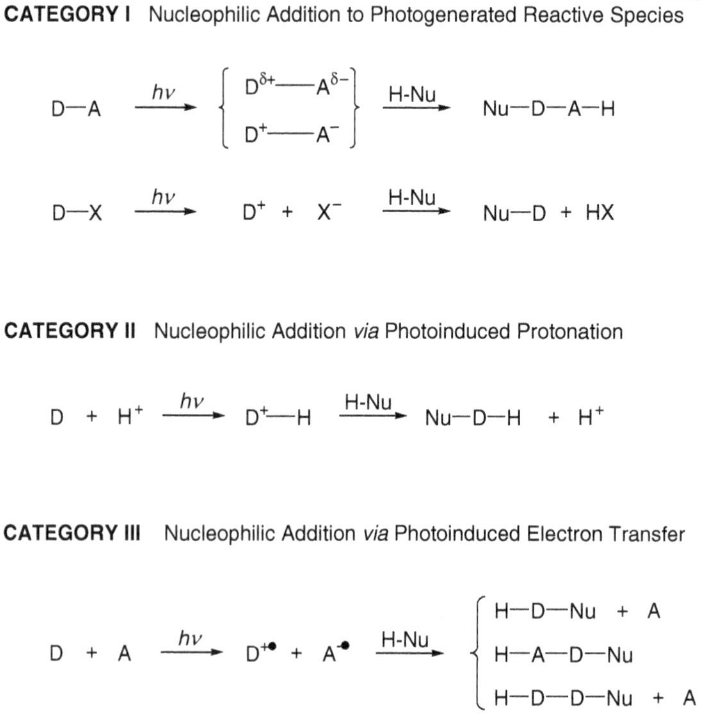

CATEGORY I   Nucleophilic Addition to Photogenerated Reactive Species

$$D-A \xrightarrow{hv} \left\{ \begin{array}{c} D^{\delta+}\!\!-\!\!A^{\delta-} \\ D^{+}\!\!-\!\!A^{-} \end{array} \right\} \xrightarrow{H\text{-}Nu} Nu-D-A-H$$

$$D-X \xrightarrow{hv} D^{+} + X^{-} \xrightarrow{H\text{-}Nu} Nu-D + HX$$

CATEGORY II   Nucleophilic Addition *via* Photoinduced Protonation

$$D + H^{+} \xrightarrow{hv} D^{+}\!\!-\!\!H \xrightarrow{H\text{-}Nu} Nu-D-H + H^{+}$$

CATEGORY III   Nucleophilic Addition *via* Photoinduced Electron Transfer

$$D + A \xrightarrow{hv} D^{+\bullet} + A^{-\bullet} \xrightarrow{H\text{-}Nu} \left\{ \begin{array}{l} H-D-Nu + A \\ H-A-D-Nu \\ H-D-D-Nu + A \end{array} \right.$$

SCHEME 6.4   Classification of PPA reactions.

6.2.2   PHOTOAMINATION VIA CATEGORY I

Direct irradiation of diphenylacetylene with R_2NH gave alkylaminostilbene and alkylamino-1,2-diphenylethane [16]. Photoamination of 1,2-diphenylcyclopropane with R_2NH gave the ring-opened product, 1-amino-1,3-diphenylpropane [17]. The photosubstitution of 1-methoxyanthraquinone with NH_3 gave 1-alkylaminoanthraquinone [18]. Photoamination of 1-hydroxyanthraquinone with n-BuNH$_2$ gave 1-hydroxy-4-(butylamino)anthraquinone and 1-hydroxy-2-(butylamino)anthraquinone [19]. Photoamination of fluorobenzene with Et_2NH gave three isomers of diethylaminofluorodihydrobenzenes (Scheme 6.5) [20]. Also, the photosubstitution of 2-fluoropyridine with t-BuNH$_2$ and Et_2NH gave 2-alkylaminopyridines [21]. Mutai and coworkers reported the photorearrangement of m-(ω-aminoalkoxy)nitrobenzene to o-(ω-hydroxyalkylamino)nitrobenzene (Scheme 6.5) [22], named photo-Smiles rearrangement. The irradiation of 2-(2-aminoethoxy)-1-nitrobenzene underwent to the intramolecular photoamination [23]. The photosubstitution of nitro group in 4-nitroanisole with hexylamine gave 4-hexylaminoanisole (Scheme 6.5) [24].

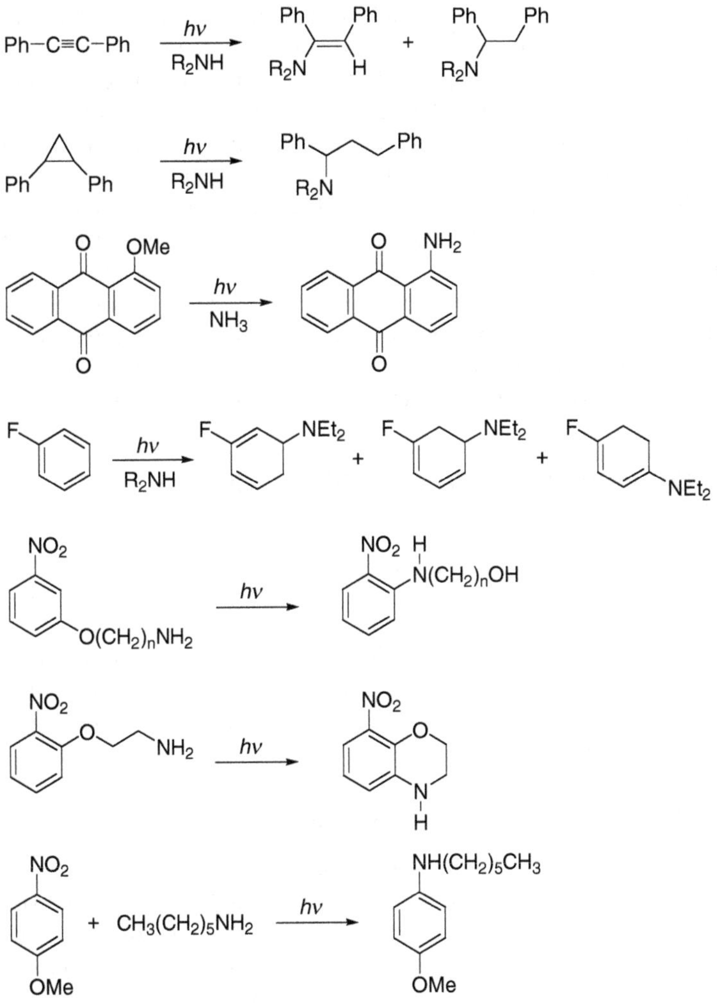

SCHEME 6.5

6.2.3 PHOTOAMINATION VIA CATEGORY II

Category II reaction is performed under acidic conditions using MeOH and AcOH as nucleophiles. Photoaminations via Category II were restricted to those of alkenylphenols, since it is hard for ammonia and amines to operate as nucleophile under acidic conditions. Irradiation of a deaerated MeCN solution containing *o*-(2-methyl-1-propenyl)phenol and RNH$_2$ gave *o*-(1-alkylamino-2-methylpropyl)phenols in relatively good yields, as shown in Scheme 6.6 [25]. However, no photoamination of O-methylated and O-acetylated derivatives with RNH_2 occurred to recover the starting materials. Therefore, a phenolic hydroxyl group is requisite for these photoaminations. Similar photoamination was applied to (E)-o-1-propenylphenols and (Z)-o-(3-hydroxy-3-methyl-1-butenyl)phenols. The fluorescence spectra of o-alkenylphenols were efficiently quenched by the addition of RNH_2, resulting in new emission at longer wavelength. Similar spectral change in the fluorescence spectra were observed in the presence of NaOH, showing that the new emission was due to the excited singlet state of phenolate anion, which is formed by the deprotonation of the phenolic hydroxyl group of o-alkenylphenols in the excited singlet state (Scheme 6.7). The decay of phenolate anion takes place via the proton transfer from the ammonium ion to the alkenyl group to form the o-quinodimethane or the zwitter ion. Therefore, the amination proceeds by a nucleophilic addition of RNH_2 at the benzylic cation center of zwitter ion.

6.3 PHOTOAMINATION VIA CATEGORY III: GENERAL ASPECTS

6.3.1 GENERATION OF CATION RADICAL

Category III reactions are one of the most fundamental photochemical processes, thus offering a potentially useful procedure for the introduction of functional group to the substrates. The photochemical generation of the cation radicals of electron-donating substrates (D) via the photoinduced electron transfer (PET) can be achieved by the following methods.

6.3.1.1 Direct Photoinduced Electron Transfer

In general, D was directly photoexcited in the presence of electron acceptor (A) to induce the electron transfer from the excited state of D to the ground state of A. m- and p-dicyanobenzenes (m- and p-DCB) which does not disturb the absorption of D in near-UV region and are effective for the reaction were selected as A for the photoamination (Scheme 6.8). Methyl p-cyanobenzoate (MCB) and the related benzoate compounds were unfavorable for photoamination due to the occurrence of solvolysis under reaction conditions.

6.3.1.2 Electron-Transfer Photosensitization

When an electron acceptor that can absorb incident light was used, i.e., 1-naphthonitrile (CNN), 1,4-dicyanonaphthalene (DCN), 9-cyanophenanthrene (CNP), 9,10-dicyanoanthracene (DCA), and N-methylphthalimide (MPI), an electron transfer from D to the excited state of A occurs to give the cation radical of D (Scheme 6.9A). This is a case where the electron acceptor plays a photosensitizer, which is absorbing an incident light to activate D. Electron-transfer photosensitization is well known to be effective for the photoaddition of weak nucleophiles such as alcohols and water.

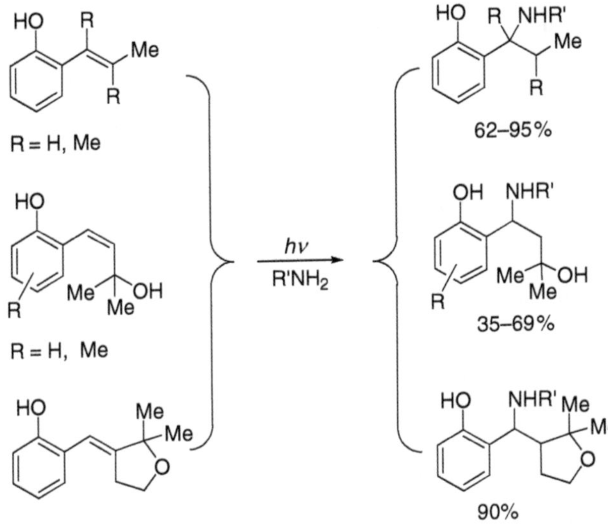

R'NH$_2$ = Et$_2$NH, MeNH$_2$, i-PrNH$_2$, C$_6$H$_{11}$NH$_2$, Et(Me)CHNH$_2$,
CH$_2$=CHCH$_2$NH$_2$, HOCH$_2$CH$_2$NH$_2$

SCHEME 6.6 Photoamination via Category II.

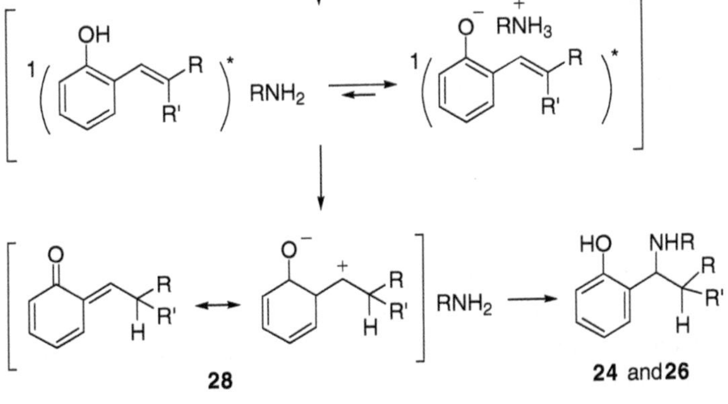

SCHEME 6.7

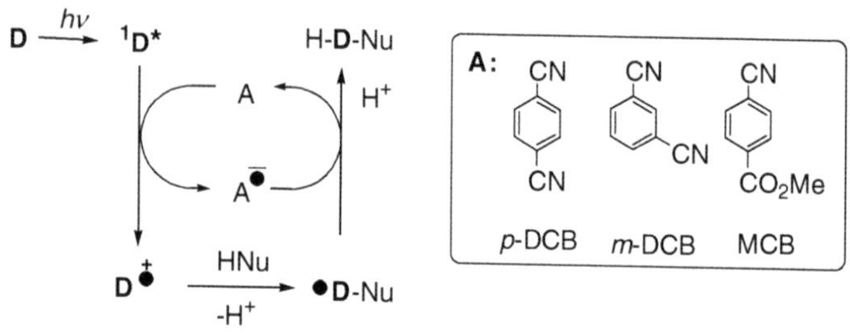

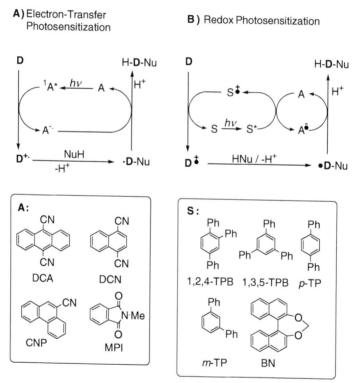

$$D \xrightarrow{h\nu} {}^1D^* \qquad \text{H-D-Nu}$$

SCHEME 6.8 Generation of cation radical via direct photoinduced electron transfer.

SCHEME 6.9 Generation of cation radical by photosensitization.

However, the electron-transfer photosensitization is not effective for photoamination except for a few cases, since the sensitizers are quenched more effectively by the electron-donating ammonia and amines than D, none resulting in the cation radical of D. One exception was reported by Pandey and coworkers, who applied the DCN-photosensitization to the cyclization of 1-(2-aminoethyl)-3,4-dimethoxybenzene to the corresponding dihydroindole [26].

6.3.1.3 Redox Photosensitization

Redox-photosensitization requires a sensitizer (S) that can effectively absorb an incident light and allow it to be quenched efficiently by A but not by D. Under the optimized conditions, the photoexcitation of S induces selectively the electron transfer from the excited state of S to A to give S^+ and A^+. The hole transfer from the cation radical of S to D generates the D^+ (Scheme 6.9B). For the efficient reaction, it is requisite that S^+ are not consumed by the nucleophilic reaction and/or reduction with the nucleophiles under reaction conditions. In 1977, Pac and Sakurai developed the redox-photosensitization using phenanthrene as photosensitizer [27], which was applied to the nucleophilic addition of MeOH to furan, indene, and 1,1-diphenylethene [28]. The phenanthrene radical cation that was generated by electron transfer from the excited singlet state of phenanthrene to p-dicyanobenzene causes the hole transfer to D, resulting in the cation radicals of D (D^+). The D^+ underwent the nucleophilic addition of MeOH. Redox-photosensitization has been applied to many kinds of the photoinduced nucleophilic addition and has enhanced the reaction efficiency.

6.3.2 Types of Reaction Products in Category III

Category III involves the electron-donating substrates (D), electron acceptors (A), nucleophiles (NuH or Nu^-), and additives in the reaction solvents. Therefore, the reaction systems are versatile in the combination of D, A, NuH (Nu^-), and additives. The cation radicals of D are very reactive species to undergo many kinds of organic

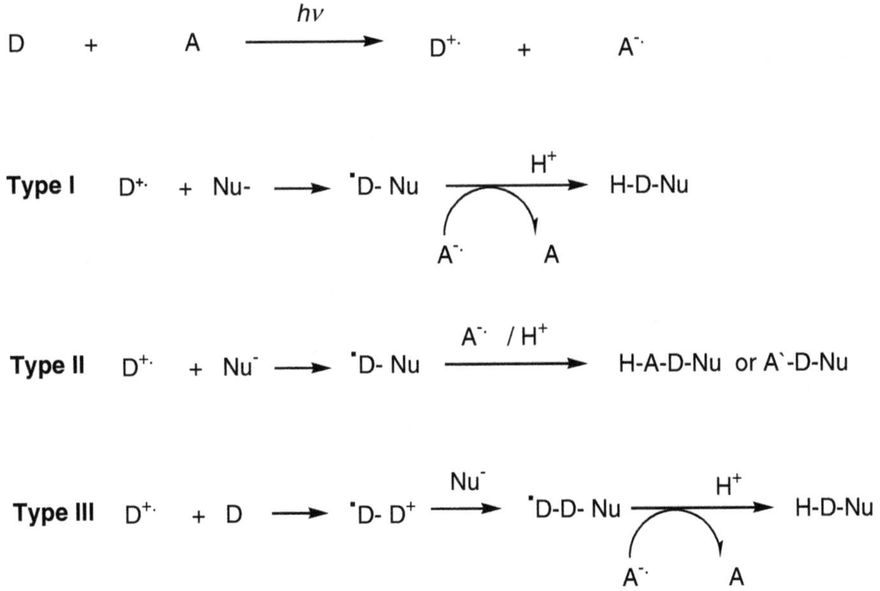

SCHEME 6.10 Product types of Category III reaction.

reaction involving PPA reaction and other type reactions (Scheme 6.10). In the presence of a certain HNu (Nu^-), the cation radical of D ($D^{+\cdot}$) is attacked by NuH to give the cation radical or radical of adducts ($\cdot D\text{-}Nu$). The cation radical or radical of the adducts is reduced by the anion radicals of A ($A^{-\cdot}$) and then protonated to afford the products ($H\text{-}D\text{-}Nu$). This is named *Type I* product. Since $D^{+\cdot}$ and $A^{-\cdot}$ have character as radical species as well as ionic species, a competitive radical coupling reaction of $D^{+\cdot}$ with $A^{-\cdot}$ may occur. Especially when it is hard to reduce the $\cdot D\text{-}Nu$ by A^-, a radical coupling between $\cdot D\text{-}Nu$ and $A^{-\cdot}$ predominantly occurs to give the coupling products ($H\text{-}A\text{-}D\text{-}Nu$ or $A'\text{-}D\text{-}Nu$; *Type II* product) where A' denotes the moiety produced by the elimination of the substituents such as CN^- from $A^{-\cdot}$. Other competitive pathway is the dimerization of $D^{+\cdot}$ with D followed by the nucleophilic addition and protonation to form $H\text{-}D\text{-}D\text{-}Nu$ (*Type III* product). Thus, the product types in Category III are divided into *Type I, II,* and *III,* as shown in Scheme 6.10.

As described above, the photoaddition of MeOH to 1,1-diphenylethene in the presence of methyl p-cyanobenzoate gave 2,2-diphenylethyl methyl ether, the *Type I* product [13]. Hixson reported the photoinduced intramolecular charge-transfer of 3-(p-cyanophenyl)-1-phenylpropene followed by the MeOH-addition [29]. On the other hand, McCullough et al. reported that the photoreaction of 2-naphthonitrile with 2,3-dimethyl-2-butene in the presence of MeOH gave the *Type II* products [30]. In 1978, Mazzocchi et al. [31] and Kubo and Maruyama [32] published the first reports on the PPA reaction using phthalimides which formed the *Type II* products. Mariano et al. reported the PPA reaction using pyrrolinium salts as A, leading to *Type II* product [33]. *Type III* product, for example, was found in the CNN-photosensitized reaction of indene with H_2O [34].

6.3.3 Substrates

It is well known that the nucleophilic addition of alcohols occurred to the localized cation radicals but not the delocalized cation radical [35]. On the other hand, the nucleophilic addition of ammonia and amines occurred for not only the localized cation radicals but also the delocalized cation radicals because of strong nucleophilicity, showing that photoamination has much versatility. Therefore, we have extensively investigated photoamination of arenes (1), alkenylnaphthalenes (2), stilbenes (3), 1,1-diarylalkenes (4), arylalkadienes (5), styrenes (6), cyclopropanes (7), alkene and alkadienes (8), and other substrates (9). Since the Category III reaction involves the electron transfer process, oxidation potentials of the substrates are important to determine the reaction efficiency. Table 6.2 lists the substrates (1–9) that have been investigated along with their oxidation potentials as well as the optimum yields of photoamination with NH_3.

TABLE 6.2
Oxidation Potentials and Fluorescence Lifetimes of the Substrates and Yields for Photoamination

Substrates	$E_{1/2}^{ox}$ (V)[a]	τ_F (ns)[b]	Yield (%)[c]	Ref.
(A) Arenes				
phenanthrene (1a)	1.29	53	84	36
9-methoxyphenanthrene (1b)	0.97	25	100	38
9-ethoxyphenanthrene (1c)	0.93	25	95	38
9-(3-aminopropoxy)phenanthrene (1d)	0.72	26	59	38
2-methoxyphenanthrene (1e)	1.13	28	67	37
3-methoxyphenanthrene (1f)	0.97	25	84	37
3,6-dimethoxyphenanthrene (1g)	0.84	20	40	37
2,3-dimethoxyphenanthrene (1h)	0.93	23	0	37
2,3,4-trimethoxyphenanthrene (1i)	0.97	16	0	37
anthracene (1j)	0.93	5	88	36
naphthalene (1k)	1.22	105	61	36
1-methylnaphthalene (1l)	1.22	—	67	36
2-methylnaphthalene (1m)	1.20	38	55	36
2,3-dimethylnaphthalene (1n)	1.20	—	81	36
2,6-dimethylnaphthalene (1o)	—	—	—	36
2-acetylnaphthalene ethyleneglycol acetal (1p)	—	—	69	36
1-methoxynaphthalene (1q)	0.93	13	62	40
2-methoxynaphthalene (1r)	1.07	15	75	40
2-ethoxynaphthalene (1s)	1.08	—	62	40
2-isobutoxynaphthalene (1t)	1.04	—	69[d]	40
2-(benzyloxy)naphthalene (1u)	1.09	—	64[d]	40
1-chloronaphthalene (1v)	—		35	36
1,3-dimethoxybenzene (1w)	—	—	58	36
biphenyl(1x)	1.45	—	56	36
(B) Alkenylnaphthalenes				
1-(2-methyl-1-propenyl)naphthalene (2a)	1.08	5	98	41
(*E*)-1-styrylnaphthalene (2b)	0.95	3	40	41
2-(2-methyl-1-propenyl)naphthalene (2c)	0.91	33	64	41
(*E*)-2-styrylnaphthalene (2d)	1.01	13	29	41
2-methoxy-6-(2-methyl-1-propenyl)naphthalene (2e)	0.84	11	63	41
2-methoxy-1-(2-methyl-1-propenyl)naphthalene (2f)	0.90	2	0	41
2-(2-cyanostyryl)naphthalene (2g)	1.40	10	3	41
2-(2,2-dicyanoethenyl)naphthalene (2h)	1.67	1	0	41
2-(2-cyanostyryl)-6-methoxynaphthalene (2i)	1.04	9	0	41
2-(2,2-dicyanoethenyl)-6-methoxynaphthalene (2j)	1.29	2	0	41

TABLE 6.2 (CONT.)

Substrates	$E_{1/2}^{ox}$ (V)[a]	τ_F (ns)[b]	Yield (%)[c]	Ref.
(C) Stilbenes				
(E)-stilbene (3a)	1.10	1.5	91	42
(E)-2-methoxystilbene (3b)	0.91	—	59	42
(E)-3-methoxystilbene (3c)	1.06	—	93	42
(E)-4-methoxystilbene (3d)	0.79	18.9	70	42
(E)-4-(benzyloxy)stilbene (3e)	0.90	12.4	53	—[e]
(E)-3,4-dimethoxystilbene (3f)	0.68	1.6	68	42
(E)-3,4-methylenedioxystilbene (3g)	0.75	0.4	27	42
(E)-3,5-dimethoxystilbene (3h)	1.09	15.4	30	42
(E)-3,4,5-trimethoxystilbene (3i)	0.72	4.7	0	42
(E)-4-methylstilbene (3j)	1.02	5.3	87	42
(E)-chlorostilbene (3k)	1.13	4.2	57	42
(E)-4,4'-dimethoxystilbene (3l)	0.64	23	46	42
(E)-3,4',5-trimethoxystilbene (3m)	0.84	3.0	39	42
6-methoxy-2-phenyl-3,4-dihydronaphthalene (3n)	0.75	—	48	—[e]
dibenzo[a,d]cycloheptene (3o)	1.07	—	79	44
dibenzo[a,d]cyclohepten-5-ol (3p)	1.30	2.2	68	44
5-methoxydibenzo[a,d]cycloheptene (3q)	1.35	2.2	80	44
5-acetoxydibenzo[a,d]cycloheptene (3r)	1.34	1.8	52[d]	44
5-methyldibenzo[a,d]cyclohepten-5-ol (3s)	1.23	—	19[d]	44
5,5-dimethoxydibenzo[a,d]cycloheptene (3t)	1.27	—	89[d]	44
5-dibenzo[a,d]cycloheptenone ethyleneglycol ketal (3u)	1.30	—	63[d]	44
5-methylenedibenzo[a,d]cycloheptene (3v)	1.06	—	36[d]	44
(E,E)-1,2-distyrylbenzene (3w)	1.04	12	67	45
(E,E)-2,3-distyrylnaphthalene (3x)	1.04	—	54	45
(D) 1,1-diarylalkenes				
1,1-diphenylethene (4a)	1.26	—	18	46
1,1-diphenyl-1-propene (4b)	1.13	—	44	46
1-phenyl-3,4-dihydronaphthalene (4c)	1.11	—	33	46
(E) Arylalkadienes				
3-methyl-1-phenyl-1,3-butadiene (5a)	1.16	—	27	47
(E,E)-2-methyl-1-phenyl-1,3-pentadiene (5b)	0.99	—	28	47
(E,E)-3-methyl-1-phenyl-1,3-pentadiene (5c)	0.98	—	37	47
(E)-4-methyl-1-phenyl-1,3-pentadiene (5d)	0.88	—	37	47
(E)-3-methyl-1-(o-methoxyphenyl)-1,3-butadiene (5e)	0.95	—	80	47
(E)-4-methyl-1-(o-methoxyphenyl)-1,3-butadiene (5f)	0.82	—	85	47
(E,E)-1,4-diphenyl-1,3-butadiene (5g)	0.87	—	92	47
(E,E)-2-methyl-1,4-diphenyl-1,3-butadiene (5h)	0.82	—	77	47
(E,E)-1-(4-methoxyphenyl)-4-phenyl-1,3-butadiene (5i)	0.64	—	65	47
(E,E)-3-methyl-1-(4-methoxyphenyl)-4-phenyl-1,3-butadiene (5j)	0.67	—	100	47
(E,E)-2-methyl-1-(4-methoxyphenyl)-4-phenyl-1,3-butadiene (5k)	0.67	—	100	47

TABLE 6.2 (CONT.)

Substrates	$E_{1/2}^{ox}$ (V)[a]	τ_F (ns)[b]	Yield (%)[c]	Ref.
(F) Styrenes				
(E)-1-(2-methoxyphenyl)-1-propene (6a)	0.86	—	75	48
(E)-1-(3-methoxyphenyl)-1-propene (6b)	1.18	—	46	48
(E)-1-(4-methoxyphenyl)-1-propene (6c)	0.93	—	91	48
(E)-1-(3,4-dimethoxyphenyl)-1-propene (6d)	0.82	—	65	48
(E)-1-(3,5-dimethoxyphenyl)-1-propene (6e)	1.10	—	57	48
(2-methyl-1-propenyl)benzene (6f)	1.59	—	39	48
6-methoxy-3,4-dihydronaphthalene (6g)	0.82	—	29	48
6-methoxy-1-methyl-3,4-dihydronaphthalene (6h)	0.69	—	35	48
6-methoxy-2-methyl-3,4-dihydronaphthalene (6i)	0.78	—	48	48
7-methoxy-3,4-dihydronaphthalene (6j)	—	—	64	48
5,7-dimethyl-3,4-dihydronaphthalene (6k)	—	—	64	—[e]
1-methyl-3,4-dihydronaphthalene (6l)	—	—	25	—[e]
(E)-1-(4-methoxyphenyl)-1-propen-3-ol (6m)	0.96	—	64	—[e]
(E)-1-(4-methoxyphenyl)-2-methyl-1-propen-3-ol (6n)	0.89	—	62	—[e]
(E)-3-acetoxy-1-(4-methoxyphenyl)-1-propene (6o)	1.03	—	45	—[e]
indene (6p)	1.33	—	79	50
2-methylindene (6q)	1.22	—	90	50
3,4-dihydronaphthalene (6r)	1.19	—	96	50
benzocycloheptadiene (6s)	1.18	—	58	50
(G) Cyclopropanes				
1,2-diphenylcyclopropane (7a)	1.12	—	68	50
(E)-1,2-bis(4-methoxyphenyl)cyclopropane (7b)	0.61	—	90	50
(E)-1-(4-methoxyphenyl)-2-phenylcyclopropane (7c)	0.82	—	79	50
(E)-1-(2,4-dimethoxyphenyl)-2-phenylcyclopropane (7d)	0.67	—	0	50
1-(4-methoxyphenyl)-2,2-dimethylcyclopropane (7e)	1.26	—	59	50
1-(3-methoxyphenyl)-2,2-dimethylcyclopropane (7f)	1.10	—	7	—[e]
1-(2-methoxyphenyl)-2,2-dimethylcyclopropane (7g)	0.90	—	22	—[e]
(Z)-1-(1-naphtyl)-2-phenylcyclopropane (7h)	1.23	3.4	52	—[e]
(Z)-1-(2-naphtyl)-2-phenylcyclopropane (7i)	1.11	3.4	44	—[e]
(H) Alkenes and Alkadienes				
2,3-dimethyl-2-butene (8a)	1.22	—	74	—[e]
2,5-dimethyl-2,4-hexadiene (8b)	0.86	—	97	—[e]
2,4-hexadiene (8c)	1.17	—	98	—[e]
1,2,3,4-tetamethyl-1,3-cyclopentadiene (8d)	0.58	—	95	—[e]
2,3-dihydrofuran (8e)	0.99	—	58	—[e]
2,3-dihydropyran (8f)	1.16	—	40	—[e]
(I) Others				
quadricyclane (9a)	0.65	—	96	50

[a] Half peak of oxidation potentials. [b] Fluorescence lifetimes. [c] Isolated yields based on the consumed substrates. [d] Yields for photoamination with $MeNH_2$. [e] Unpublished data.

6.4 PHOTOAMINATION VIA CATEGORY III: PRODUCTS ANALYSIS

6.4.1 PHOTOAMINATION OF ARENES (1)

Photoamination via Category III has been investigated in polycyclic arenes such as phenanthrene (1a) and its derivatives (1b–1i), anthracene (1j), and naphthalene derivatives (1k–1v) [36-37]. A typical example was photoamination of 1a with NH_3, which was performed by the irradiation of an ammonia-saturated MeCN-H_2O solution containing 1a and m-DCB for 17 h to give the *Type I* aminated product, 9-amino-9,10-dihydrophenanthrene (10a), in 84% yield (Scheme 6.11). It was applied to photoamination with primary amines such as alkylamines, ethanolamine, allylamine, and glycine ethyl ester and so on, which gave efficiently N-alkyl derivatives of 10a. But photoamination with secondary alkylamines such as dimethylamine and diethylamine were inefficient [38].

Similar photoaminations were applied to a variety of methoxyphenanthrenes (1e–1i) (Scheme 6.12) [37], anthracene (1j) (Scheme 6.13), naphthalene (1k), and naphthalene derivatives (1l–1q) (Scheme 6.14) [36]. However, photoaminations of naphthalene derivatives substituted with electron-withdrawing group and benzene derivatives such as m-dimethoxybenzene (1w) and biphenyl (1x) occurred inefficiently.

Stereochemical studies were performed on photoamination of 1a, 9-alkoxyphenanthrene (1b–1c), and 1j [39]. Photoamination of 9-methoxyphenanthrene (1b) with NH_3 gave cis- and trans-9-amino-10-methoxy-9,10-dihydrophenanthrene (11a) in a ratio of 75:25 (Scheme 6.15). Photoamination of 1b with propylamine and isopropylamine and 9-ethoxyphenanthrene (1c) with t-butylamine gave exclusively cis-9-alkylamino-10-alkoxy-9,10-dihydrophenanthrene (11b and 11c). Thus the selective trans-addition of ammonia and alkylamines to arenes was suggested to arise from the protonation into the stable conformation of the aminated anion.

Photoamination of 2-alkoxynaphthalenes (1r–1u) was applied to the preparation of 1-amino-2-tetralone derivatives (13). Photoamination of 1r–1u with alkylamines

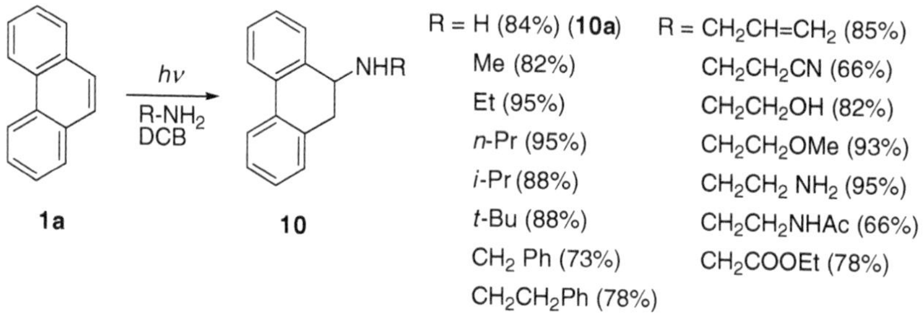

R = H (84%) **(10a)**	R = $CH_2CH=CH_2$ (85%)
Me (82%)	CH_2CH_2CN (66%)
Et (95%)	CH_2CH_2OH (82%)
n-Pr (95%)	CH_2CH_2OMe (93%)
i-Pr (88%)	$CH_2CH_2NH_2$ (95%)
t-Bu (88%)	CH_2CH_2NHAc (66%)
CH_2Ph (73%)	CH_2COOEt (78%)
CH_2CH_2Ph (78%)	

SCHEME 6.11

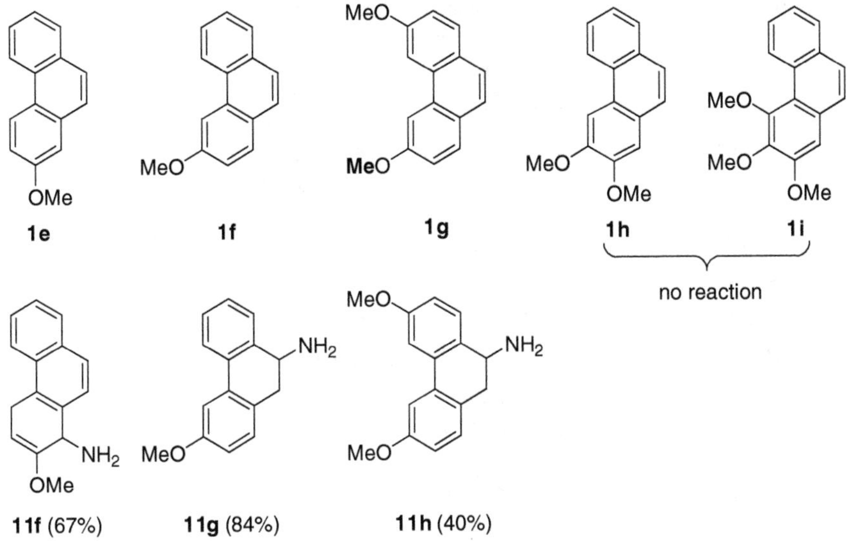

SCHEME 6.12

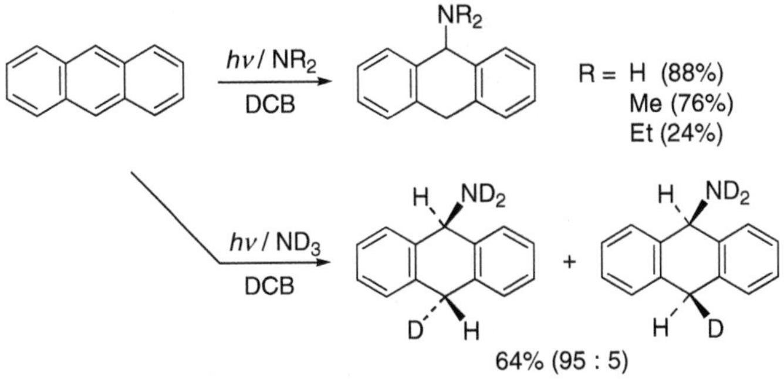

SCHEME 6.13

gave 1-alkylamino-2-alkoxy-1,4-dihydronaphthalenes (12). Then, 1-amino-2-tetralones (13) were prepared by treatment of the acetamides of 12 with BF$_3$·OEt$_2$ along with the formation of the isomerized products of 12 (14) (Scheme 6.16) [40].

6.4.2 PHOTOAMINATION OF ALKENYLNAPHTHALENES (2)

Photoamination of 1-(2-methyl-1-propenyl)naphthalene (2a) and 1-styrylnaphthalene (2b) occurred selectively at the alkenyl group to give 1-(2-amino-2-methylpropyl)

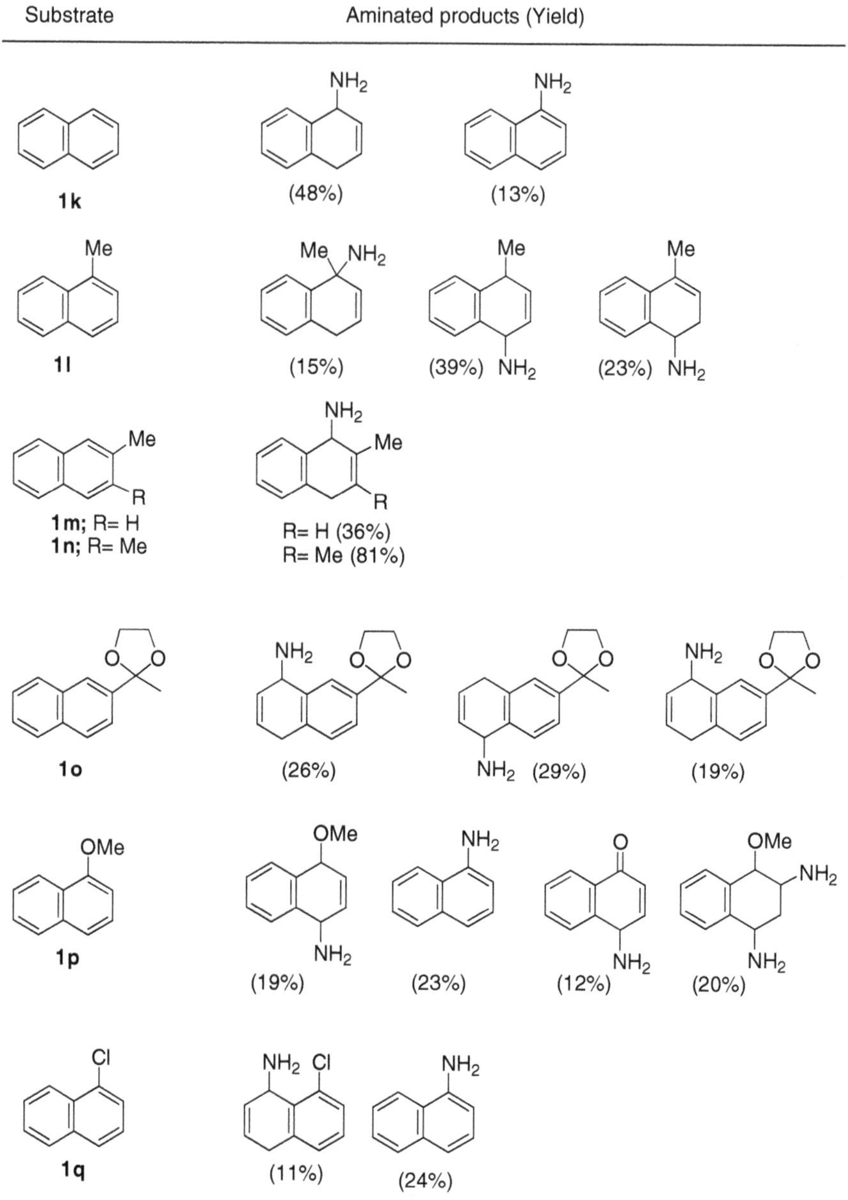

Substrate	Aminated products (Yield)			

SCHEME 6.14

naphthalene (15a) and 1-(2-amino-2-phenylethyl)naphthalene (15b), respectively [41]. Also, photoamination of 2-naphthyl isomers (2c–d) occurred regioselectively at alkenyl groups not at aryl moiety to give 15c–d (Scheme 6.17). Irradiation of 2a–b for a long time under photoamination conditions gave 1-methylnaphthalene (16a) as a

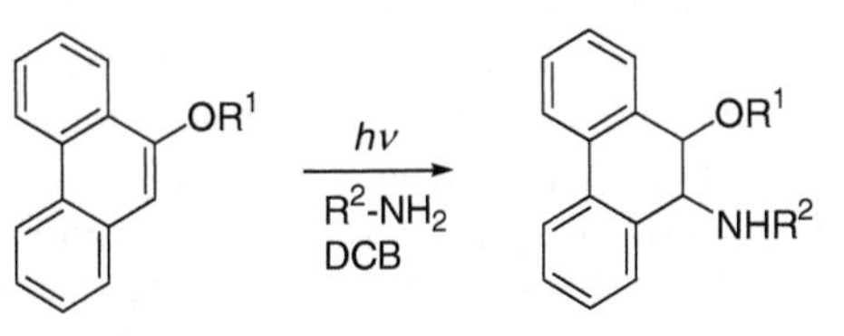

1b: R^1 = Me
1c: R^1 = Et
11

11	R^1	R^2	Yield/% (*cis* : *trans*)
11a	Me	H	100 (75 : 25)
11b	Me	*n*-Pr	97 (100 : 0)
11c	Me	*i*-Pr	99 (100 : 0)
11d	Et	H	95 (73 : 27)
11e	Et	*t*-Bu	81 (100 : 0)

SCHEME 6.15

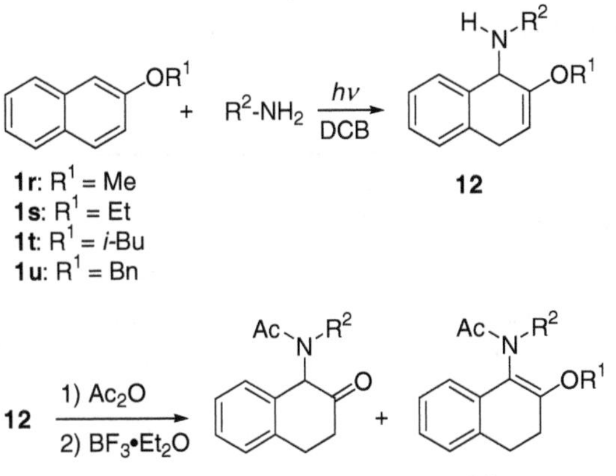

1r: R^1 = Me
1s: R^1 = Et
1t: R^1 = *i*-Bu
1u: R^1 = Bn
12, **13**, **14**

R^1	R^2	Yield (%)		
		12	**13**	**14**
Me	Me	69	92	
Me	Allyl	83	55	
Me	Me	67	80	9
Me	Et	67	49	12
Me	*n*-Pr	40		42
Me	*i*-Pr	56		86
Et	Me	59	75	12
i-Bu	Me	64	47	40
Bn	Me	60	77	
Bn	Et	76	49	42
Bn	*i*-Pr	71	40	

SCHEME 6.16

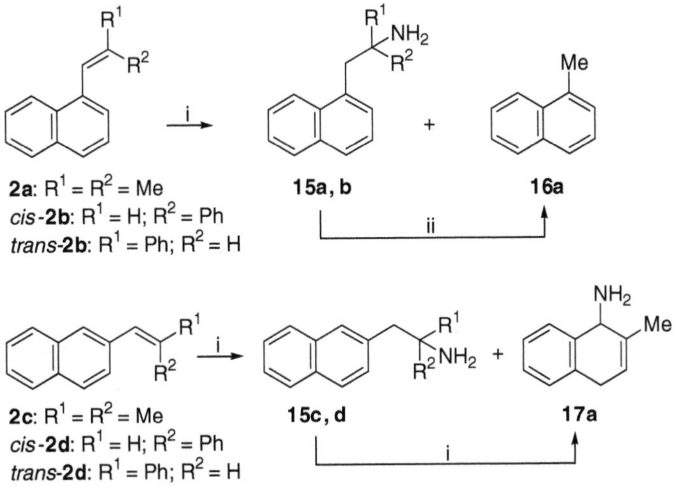

2a: $R^1 = R^2$ = Me
cis-**2b**: R^1 = H; R^2 = Ph
trans-**2b**: R^1 = Ph; R^2 = H
15a, b **16a**

2c: $R^1 = R^2$ = Me
cis-**2d**: R^1 = H; R^2 = Ph
trans-**2d**: R^1 = Ph; R^2 = H
15c, d **17a**

SCHEME 6.17 Reagents. i: *hν*/NH_3/*p*-DCB; ii: *hν*/*p*-DCB.

consequence of β-fission of the amino groups from 15a–b via electron transfer. In the case of 2c, irradiation for a long time gave 1-amino-2-methyl-1,4-dihydronaphthalene (17a), which would be produced by photoamination of 2-methylnaphthalene formed by the decomposition of 2c. The products and yields are shown in Table 6.3.

Photoamination of 2e occurred at the alkenyl group to give 15e along with the formations of 16b and 17b as secondary products (Scheme 6.18). On the other hand, no photoamination of 2-methoxy-1-(2-methyl-1-propenyl)naphthalene (2f) occurred at either the alkenyl group or the naphthalene ring, so almost 2f was recovered. In this case, the localization of the positive charge on aromatic ring makes the reactivity of olefinic moiety lower. Photoamination of 2g–j, which was substituted by a cyano group at the alkenyl groups, afforded little or no aminated products because of no occurrence of the photoinduced electron transfer from 2g–j to p-DCB.

TABLE 6.3
Photoamination of Alkenylnaphthalenes (2) with NH_3[a]

			Recovery (%)	
2	**t / h[b]**	**Products (Yield / %)[c]**	**2**	**p-DCB**
2a	4	**15a** (45)	54	73
cis-**2b**	7	**15b** (67)	0	63
trans-**2b**	7	**15b** (40)	0	68
2c	1	**15c** (64)	0	84
cis-**2d**	7	**15d** (54)	0	94
trans-**2d**	3	**15d** (29)	0	86

[a] For an ammonia-saturated $MeCN-H_2O$ (8:2, 70 ml) solution containing 2 (3.5 mmol) and p-DCB (3.5 mmol). [b] Irradiation time. [c] Isolated yields based on 2 used.

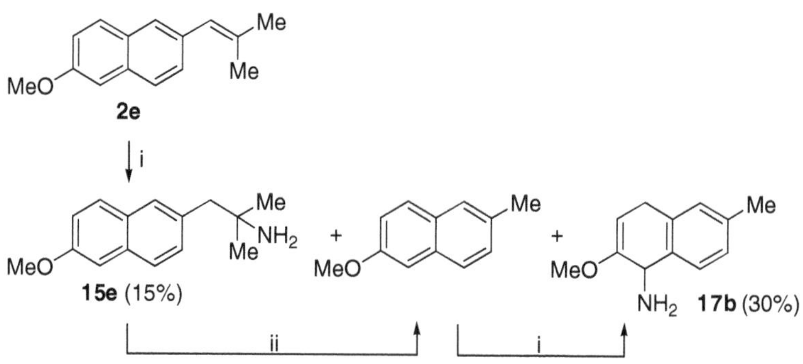

SCHEME 6.18　　Reagents. i: $hv/NH_3/p$-DCB; ii: hv/p-DCB.

6.4.3 PHOTOAMINATION OF STILBENES (3)

Photoaminations of 1,2-diarylethene derivatives (3a–k) with NH_3 were carried out by irradiation of an ammonia-saturated MeCN-H_2O or MeCN-benzene-H_2O solution containing 3 and p-DCB (Scheme 6.19 and Table 6.4). Regiochemistry was dependent on the substituent on the aryl group. Photoamination of 3d–g having the alkoxy substituents on p-position gave 1-amino-2-aryl-1-phenylethane (18) selectively. On the contrary, photoaminations of 1,2-diarylethenes (3b,c,h,j,k) having p-methyl, p-chloro, and o- and m-methoxy gave 1-amino-1-phenyl-2-arylethane (18) and 1-amino-2-phenyl-1-arylethane (19). But no photoamination of 3,4,5-trimethoxystilbene (3i) occurred, probably because the positive charge develops on the aromatic ring. Moreover, it should be noted that photoamination proceeds more efficiently in MeCN-benzene-H_2O (7:2:1) than in MeCN-H_2O (9:1) [42]. As stilbene analogs, photoamination of styrylthiophene occurred at olefinic moiety [43].

Photoaminations of dibenzo[a,d]cycloheptene (3o), and dibenzo[a,d]cyclohepten-5-ol (3p) with RNH_2 gave 5-substituted 10-alkylamino-10,11-dihydro-5H-dibenzo[a,d]

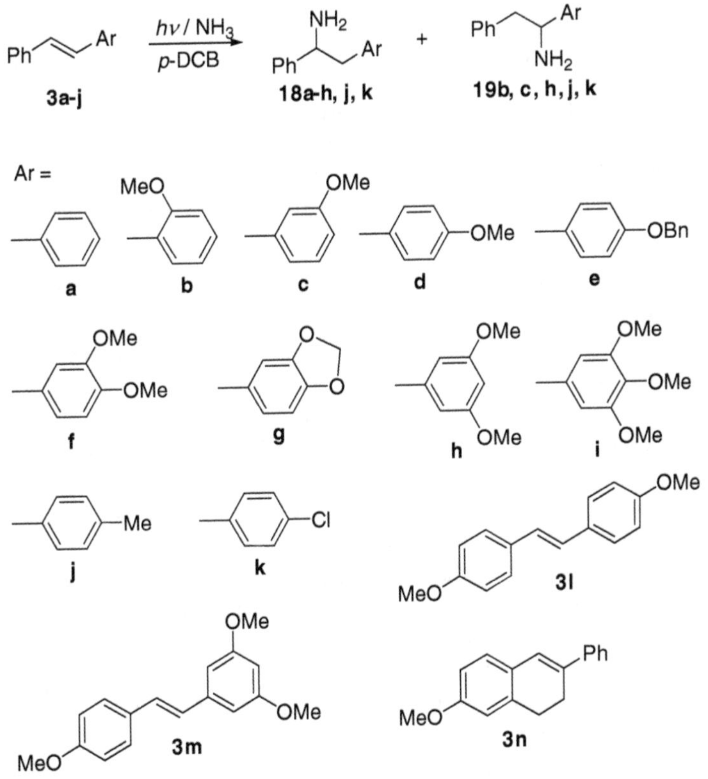

SCHEME 6.19

TABLE 6.4

Photoamination of Stilbenes (3) with NH_3[a]

3	Solvent[b]	t/h[c]	Yield (%)[d] (18 : 19)	Recovery/% 3 (Z : E)	p-DCB
3a	9:0:1	20	46	14 (9:1)	97
3a	7:2:1	31	88	3 (1:1)	96
3b	9:0:1	7	59 (1:0.4)	0	97
3c	8:1:1	20	91 (1:0.7)	2 (1:1)	91
3d	8:1:1	15	62 (1:0)	12 (4:1)	89
3e	8:1:1	10	53 (1:0)	0	85
3f	9:0:1	20	21 (1:0)	69 (4:1)	88
3g	8:1:1	20	21 (1:0)	21 (2:1)	100
3h	7:2:1	17	60 (1:0.7)	2 (2:1)	82
3i	9:0:1	20	0	70 (5:1)	88
3j	7:2:1	20	87 (1:0.6)	0	92
3k	7:2:1	8	54 (1:0.9)	6 (1:2)	100
3l	7:2:1	41	44	5 (2:1)	100
3m	7:2:1	20	16 (1:0)	59 (1:0)	85
3n	9:0:1	15	25 (1:0)	28	97

[a] A MeCN-benzene-H_2O solution (70 mL) containing 3 (7 mmol) and p-DCB (7 mmol) was bubbled with ammonia and then irradiated. [b] The values are the ratio of MeCN-benzene-H_2O. [c] Irradiation time. [d] Isolated yields based on 3 used.

cycloheptenes (20a–h) in a mixture of *cis* and *trans* isomers (Scheme 6.20) [44]. Photoaminations of 3p and 5-methoxy- and 5-acetoxydibenzo[*a,d*]cycloheptenes (3q and 3r) were applied to the synthesis of dibenzo[*a,d*]cycloheptenimines, as mentioned in Section 6.4.10.2.

Styryl and vinyl groups are well known as photochemically and thermally reactive functional groups. Especially, 2-vinylstilbene, (*E,E*)-1,2-distyrylbenzene (3w), and the naphthalene analog (3x) having these reactive groups at *ortho*-position of aromatic ring are highly reactive for photochemical dimerization and valence isomerization in solution, even in solid state, and in adsorbed state on silica gel. However, little is known about the cation radical of 3w, probably because the cation radical of 3w is too reactive to afford the identifiable products.

In order to elucidate the reactivity of the cation radical of 3w, an ammonia-saturated MeCN-benzene-H_2O solution containing 3w and p-DCB was irradiated giving 1-benzyl-3-phenyl-1,2,3,4-tetrahydroisoquinoline (21a; 37%) and 1-amino-3-benzyl-2-phenylindan (22a; 30%) (Scheme 6.21) [45]. The formation of the *Type III* aminated product, 22a, suggests that 3w[+·] was isomerized to the closed-type cation

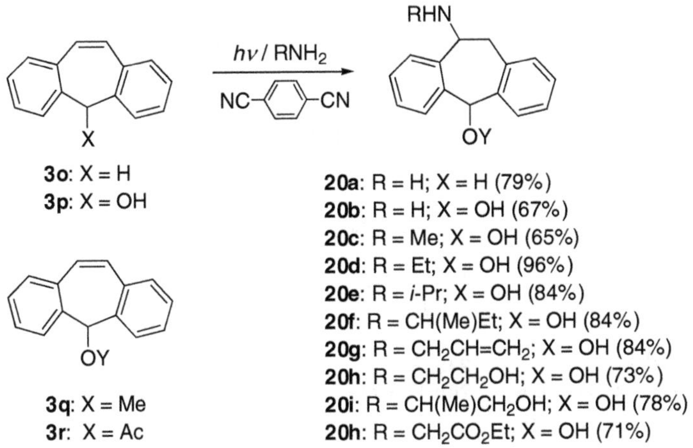
RHN
hν / RNH2
NC–C6H4–CN
X
OY
3o: X = H
3p: X = OH
OY
3q: X = Me
3r: X = Ac
20a: R = H; X = H (79%)
20b: R = H; X = OH (67%)
20c: R = Me; X = OH (65%)
20d: R = Et; X = OH (96%)
20e: R = i-Pr; X = OH (84%)
20f: R = CH(Me)Et; X = OH (84%)
20g: R = CH2CH=CH2; X = OH (84%)
20h: R = CH2CH2OH; X = OH (73%)
20i: R = CH(Me)CH2OH; X = OH (78%)
20h: R = CH2CO2Et; X = OH (71%)

SCHEME 6.20

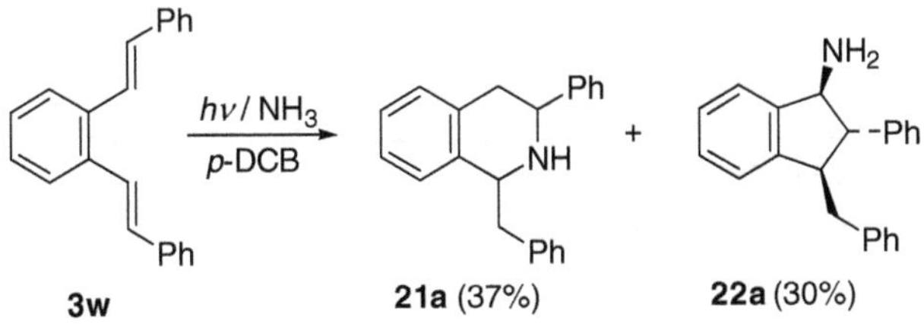
Ph
Ph
hν / NH3
p-DCB
Ph
NH
Ph
+
NH2
-Ph
Ph
3w
21a (37%)
22a (30%)

SCHEME 6.21

radical (closed-3w$^{+\cdot}$). Thus, valence isomerization of opened-3w$^{+\cdot}$ to closed-3w$^{+\cdot}$ can be elucidated by use of the nucleophilic addition with NH_3, but not by weak nucleophiles such as MeOH and H_2O.

6.4.4 Photoamination of 1,1-Diarylalkenes (4)

Photoamination of 1,1-diphenylethene (4a) and 1,1-diphenyl-1-propene (4b) with RNH_2 in the presence of *p*-DCB gave the corresponding *Type I* aminated products (23) (Scheme 6.22) [46]. The yields of 23 were poor along with the formation of the minor products (Table 6.5). It was noteworthy that photoaminations of 4a and 4b with *t*-butylamine gave mainly the acetonitrile-incorporated products, 4,4-diphenylbutanenitrile (24a) and 4,4-diphenyl-3-methylbutanenitrile (24b). Photoamination of 1-phenyl-3,4-dihydronaphthalene (4c) with RNH_2 gave 2-alkylamino-1-phenyl-1,2,3,4-tetrahydronaphthalene in a mixture of *cis* and *trans isomers*.

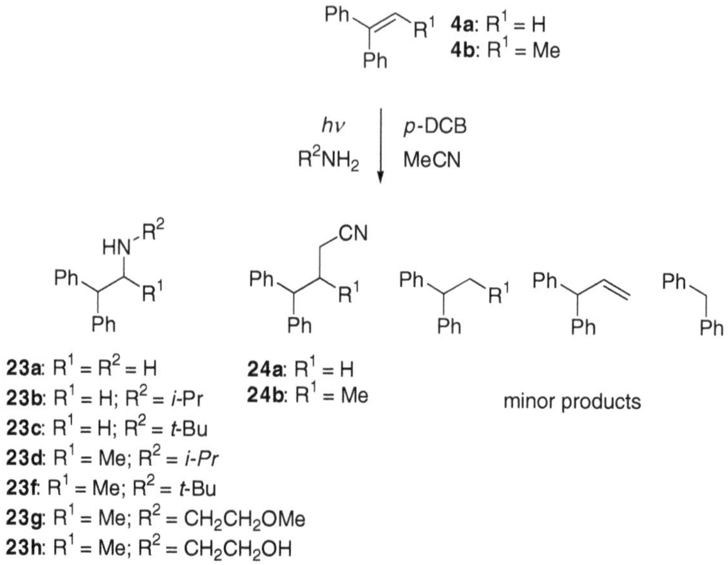

4a: R^1 = H
4b: R^1 = Me

hv | p-DCB
R^2NH_2 | MeCN

23a: $R^1 = R^2$ = H
23b: R^1 = H; R^2 = i-Pr
23c: R^1 = H; R^2 = t-Bu
23d: R^1 = Me; R^2 = i-Pr
23f: R^1 = Me; R^2 = t-Bu
23g: R^1 = Me; R^2 = CH_2CH_2OMe
23h: R^1 = Me; R^2 = CH_2CH_2OH

24a: R^1 = H
24b: R^1 = Me

minor products

SCHEME 6.22

TABLE 6.5
Photoamination of 4

		Products (Yield/%)[a]	
4	**RNH$_2$**	**23**	**24**
4a	NH$_3$	**23a** (18)	**24a** (15)
	i-PrNH$_2$	**23b** (48)	**24a** (27)
	t-BuNH$_2$	**23a** (22)	**24a** (60)
4b	NH$_3$	**23d** (44)	**24a** (7)
	i-PrNH$_2$	**23e** (65)	**24a** (3)
	t-BuNH$_2$	**23f** (0)	**24a** (72)
	MeOCH$_2$CH$_2$NH$_2$	**23g** (55)	**24a** (0)
	HOCH$_2$CH$_2$NH$_2$	**23h** (58)	**24a** (7)

[a] Isolated yield based on the consumed 4.

6.4.5 PHOTOAMINATION OF ARYLALKADIENES (5)

Irradiation of ammonia-saturated MeCN-H_2O solution containing 3-methyl-1-phenyl-1,3-butadiene (5a), 1-phenyl-1,3-pentadienes (5b–d), 1-(o-methoxyphenyl)-3-methyl-1,3-butadiene (5e), and 1-(o-methoxyphenyl)-4-methyl-1,3-pentadiene (5f) in the presence of *p*-DCB gave exclusively the corresponding 1,4-addition products, 4-amino-1-aryl-2-alkenes (25a–f), in 26 to 85% yields (Scheme 6.23) [47]. Moreover, photoamination of 1,4-diphenyl-1,3-butadiene (5g) gave 1-amino-1,4-diphenyl-2-butene (25g), while photoamination of 2-methyl-1,4-diphenyl-1,3-butadiene (5h) gave a mixture of 1-amino-2-methyl-1,4-diphenyl-2-butene (25h; 58%) and 1-amino-3-methyl-1,4-diphenyl-2-butene (25h'; 11%) (Scheme 6.24). Also, photoamination of 1-(*p*-methoxyphenyl)-4-phenyl-1,3-butadiene (5i) and the methyl-substituted isomers (5j–k) were investigated. Photoamination of 5i–j gave exclusively 25i–j, while photoamination of 5k gave a mixture 25k' (58%) and 25k (13%).

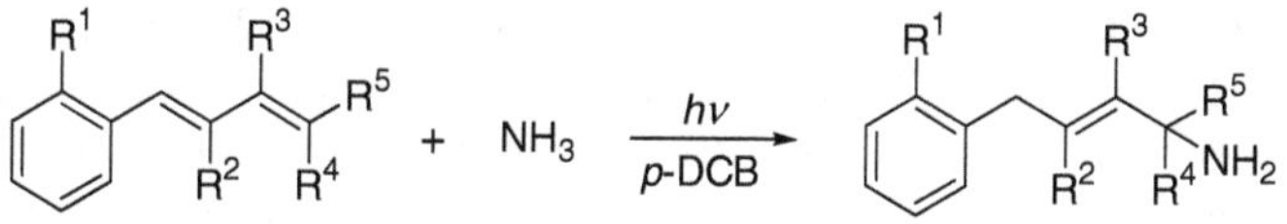

	R^1	R^2	R^3	R^4	R^5	Yield/%
a	H	H	Me	H	H	27
b	H	Me	H	H	Me	28
c	H	H	Me	Me	H	37
d	H	H	H	Me	Me	37
e	OMe	H	Me	H	H	80
f	OMe	H	H	Me	Me	85

SCHEME 6.23

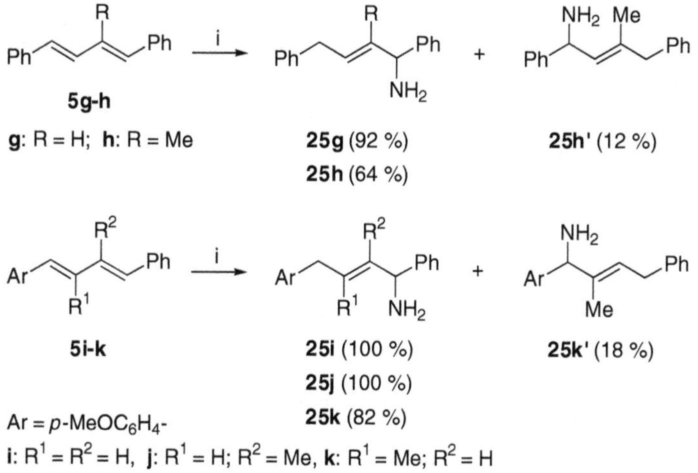

g: R = H; h: R = Me

25g (92 %)   25h' (12 %)
25h (64 %)

Ar = *p*-MeOC$_6$H$_4$-

25i (100 %)   25k' (18 %)
25j (100 %)
25k (82 %)

i: R^1 = R^2 = H, j: R^1 = H; R^2 = Me, k: R^1 = Me; R^2 = H

SCHEME 6.24   Reagents. i: *hv*/NH$_3$/*p*-DCB.

6.4.6 PHOTOAMINATION OF STYRENES (6)

Irradiation of an ammonia-saturated MeCN-H$_2$O solution containing *trans*-anethole (6a) and *m*-DCB gave 2-amino-1-(4-methoxyphenyl)propane (26a) as the sole product (Scheme 6.25) [48]. The yield of 26a was remarkably improved by the addition of 1,3,5-triphenylbenzene (TPB) or *m*-terphenyl (*m*-TP) (Table 6.6). After the photoreaction, these polyphenylbenzenes were recovered in high yields without the reaction with NH$_3$.

The similar photoamination of 1-aryl-1-propenes (6b–e), 3,4-dihydronaphthalenes (6g–l) and 1-(4-methoxyphenyl)-1-propen-3-ol (6m) and their derivatives (6n-o) with NH$_3$ were performed in the presence of 1,3,5-TPB and *m*-DCB, as shown in Scheme 6.26. Photoamination of *trans*-1-(3,5-dimethoxyphenyl)propene (6f) with *i*-PrNH$_2$ occurred at aromatic ring to give *trans*-1-(2-isopropylamino-3,5-dimethoxyphenyl)propene (27) (Scheme 6.27). No indication was obtained for the significant formation of other products occurring from the addition of the amine to the propenyl group. The positive charge in 6e$^{+\cdot}$ develops mainly over aromatic nuclei where two methoxyl groups may highly stabilize the positive charge.

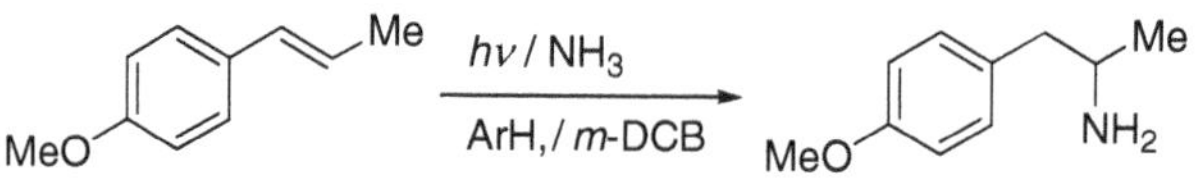

SCHEME 6.25

TABLE 6.6
Photoamination of Styrene Derivatives (6) with NH$_3$[a]

6 ($E_{1/2}^{ox}$/V)[b]	ArH[c]	Yield of 26a (%)[d]	Recovery	
			6	*m*-DCB
6a 0.93)	None	52	5	87
	1,3,5-TPB (1.52)	91	2	82
	m-TP (1.52)	85	2	95
	1,2,4-TP (1.54)	67	13	96
	p-TP (1.51)	66	3	90
	BP (1.45)	58	0	90

[a] Irradiation of an ammonia-saturated MeCN-H$_2$O (9:1; 75 ml) solution containing 6 (2 mmol), DCB (3.75 mmol), and ArH (0.75 mmol). [b] Oxidation potentials vs. Ag/AgNO$_3$. [c] TPB = 1,3,5-triphenylbenzene; *m*-TP= *m*-terphenyl; 1,2,4-TPB = 1,2,4-triphenylbenzene; *p*-TP = *p*-terphenyl; BP = biphenyl. After the photoreaction, ArH were recovered in > 68% yields. [d] Isolated yields based on consumed 6.

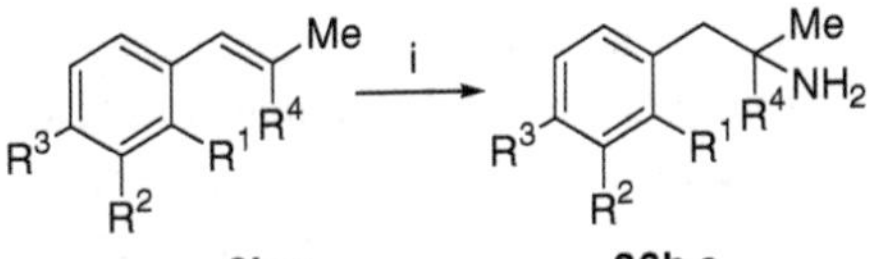

6b-e **26b-e**

26b: R¹ = OMe; R² = R³ = R⁴ = H (46%)
26c: R² = OMe; R¹ = R³ = R⁴ =H (75%)
26d: R² = R³= OMe; R¹ =R⁴ = H (65%)
26e: R¹ = R² = R³ = H; R⁴ = OMe (39%)

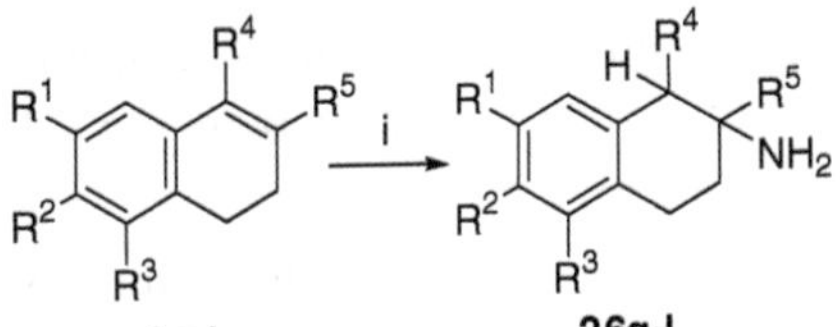

6g-l **26g-l**

26g: R¹ = R³ = R⁴ = R⁵ = H; R² = OMe (29%)
26h: R¹ = R³= R⁵= H; R² = OMe; R⁴ = Me (35%)
26i: R¹ = R³ = R⁴ = H; R² = OMe; R⁵ = Me (48%)
26j: R² = R³ = R⁴ = R⁵ = H; R¹ = OMe (64%)
26k: R¹ = R³ = Me; R² = R⁴ = R⁵ = H (64%)
26l: R¹ = R² = R³ = R⁵ = H; R⁴ = Me (25%)

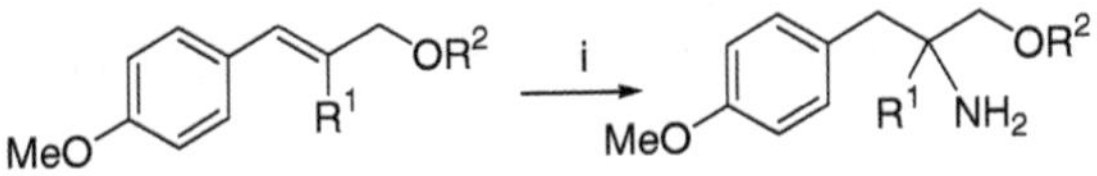

6m-o **26m-o**

26m: R¹ = R² = H (64%)
26n: R¹ = Me; R² = H (62%)
26o: R¹ = H, R² = Ac (45%)

SCHEME 6.26 Reagents. i: $h\nu/NH_3/m$-DCB.

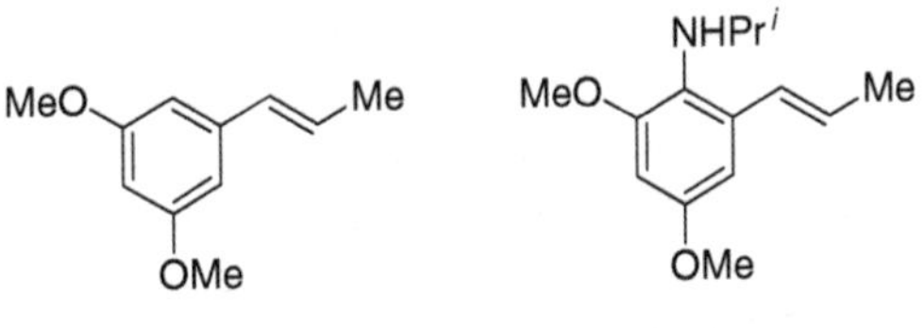

6f **27**

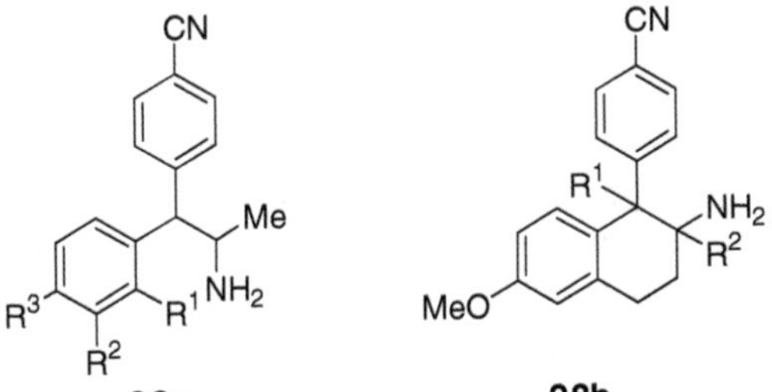

28a **28b**

SCHEME 6.27

Photoamination of 6 proceeds via the nucleophilic addition to the radical ion pairs between 6 and m-DCB followed by one-electron reduction of the aminated radicals with m-DCB$^-$ and protonation to give 26 according to Scheme 6.8. However, the highly reactive localized cation radicals of 6, in general, tends to cause side reactions involving dimerization, deprotonation, and isomerization, resulting in the amination in lower yields. It is well known that aromatic hydrocarbons (ArH) work as π-donor that can interact with a cation radical [49]. The additive effect of 1,3,5-TPB or m-TP appears to come from the formation of π-complex with 6$^{+\cdot}$. The π-complex formation with ArH would lower the reactivity of the cation radicals and suppress the side reactions, resulting in the effective photoamination of 6.

When p-DCB was used as an electron acceptor in place of m-DCB, photoamination of 6a–e and 6g–l gave mainly *Type II* aminated products, 28a and 28b, respectively, in which the cyanophenyl group was incorporated. The formation of 28 can be easily interpreted in terms of a coupling reaction between p-DCB$^-$ and the aminated radicals. The facile radical coupling reactions can be attributed to slow reduction of the aminated radicals by p-DCB$^-$, owing to strong electron-donating ability of methoxyl group.

The unsubstituted styrene derivatives such as indene (6p), 2-methylindene (6q), 3,4-dihydronaphthalene (6r), and 1,2-benzo-1,3-cyloheptadiene (6s) have weak absorption in near UV region, showing that direct irradiation could not induce the efficient photoamination. Therefore, redox-photosensitization was applied. Our efforts were paid to find arenes as sensitizers which are inert toward NH$_3$ and amines, because many arenes readily react with NH$_3$ under the conditions of photoamination. Moreover, the redox-photosensitizers are required to have relatively higher oxidation potentials than the substrates. The polyphenylbenzenes have relatively high oxidation potentials near 1.50 V and are inert toward NH$_3$ under photoamination conditions. However, many polyphenylbenzenes are poorly soluble in aqueous MeCN solutions. As the sensitizers that were moderately soluble in an aqueous MeCN solution, therefore, we selected 1,2,4-triphenylbenzene (1,2,4-TPB), 1,3,5-triphenylbenzene (1,3,5-TPB), m-terphenyl (m-TP), and p-terphenyl (p-TP) (Scheme 6.9B) [48]. Also 2,2'-methylenedioxy-1,1'-binaphthalene (BN) ($E_{1/2}^{ox} = 1.20$ V) was used as redox-sensitizer having lower oxidation potential.

In order to find a more effective sensitizer among the polyphenylbenzenes, the control experiments were performed for photoamination of 6p with NH$_3$. Irradiation of an ammonia-saturated MeCN-H$_2$O (19:1, 50 ml) containing 6p (2 mmol), m-DCB (3.75 mmol), and the polyphenylbenzene (0.75 mmol) for 5 to 13 h gave 29a (Scheme 6.28). When 1,2,4-TPB and p-TP were used as the sensitizer, photoamination of 1a proceeded in good yields. However, p-TP deposited during the irradiation because of the poor solubility in MeCN-H$_2$O solution. Therefore, we selected 1,2,4-TPB as the sensitizer for photoamination. The redox-photosensitized amination of 6p–s are shown in Scheme 6.28.

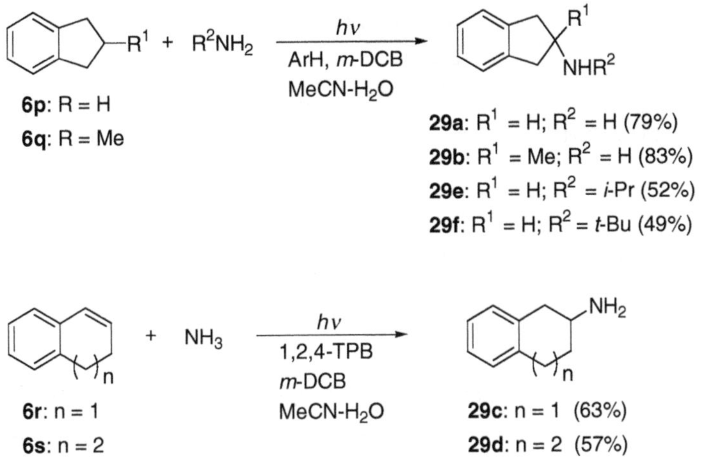

6p: R = H

6q: R = Me

29a: R^1 = H; R^2 = H (79%)

29b: R^1 = Me; R^2 = H (83%)

29e: R^1 = H; R^2 = i-Pr (52%)

29f: R^1 = H; R^2 = t-Bu (49%)

6r: n = 1

6s: n = 2

29c: n = 1 (63%)

29d: n = 2 (57%)

SCHEME 6.28

6.4.7 PHOTOAMINATION OF CYCLOPROPANES (7)

The redox-photosensitization was applied to the amination of arylcyclopropanes (7a–e) (Scheme 6.29) [50,51]. The amination of 1,2-diphenylcyclopropane (7a) photosensitized by a pair of 1,2,4-TPB and m-DCB gave 1-amino-1,3-diphenylpropane (30a) in higher yield than either cases of the BN-photosensitization or photoamination without the sensitizer. Especially, the direct irradiation of 7a with NH$_3$ in the absence of m-DCB did not form 30a, although it has been reported that the direct irradiation of 7a with alkylamines (e.g., C$_6$H$_{11}$NH$_2$, piperidine, n-BuNH$_2$) gave the aminated products [17]. When BN and 1,2,4-TPB were used as sensitizers, the yields of 1-amino-1,2-di(p-methoxyphenyl)propane (30b) from photoamination of trans-1,2-di(p-methoxyphenyl)cyclopropane (7b) were 90 and 54% yields, respectively. In the case of photoamination of unsymmetric trans-1-(4-methoxyphenyl)-2-phenylcyclopropane (7c), 1-amino-1-(p-methoxyphenyl)-3-phenylpropane (30c) and its regioisomer (30c′) were formed. However, trans-1-(2,4-dimethoxyphenyl)-2-phenylcyclopropane (7d) was not aminated. Photoamination of 1-(p-methoxyphenyl)-2,2-dimethylcyclopropane (7e) was carried out. In the case of the photosensitization by a pair of BN or 1,2,4-TPB and p-DCB, the p-cyanophenyl group-incorporated aminated product, 31a, was mainly obtained along with the formation of 30e, while direct photoamination without the sensitizer gave 31a in 2% yield. These aminated products from arylcyclopropanes (7a–e) are shown in Scheme 6.29 with the optimum yields. In cases of 7a–e, photoamination were carried out in the presence of Et$_4$NBF$_4$. Without Et$_4$NBF$_4$, the yield of aminated products decreased, for example, the yields of 30c–30c′ were 32%. Also, photoamination in the absence of DCB were very inefficient; for example, photoamination of 7c in the presence of 1,2,4-TPB and in the absence of DCB gave a small amount of 30c–30c′ (6%).

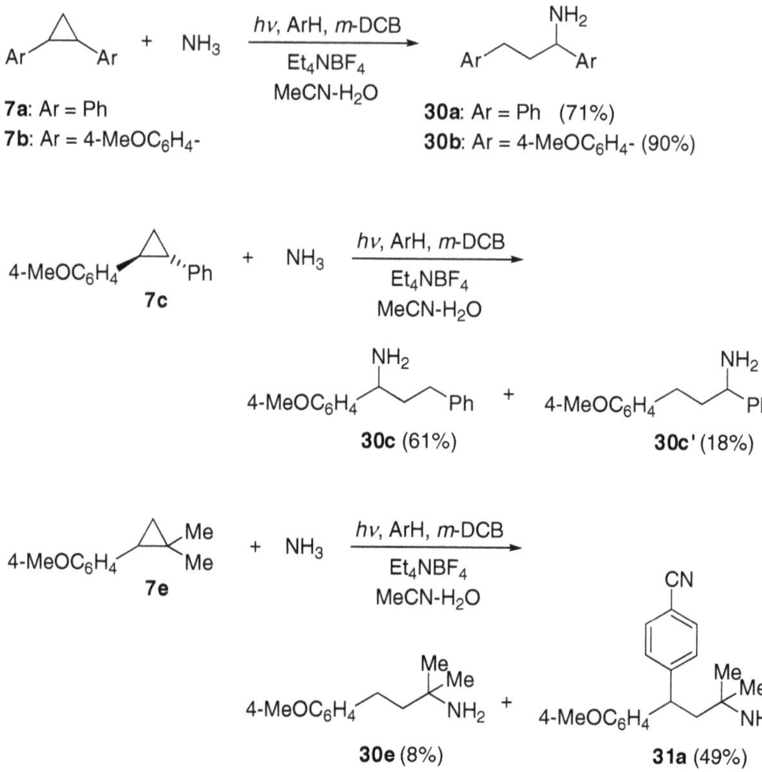

SCHEME 6.29

6.4.8 PHOTOAMINATION OF ALKENES AND ALKADIENES

The redox-photosensitized amination was applied to the amination of simple alkenes and 1,3-alkadienes (8) with NH_3 (Scheme 6.30) [52]. Photoamination of 2,3-dimethyl-2-butene (8a) gave the *Type II* aminated products, 32a and 32a'. It is noteworthy that photoamination of 8 occurred accompanied by the incorporation of 4-cyanophenyl groups. In the cases of 2,4-dimethyl-2,4-hexadiene (8b) and 2,4-hexadiene (8c), the aminated compounds (32b–i) in which 1,4-addition of the amino and 4-cyanophenyl groups occurred were obtained in fairly good yields. The similar photoamination of 1,2,3,4-tetramethyl-1,3-cyclopentadiene (8d) gave selectively 1-amino-4-(4-cyanophenyl)-1,2,3,4-tetramethyl-2-cyclopentene (32j) in 95%. Similar photoamination can be applied to 2,3-dihydrofuran (8e) and 2,3-dihydropyran (8f).

6.4.9 PHOTOAMINATION OF OTHER SUBSTRATES

The photosensitization by a pair of BN and *p*-DCB was effective for the amination of quadricyclane (9a) with NH_3 that gave 7-amino-5-(*p*-cyanophenyl)bicyclo[2, 2,1]hept-2-ene (33) in 62% yield (Scheme 6.31), while the photosensitization by a

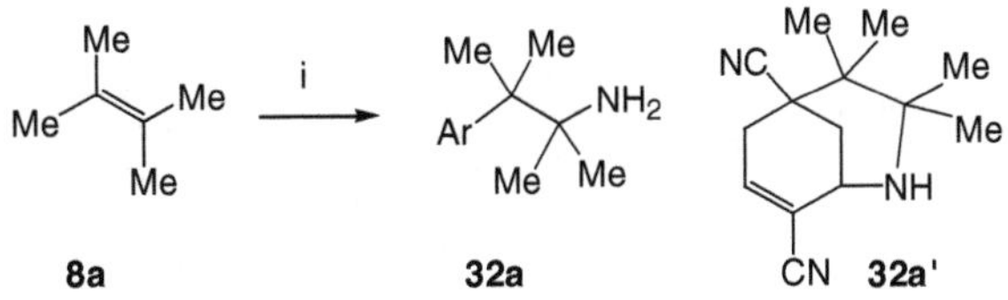

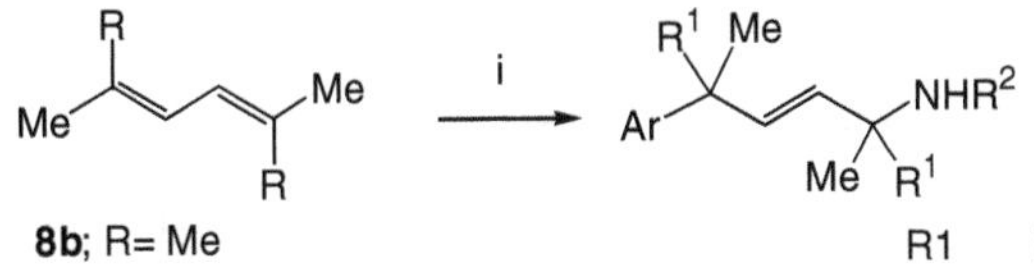

	R1	R2	Yield(%)
32b	Me	H	79
32c	H	H	98
32d	Me	i-Pr	70
32e	Me	t-Bu	62
32f	Me	(CH$_2$)$_2$OH	62
32g	H	i-Pr	53
32h	H	t-Bu	42
32i	H	i-Bu	67

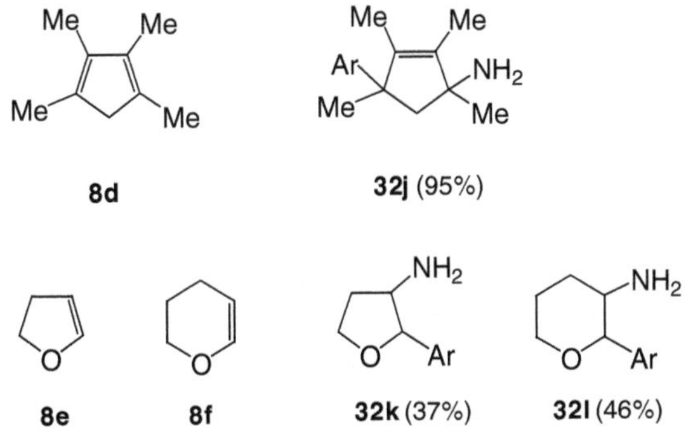

SCHEME 6.30 Reagents. i: hv/RNH$_2$/1,2,4-TPB/p-DCB; Ar=4-NC-C$_6$H$_4$-.

pair of 1,2,4-TPB and p-DCB was ineffective for the formation of 33 at all [50]. The structure of the acetamide of 33 was confirmed by an X-ray crystallographic analysis. Roth et al. has reported the detailed product analysis for the p-DCB-photosensitized nucleophilic addition of MeOH to 9a which gave several types of the MeOH-adducts [53]. On the other hand, photoamination of 9a gave 33 as a sole product. Nucleophilic addition of NH$_3$ to the cation radical of 9a might be faster than the competitive side reaction such as isomerization because of strong nucleophilicity of NH$_3$. Also, the coupling of the aminated product with the p-cyanophenyl radical were faster compared

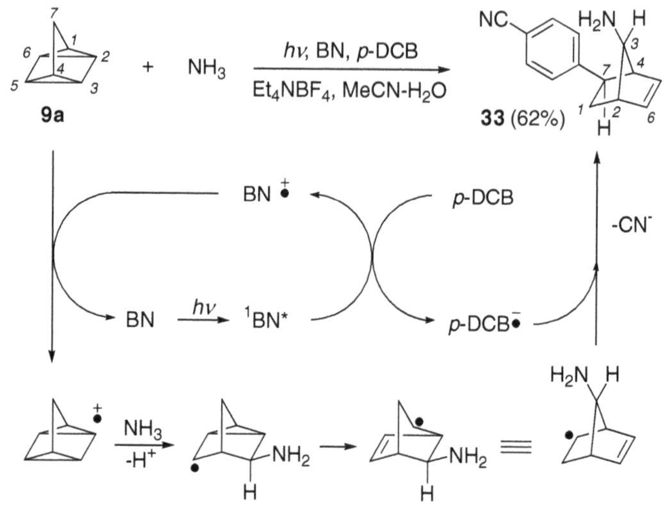

SCHEME 6.31

with the case of the *p*-DCB-photosensitized addition of MeOH, probably because the redox-photosensitization enhanced the concentration of the anion radical of *p*-DCB, thus providing the efficient formation of 33.

6.4.10 Synthesis of Heterocycles by Photoamination

6.4.10.1 Isoquinolines

We applied photoamination of stilbene (3a) and *p*-methoxystilbene (3b) to the construction of isoquinoline moiety [54]. Photoamination of 3a–b with amino alcohols, allylamine, and aminoacetaldehyde diethyl acetal followed by the methylation with HCO_2H/H_2CO gave *N*-(1,2-diarylethyl)-amino alcohols (18l and 18n), *N*-(1,2-diarylethyl)allylamine (18m and 18o), and *N*-(1,2-diarylethyl)aminoace taldehyde diethyl acetals (18p and 18q), respectively (Scheme 6.32). The 18l and 18m were treated with trifluoromethanesulfonic acid (TFSA) to give the 2,4-dimethyl-1-benzyl-1,2,3,4-tetrahydroisoquinolines (33a) in 60 and 94% yields, respectively. Treatment of 18n and 18o with TFSA gave 2,4-dimethyl-1-(*p*-acetoxybenzyl)-1,2,3,4-tetrahydroisoquinoline (34b) after the acetylation with acetic anhydride in 89 and 80% yields, respectively. The cyclization of benzylamino acetals (18p) was performed with gaseous BF_3 in dichloromethane to give the isopavines (35a; 89%). But HCl, H_2SO_4, and $BF_3.OEt_2$ were not effective for the cyclization of these benzylamino acetals. Moreover, the treatment of 18g with BF_3 gave 35b (28%).

Also, the similar treatment of phenanthrene (1a) gave *N*-substituted *N*-methyl-9-amino-9,10-dihydrophenanthrenes (10p–r) (Scheme 6.33). Cyclization of 10p and 10r

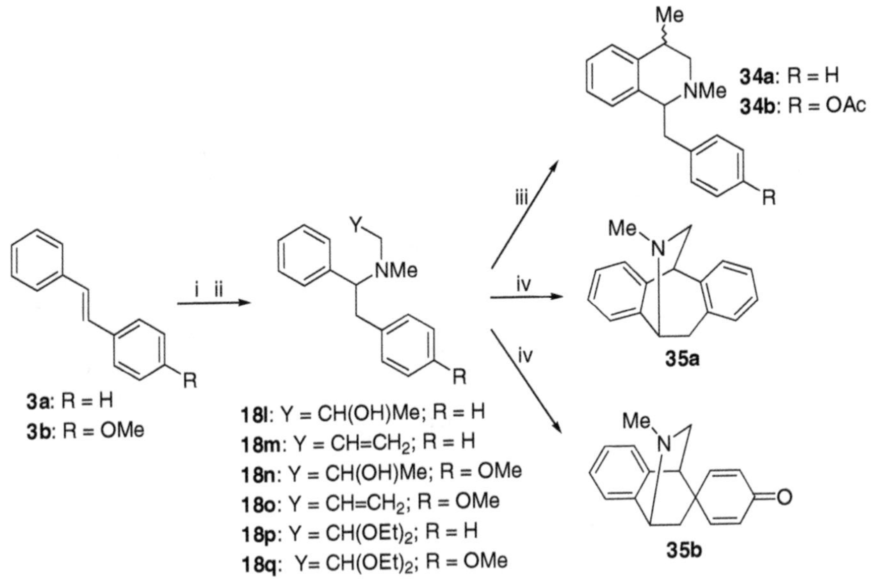

Me

34a: R = H
34b: R = OAc

iii

iv

iv

3a: R = H
3b: R = OMe

18l: Y = CH(OH)Me; R = H
18m: Y = CH=CH₂; R = H
18n: Y = CH(OH)Me; R = OMe
18o: Y = CH=CH₂; R = OMe
18p: Y = CH(OEt)₂; R = H
18q: Y = CH(OEt)₂; R = OMe

35a

35b

SCHEME 6.32 Reagents. i: hv/H₂NCH₂-Y/DCB; ii: HCO₂H / HCHO; iii: CF₃SO₃H; iv: BF₃.

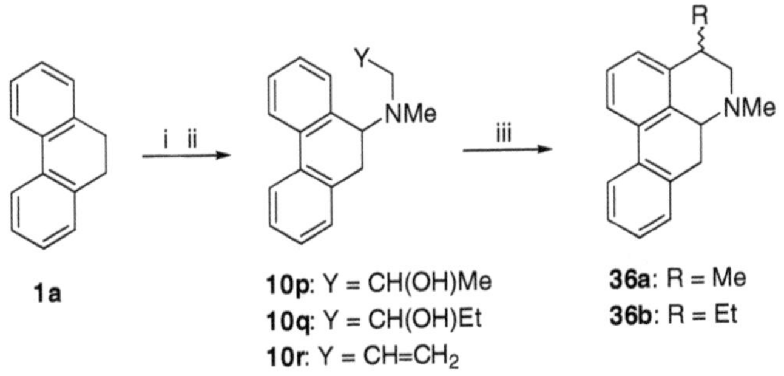

1a

i ii

iii

10p: Y = CH(OH)Me
10q: Y = CH(OH)Et
10r: Y = CH=CH₂

36a: R = Me
36b: R = Et

SCHEME 6.33 Reagents. i: hv/H₂NCH₂-Y/p-DCB; ii: HCO₂H / HCHO; iii: CF₃SO₃H.

with TFSA gave the aporphine derivative (36a) in 41 and 62% yields, respectively. The treatment of 10q with TFSA gave 36b (36%).

6.4.10.2 Dibenzo[a,d]cycloheptenimine

Dibenzo[a,d]cycloheptenimine was prepared by photoamination using dibenzo [a,d]cyclohepten-5-ol (3p) and 5-methoxy- and 5-acetoxydibenzo[a,d]cyclohepten e (3q and 3r) as the starting materials [44]. Synthetic routes are shown in Scheme 6.34. Photoamination of 3p–r with RNH₂ gave a mixture of *cis* and *trans* isomers

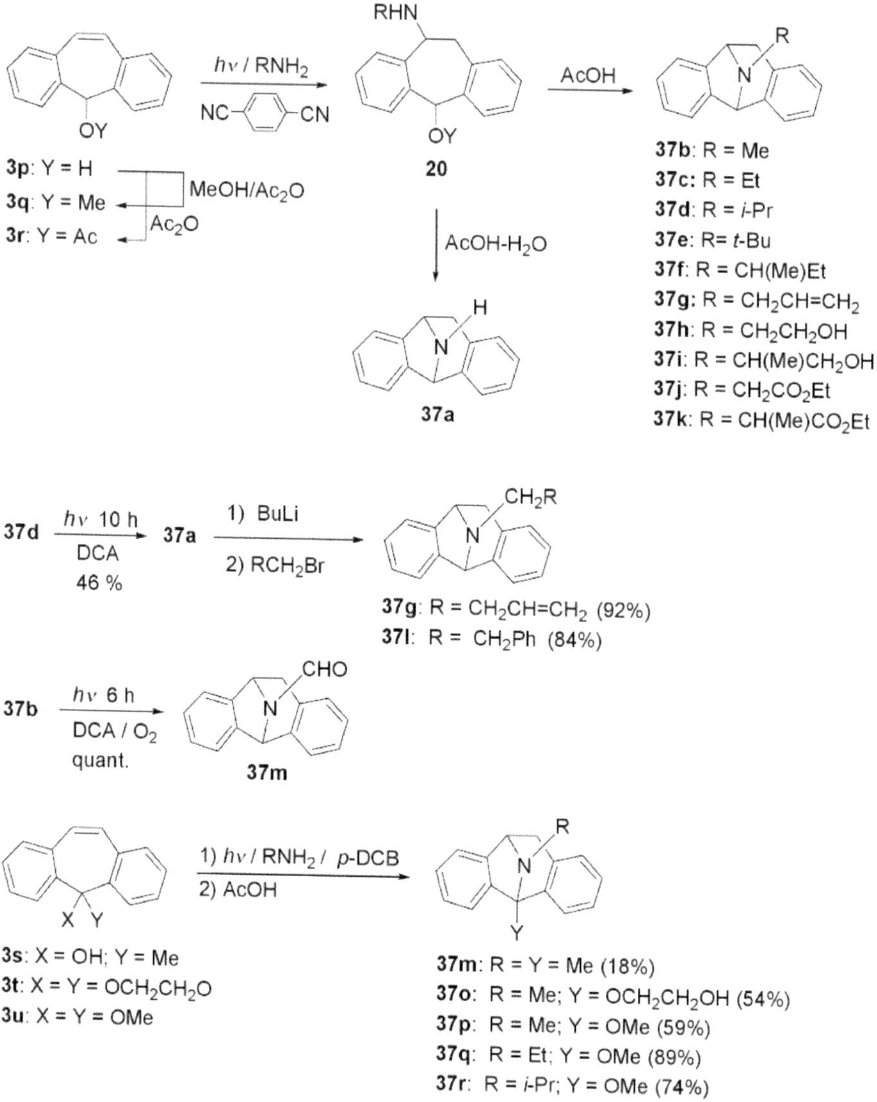

SCHEME 6.34 Synthesis and transformation of dibenzo[a,d]cycloheptenimines (37).

of 5-substituted 10-alkylamino-10,11-dihydro-5H-dibenzo[a,d]cycloheptenes (20). The reaction mixture was treated with AcOH at refluxing temperature to N-alkyl dibenzo[a,d]cycloheptenimines (37b–k). The results are summarized in Table 6.7. The transformation of N-unsubstituted analog of 20 (R=H) to 37a was performed by

TABLE 6.7
Synthesis of 37a-k via Photoamination of 3 Followed
by the Cyclization Reaction[a]

3	37 (Yield/%)	3	37 (Yield/%)
3p	37a (19)	3p	37f (62)
3q	37a (80)	3p	37g (32)
3p	37b (49)	3p	37h (41)
3q	37b (81)	3q	37h (63)
3r	37b (48)	3p	37i (53)
3p	37c (62)	3p	37j (28)
3p	37d (69)	3r	37j (44)
3q	37d (79)	3r	37k (53)
3q	37e (85)		

[a] Photoamination was performed by irradiating an MeCN-H$_2$O (9:1; 100 ml) solution containing 3 (6 mmol), p-DCB (12 mmol), and RNH2 (30 mmol) for 8 h when photoamination proceeded up to > 71% conversion. The transannular reaction was performed by heating of 20 with AcOH at 100°C for 5 h.

heating in AcOH-H$_2$O (4:1) at 100°C. It was found that 3q is good starting material to prepare 37a.

The alkylation of 37a with allylbromide and benzylbromide in the presence of n-BuLi gave N-allyl and N-benzyl analogs, 37g and 37l, in 92 and 84% yields, respectively. Moreover, 9,10-DCA-photosensitized reaction of methyl analog (37b) and isopropyl analog (37c) in MeCN gave N-formyl analog (37m) and 37a in 100 and 46% yields, respectively.

6.4.10.3 Intramolecular Photoamination

Intramolecular photoamination was applied to the synthesis of heterocyclic compounds. Photoamination of 9-(4-amino-1-butoxy)phenanthrene (1d) gave phenanthro[9,10-b]-4-oxazepine derivatives (38) in a cis to trans isomer ratio of 65:35, along with 4-[N-(9-phenanthrylamino)]butanol (39) (Scheme 6.35) [39]. Recently, Lewis and coworkers applied photoamination to the intramolecular amination of o-(3-aminoethyl)stilbenes (40a), which gave 1-benzyl-1,2,3,4-tetrahydroisoquinoline (41a; 76%) [55]. Similarly, photoamination of o-(3-aminopropyl)stilbenes (40) afforded benzazepine derivative (41; 70%) (Scheme 6.36). Also the aporphine and azepine prepared the intramolecular photoamination of 1-(aminoalkyl)phenanthrenes [56].

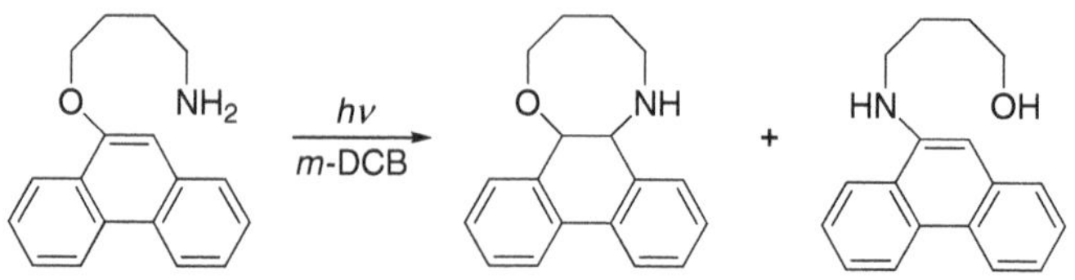

1d

38

39

41% (*cis* : *trans* = 65 : 35)   18%

SCHEME 6.35

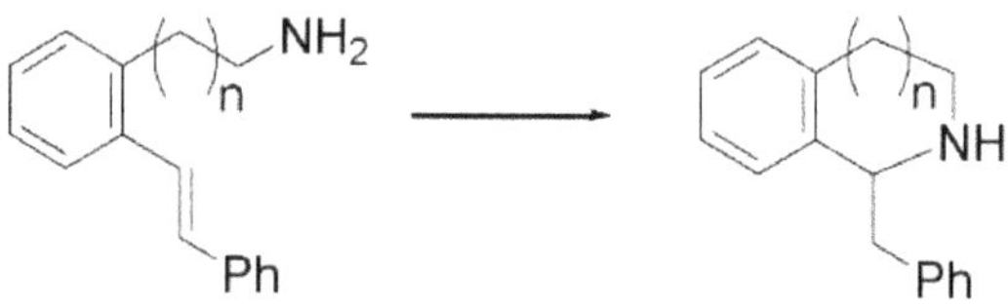

40a ; n=1   **41a**; n=1 (76%)

40b; n= 2   **41b**; n= 2 (70%)

SCHEME 6.36   Reagents i) *hv*/*p*-DCB.

6.5 PHOTOAMINATION VIA CATEGORY III: MECHANISTIC ASPECTS

The photoinduced electron transfer (Scheme 6.8) is usually confirmed by the following experimental results:

- Stern–Volmer quenching of the fluorescence of the substrates (D) by an electron acceptor (A) occurred at a nearly diffusion-controlled rate.
- The free-energy changes from the excited singlet state of D to A is calculated to be negative by a Rehm–Weller equation using the oxidation potentials and excitation energy of the substrates and reduction potentials of electron acceptors.

In the case of redox photosensitization (Scheme 6.9B), the Stern–Volmer quenching of fluorescence of the sensitizer (S) at a nearly diffusion-controlled rate and negative free-energy changes from the excited singlet state of S to A is required for the confirmation of the electron transfer from the exited singlet of S to A. The hole transfer from the resulting cation radical of S to D depends on the difference between the sensitizer and the substrates in the oxidation potentials.

6.5.1 KINETIC ANALYSIS

Photoamination of 1a with ammonia and aliphatic primary amines (RNH_2) in the presence of m-DCB in MeCN-H_2O (9:1) was kinetically analyzed by Yasuda and coworkers. The fluorescence of 1a was quenched by m-DCB at a diffusion-controlled limit, but not at all by RNH_2, and photoamination of 1a with RNH_2 did not occur in the absence of m-DCB. Therefore, an initiation process for photoamination is the photoinduced electron transfer from 1a to m-DCB. For kinetic analysis, Scheme 6.37, involving the nucleophilic addition of RNH_2 to the cationic intermediate, was postulated. As the cationic species, the cation radical of 1a ($D^{+\cdot}$) and the ion pair $[D^{+\cdot}/$

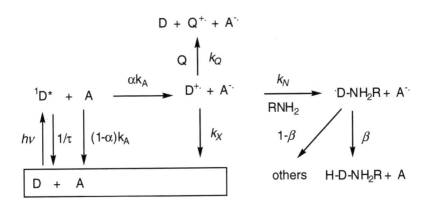

In the absence of Q

$$\Phi^{-1} = \frac{1}{\alpha\beta}\left(1 + \frac{1}{k_A\tau[A]}\right)\left(1 + \frac{k_X}{k_N[RNH_2]}\right) \qquad \text{- (1)}$$

$$k_A\tau[A] = 9.6$$

$$\Phi^{-1} = \frac{1}{\alpha\beta}\left(1 + \frac{k_X}{k_N[RNH_2]}\right) \qquad \text{- (2)}$$

In the presence of Q

$$\Phi^{-1} = \frac{1}{\alpha\beta}\left(1 + \frac{k_X}{k_N[RNH_2]} + \frac{k_Q[Q]}{k_N[RNH_2]}\right) \qquad \text{- (3)}$$

SCHEME 6.37

A^-] between the cation radical of 1a and the anion radical of m-DCB were proposed. In the absence of quencher, the quantum yield (Φ) for the formation of H-ArH-NHR is expressed by Equation 1 in Scheme 6.37. 1,4-Dimethoxybenzene and 1,3,5-trimethoxybenzene were used as a quencher (Q) that can quench the cationic species at a nearly diffusion-controlled rate (1×10^{10} dm^3 mol^{-1}s^{-1}). In the presence of Q, the quantum yield (Φ^Q) is expressed by Equation 3 in Scheme 6.37. As the results, the rate constants (k_N) for the nucleophilic addition of RNH$_2$ were obtained, as shown in Table 6.8. Hammet plots of $\ln k_N$ vs Taft σ* gave a linear correlation with a large negative slope (−2.1), showing that a substantial positive charge should be populated on the nitrogen atom of RNH$_2$ in the transition state (Figure 6.1).

Lemmetyinen and coworkers have reported the mechanistic analysis for photoamination of 1a with ethylenediamine in the presence of p-DCB in MeCN-H$_2$O (9:1) by laser flash photolysis [57]. During the irradiation, the short-lived transient appeared at 420 nm, followed by the formation of long-lived transient (420 nm) due to the cation radical of 1a. The short-lived transient absorption is due to the exciplex between 1a and p-DCB and/or ion pair between the cation radical of 1a and the anion radical of p-DCB. By the addition of ethylenediamine, the absorption of exciplex was affected but the absorption of the cation radical of 1a was not affected. Therefore, the nucleophilic attack of the amine occurred on the exciplex between 1a and p-DCB. They have concluded that nucleophilic attack of the amine to the cation radical of 1a seems to be impossible.

6.5.2 REGIOCHEMISTRY AND PM3-CALCULATION

As shown in Scheme 6.8, photoamination proceeds via the nucleophilic addition of NH$_3$ to the cation radical ($D^{+\cdot}$) to give the aminated cation radicals that are deprotonated to give aminated radicals ($\cdot$D-NH$_2$). It is reduced with the DCB$^{-\cdot}$ and

---

TABLE 6.8
Kinetic Data for Photoamination

RNH$_2$	Taft σ*	k_N (10^8 mol dm^{-3} s^{-1})
HNH$_2$	0.49	0.3
NCCH$_2$CH$_2$NH$_2$	0.46	1.0
HOCH$_2$CH$_2$NH$_2$	0.20	0.6
MeOCH$_2$CH$_2$NH$_2$	0.19	0.7
CH$_2$=CHCH$_2$NH$_2$	0.13	1.5
MeNH$_2$	0.00	2.0
EtNH$_2$	−0.10	4.0
n-PrNH$_2$	−0.115	4.0
i-PrNH$_2$	0.19	7.9
t-BuNH$_2$	−0.30	8.9

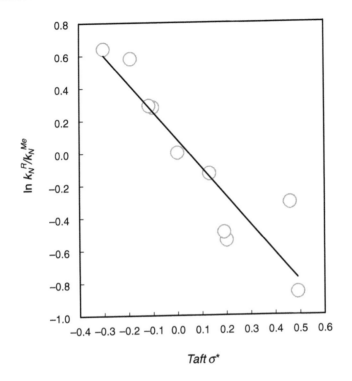

FIGURE 6.1   Hammet plots of lnk_N vs. Taft σ^*.

protonated to give the final products. Thus photoamination proceeds via three kinds of the intermediates: $D^{+\cdot}$, $\cdot D\text{-}NH_2$, and the aminated anion ($^-D\text{-}NH_2$). The rates of the nucleophilic addition should depend on the positive charge density of the reaction site in $D^{+\cdot}$ and the stabilities of $\cdot D\text{-}NH_2$ and/or $^-D\text{-}NH_2$. Therefore, we deduced these values by semi-empirical PM3-UHF/RHF or *ab initio* calculation of the positive charge distribution of $D^{+\cdot}$ and the heat of formation of $\cdot D\text{-}NH_2$.

In the case of the cation radical of 1,2-distyrylbenzene (3w), the PM3 calculation indicated that the negative and positive charges appeared on α- and β-positions, respectively, suggesting that the nucleophilic addition of NH_3 to the opened 3w$^{+\cdot}$ occurred selectively on β-position (Scheme 6.38) [45]. Moreover, it is suggested that the formation of 22a proceeded via the isomerization of the opened-3w$^{+\cdot}$ to the closed 3w$^{+\cdot}$ followed by the nucleophilic addition of NH_3.

In the case of the cation radical of the arylalkadiene (5$^+$), the regiochemistry can be related with the distribution of positive charge of 5$^+$ as well as the stabilities of $\cdot D\text{-}NH_2$ [47]. The nucleophilic attack of NH_3 occurred at the positive sites of 5$^+$ and proceeded via the most stable $\cdot D\text{-}NH_2$. If there is no difference in the stabilities of $\cdot D\text{-}NH_2$, the amination pathway is determined by the stability of $^-D\text{-}NH_2$.

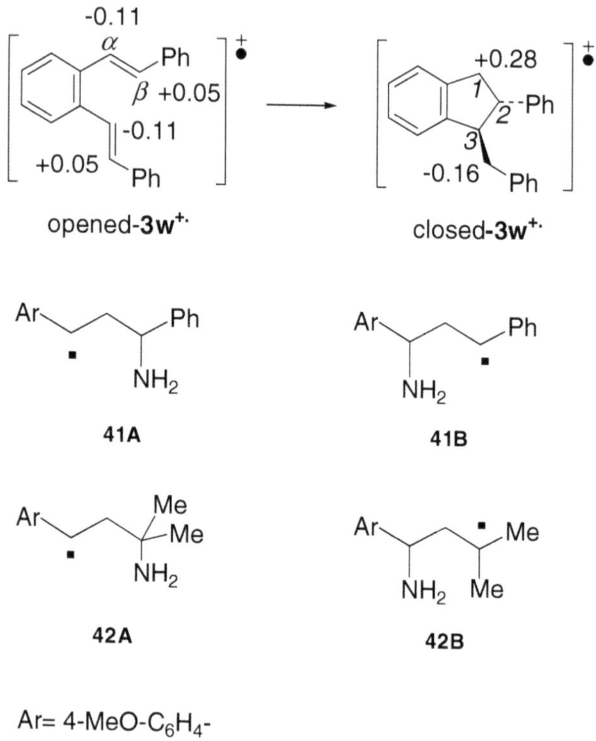

opened-**3w**$^{+\cdot}$ closed-**3w**$^{+\cdot}$

41A   **41B**

42A   **42B**

Ar= 4-MeO-C$_6$H$_4$-

SCHEME 6.38

Another example is the regiochemistry of arylcyclopropanes (7). The calculation of the charge distribution of the cation radicals of 7 were performed for the ring-closed cation radicals, since Dinnocenzo et al. has reported that the photoinduced nucleophilic addition of MeOH to the arylcyclopropanes which proceeded via the nucleophilic attack to the cation radicals of the ring-closed cyclopropanes. [58]. The positive charge develops over the methoxy-substituted aryl group except for the case of 7a. In the case of 7d having two methoxy groups, the positive charge localized more highly on the aryl group compared with the cases of the other cyclopropanes. This is a reason that the cation radical of 7d did not allow the nucleophilic attack of NH$_3$ to cyclopropane moiety, leading to no aminated products. The nucleophilic addition of NH$_3$ at the ring-closed cation radicals of 7 gave the aminated radicals to give the aminated products (30) after the reduction by A$^{-\cdot}$ and the protonation. Therefore, the isomer ratio of the aminated products (30) would depend on the relative stabilities of the aminated radicals ($\cdot$D-Nu, Nu=NHR in Scheme 6.1). In the case of photoamination of 7e, the stabilities of the aminated radicals (41A and/or 41B) were estimated by the heat of the formation calculated by *ab initio* method. Since the 41A radical was *ca.* 127 kJ mol^{-1} stable than 41B, photoamination occurred selectively via the 41A intermediate, leading to the formation of 30e and 31a (Scheme 6.38). In the case of 7c, on the other hand, the heats of the formations of the aminated radicals (42A and 42B) were nearly equal to each other, thus giving a mixture of 30c and 30c'. In the cases of 7a–d, photoamination proceeded according to Scheme 6.9B and the regiochemistry of 30a–d were in accord with the stabilities of the aminated radicals, as shown in Table 6.4.

6.5.3 CONSIDERATION BASED ON THE OXIDATION POTENTIALS

The hole transfer from the cation radical of sensitizer ($S^{+\cdot}$) to the substrates is a key pathway for the efficient photoamination [50]. The efficiency of the hole transfer should depend on the difference in $E_{1/2}^{ox}$ between S and the substrates (Scheme 6.39). In the case of the redox-photosensitized amination of the substrates whose $E_{1/2}^{ox}$ are lower than those of S, the efficient hole transfer would occur. However, the hole transfer from $S^{+\cdot}$ to the substrates whose $E_{1/2}^{ox}$ are much higher than that of S would not occur. Moreover, 1,2,4-TPB-photosensitized amination of 9a was inefficient,

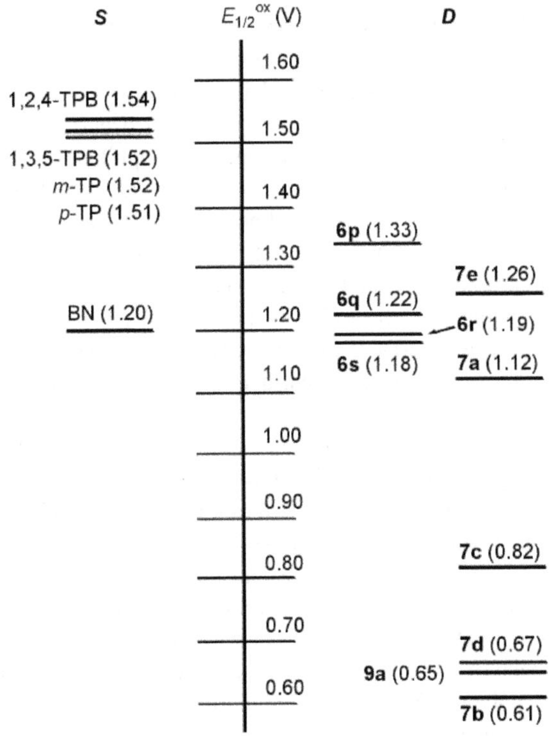

S	$E_{1/2}^{ox}$ (V)	D
	1.60	
1,2,4-TPB (1.54)		
	1.50	
1,3,5-TPB (1.52)		
m-TP (1.52)	1.40	
p-TP (1.51)		
		6p (1.33)
	1.30	
		7e (1.26)
		6q (1.22)
BN (1.20)	1.20	6r (1.19)
	1.10	6s (1.18) 7a (1.12)
	1.00	
	0.90	
	0.80	7c (0.82)
	0.70	
		7d (0.67)
		9a (0.65)
	0.60	
		7b (0.61)

SCHEME 6.39 Relationship between the oxidation potentials of the substrates (D) and those of the sensitizer (S).

although $E_{1/2}^{ox}$ of 9a was much lower than that of 1,2,4-TPB. Probably this process lies on Marcus inverted region, because the hole transfer process from the cation radical of 1,2,4-TPB to 9a is much exergonic process (0.89 eV = 85.8 kJ mol^{-1}), this process lies on Murcus inverted region.

6.6 SCOPE AND LIMITATION

Scheme 6.40 summarizes the relation between the substrates investigated and the numbers (n) of π-electrons of the substrates. Photoamination can be performed for a variety of hydrocarbons which form not only localized cation radicals, but also delocalized cation radicals. The absorptions of incident light by the substrates are strongly related to their number of π-electrons. The substrates having less than 8 π-electrons have weak absorption near UV regions, requiring the photosensitized reaction. The photosensitization with DCA, CNN, and DCN, however, were not effective for photoamination with RNH$_2$, since the excited states of the sensitizers are efficiently quenched by RNH$_2$. Therefore, we have developed the redox-photosensitized amination using 1,2,4-TPB and BN as redox-sensitizer. These sensitizers can sensitize the substrates whose $E_{1/2}^{ox}$ were *ca.* 0.06 V higher and *ca.* 0.9 V lower than those of the sensitizer. On the other hand, the heterocycles such as pyridines, thiophenes, and furans do not allow photoamination.

With ammonia and primary alkylamines having functional groups such as hydroxy, vinyl, ester, cyano, and acetal groups, photoamination occurred without other side reactions, since the amino group has the stronger nucleophilicity and photoamination can be performed under mild conditions. However, the secondary amines whose oxidation potentials are relatively low could not be used because of the occurrence of the electron exchange of the cation radicals of the substrates with the amines.

As an electron acceptor, we usually used *m*- and *p*-DCB having no absorption at over 300 nm not to prevent the light-absorption by the substrates. The photoinduced electron transfer occurs preferably to *p*-DCB than *m*-DCB while the resulting anion radical of *m*-DCB plays as a stronger reducing agent than that of *p*-DCB because the reduction potential of *m*-DCB is more negative than that of *p*-DCB ($E_{1/2}^{red}$ = −1.93 V for *p*-DCB and = −2.15 V for *m*-DCB). Moreover, the anion radical of *p*-DCB has potentially strong coupling ability with the radical site of the cation radicals of the substrates, giving the *p*-cyanophenyl group-incorporated *Type II* aminated products [59].

As the reaction solvents, MeCN-water (9:1 or 19:1) are best. MeCN is a highly polar solvent: $(\varepsilon - 1)/(2\varepsilon + 1) = 0.48$, where dielectric constant (ε) is 35.94 and a Dimroth-Reichardt value $E_T(30) = 45$ kcal/mol [60]. Less polar solvents such as benzene are poor solvents. DMSO, DMF, and MeOH are ineffective solvents than MeCN. The presence of water is requisite to dissolve ammonia gas and the volatile amines in reaction solvents. Moreover, photoamination in MeCN-water is clean and efficient than that in water-free MeCN. Photoamination in solvents with relatively low polarity such as 1,2-dimethoxyethane, tetrahydrofuran, and 1,4-dioxane was achieved by an addition of Bu$_4$NBF$_4$. Photoamination of 1a with propylamine in the

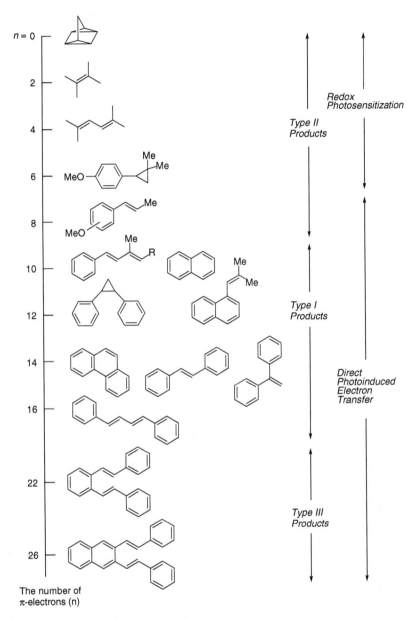

SCHEME 6.40 Substrates for photoamination.

presence of *p*-DCB in these solvents proceeds via the salt-induced charge separation of an exciplex between 1a and *p*-DCB and the subsequent nucleophilic addition of the amine [61].

TABLE 6.9
Scope and Limitation of Photoamination

	Favorable	Unfavorable
Substrates	See Scheme 6.39	Furan, thiophene, indole, and their derivatives
Amines	Ammonia	Secondary amine
	Primary amines	Tertiary amines
Electron acceptors	DCB	DCN, DCNA, DCP, MPI, MBA
Solvents	MeCN-H$_2$O	Nonpolar solvents
	MeCN	DMSO
	Dioxane[a]	DMF
	THF[a]	MeOH, EtOH
	1,2-Dimethoxyethane[a]	

[a] Involving Et$_4$NBF$_4$.

Salt effects on the photoinduced electron transfer have been well known to prevent back-electron transfer between the cation radical of substrates and the anion radical of electron acceptor [62]. In photoamination of 7a–e, the addition of Et$_4$NBF$_4$ were effective in preventing the competitive back electron transfer in the cases of 7a–e [50].

The aminated products formed by photoamination of arenes and stilbenes were the dihydroarenes and 1-amino-1,2-diarylethanes, respectively. These compounds were not so stable for thermolysis [63]. For example, the glc analysis of the aminated products showed no peaks in some cases because of the thermal decomposition at the injection temperature. Therefore, it is necessary to select suitable reagents that can be used under mild conditions for the cyclization and further reaction.

6.7   CONCLUSION

In general, many organic syntheses have been achieved by the combination of the reactive species such as reactive substrates, reactive reagents, and/or reactive catalysts. In many cases, therefore, large amounts of materials that are not incorporated to final products remain as by-products, which will be emitted to environmental field after the appropriate treatments. Photoamination diminishes remarkably the amounts of reagents compared with the usual amination, since photons are clean "reagents." Moreover, photoamination has provided a useful synthetic tool to introduce the amino group to C-C double bonds under mild conditions. For example, 2-aminoindan, which has biological activities and is used as an intermediate for synthesis of medicine, has been prepared in four steps from indene, while photoamination of indene gave 2-aminoindan in one step without special reagents such as acids and bases [50].

Thus, the present photoamination provides an environmentally friendly synthetic process for transformation of a variety of substrates into the corresponding aminated compounds.

ACKNOWLEDGMENTS

It is a great pleasure to acknowledge the contributions of many students whose names appear in the various references. We thank Chyongjin Pac for his valuable discussions. Financial support has been partially provided by the Ministry of Education, Culture, Sports, Science, and Technology (MEXT) of the Japanese government.

REFERENCES

1. L. Lehninger, D. L. Nelson, and M. M. Cox, *Principle of Biochemistry*, Worth Publishers, New York, p. 688 (1997).
2. M. S. Gibson, in *The Chemistry of the Amino Group*, S. Patai, Ed., Interscience, New York, p. 37 (1968).
3. D. Bryce-Smith, A. Gilbert, and G. Klunklin, *J. Chem. Soc., Chem. Commun.*, 1973, 330.
4. N. C. Yang and J. Libman, *J. Am. Chem. Soc.*, 95, 5783 (1973).
5. F. D. Lewis and T.-I. Ho, *J. Am. Chem. Soc.*, 99, 7991 (1977).
6. S. S. Hixson, *J. Am. Chem. Soc.*, 96, 4866 (1974); H. Tomioka and H. Miyagawa, *J. Chem. Soc., Chem. Commun.*, 1988, 1183.
7. F. D. Lewis, D. Bassani, E. L. Burch, B. E. Cohen, J. A. Engelman, G. D. Reddy, S. Schneider, W. Jaeger, P. Gedeck, and M. Gahr, *J. Am. Chem. Soc.*, 117, 660 (1995).
8. A. Sugimoto, R. Hiraoka, H. Inoue, and T. Adachi, *J. Chem. Soc. Perkin Trans. 1*, 1992, 1559; A. Sugimoto, N. Fukada, T. Adachi, and H. Inoue, *J. Chem. Soc. Perkin Trans. 1*, 1995, 1597.
9. U. C. Yoon, P. S. Mariano, R. S. Givens, and B. W. Atwater, *Advances in Electron Transfer Chemistry*, P. S. Mariano, Ed., JAI Press, Greenwich, CT, Vol. 4, p. 117 (1994); U. C. Yoon and P. S. Mariano, *Acc. Chem. Res.*, 25, 233 (1992).
10. F. D. Lewis and P. E. Correa, *J. Am. Chem. Soc.*, 103, 7347 (1981).
11. M. Yasuda, C. Pac, and H. Sakurai, *Bull. Chem. Soc. Jpn.*, 54, 2352 (1981).
12. F. D. Lewis, *Advances in Electron Transfer Chemistry*, P. S. Mariano, Ed., JAI Press, Greenwich, CT, Vol. 5, p. 1 (1996).
13. R. A. Neunteufel and D. R. Arnold, *J. Am. Chem. Soc.*, 95, 4080 (1973).
14. M. Yasuda, T. Yamashita, T. Matsumoto, K. Shima, and C. Pac, *J. Org. Chem.*, 50, 3667 (1985).
15. M. Yasuda and K. Mizuno, *Handbook of Photochemistry and Photobiology*, H. S. Nalwa, Ed., American Scientific, New York, Vol 2. 393 (2003).
16. M. Kawanishi and K. Matsunaga, *J. Chem. Soc., Chem. Commun.*, 1972, 313.
17. S. S. Hixson, *J. Chem. Soc., Chem. Commun.*, 1972, 1170; S. S. Hixson, *J. Am. Chem. Soc.*, 96, 4866 (1974).
18. J. Griffiths and C. Hawkins, *J. Chem. Soc., Chem. Commun.*, 1973, 111.

19. M. Tajima, K. Kato, K. Matsunaga, and H. Inoue, *J. Photochem. Photobiol. A: Chemistry*, 140, 127 (2001).

20. D. Bryce-Smith, A. Gilbert, and S. Krestonosich, *J. Chem. Soc., Chem. Commun.*, 1976, 405.

21. D. Bryce-Smith, A. Gilbert, and S. Krestonosich, *Tetrahedron Lett.*, 1977, 385.

22. G. G. Wubbels and W. D. Cotter, *Tetrahedron Lett.*, 30, 6477 (1989).

23. G. G. Wubbels, A. M. Halverson, J. D. Oxman, and V. H. DeBruyn, *J. Org. Chem.*, 50, 4499 (1985).

24. A. Cantos, J. Marquet, and M. Moreno-Manas, *Tetrahedron Lett.*, 30, 2423 (1989)

25. M. Yasuda, T. Sone, K. Tanabe, and K. Shima, *J. Chem. Soc., Perkin Trans. 1*, 1995, 459; M. Yasuda, T. Sone, K. Tanabe, and K. Shima, *Chem. Lett.*, 1994, 453.

26. G. Pandey, M. Sridhar, and U. Bhalerao, *Tetrahedron Lett.*, 31, 3573 (1990).

27. C. Pac, A. Nakasone, and H. Sakurai, *J. Am. Chem. Soc.*, 99, 5806 (1977).

28. T. Majima, C. Pac, A. Nakasone, and H. Sakurai, *J. Am. Chem. Soc.*, 103, 4499 (1981).

29. S. S. Hixson, *Tetrahedron Lett.*, 1971, 4211.

30. J. J. McCullough and W. S. Wu, *J. Chem. Soc., Chem. Commun.*, 1972, 1136.

31. P. H. Mazzocchi, S. Minamikawa, and P. Wilson, *Tetrahedron Lett.*, 1978, 4361.

32. K. Maruyama and Y. Kubo, *J. Am. Chem. Soc.*, 100, 7772 (1978).

33. P. S. Mariano, J. L. Stavinoha, G. Pepe, and E. F. Meter, Jr, *J. Am. Chem. Soc.*, 100, 7114 (1978).

34. M. Yauda, C. Pac, and H. Sakurai, *Bull. Chem. Soc. Jpn.*, 53, 502 (1980).

35. F. D. Lewis, *Adv. Photochem.*, 13, 4404 (1986).

36. M. Yasuda, T. Yamashita, T. Matsumoto, K. Shima, and C. Pac., *J. Org. Chem.*, 50, 3667 (1985); M. Yasuda, T. Yamashita, K. Shima, and C. Pac., *J. Org. Chem.*, 52, 753 (1987).

37. M. Yasuda, Y. Watanabe, K. Tanabe, and K. Shima, *J. Photochem. Photobiol. A: Chemistry*, 79, 61 (1994).

38. M. Yasuda, Y. Matsuzaki, K. Shima, and C. Pac., *J. Chem. Soc., Perkin Trans. 2*, 1988, 745.

39. M. Yasuda, K. Shiomori, S. Hamasuna, K. Shima, and T. Yamashita, *J. Chem. Soc. Perkin Trans. 2*, 1993, 305.

40. T. Yamashita, K. Yamano, M. Yasuda, and K. Shima, *Chem. Lett.*, 1993, 637; T. Yamashita, K. Tanabe, K. Yamano, M. Yasuda, and K. Shima, *Bull. Chem. Soc. Jpn.*, 67, 246 (1994).

41. M. Yasuda, R. Kojima, R. Ohira, T. Shiragami, and K. Shima, *Bull. Chem. Soc. Jpn.*, 71, 1655 (1998).

42. M. Yasuda, T. Isami, J. Kubo, M. Mizutani, T. Yamashita, and K. Shima, *J. Org. Chem.*, 57, 1351 (1992).

43. J.-H. Ho and T.-I. Ho, *J. Chinese Chem. Soc. Taipei*, 50, 109 (2003).

44. M. Yasuda, T. Wakisaka, R. Kojima, K. Tanabe, and K. Shima, *Bull. Chem. Soc. Jpn.*, 68, 3169 (1995).

45. R. Kojima, T. Shiragami, K. Shima, M. Yasuda, and T. Majima, *Chem. Lett.*, 1997, 1241.

46. T. Yamashita, K. Shiomori, M. Yasuda, and K. Shima, *Bull. Chem. Soc. Jpn.*, 64, 366 (1991).

47. R. Kojima, T. Yamashita, K. Tanabe, T. Shiragami, M. Yasuda, and K. Shima, *J.*

Chem. Soc., Perkin Trans 1, 1997, 217.
48. T. Yamashita, M. Yasuda, T. Isami, S. Nakano, K. Tanabe, and K. Shima, *Tetrahedron Lett.*, 34, 5131 (1993); T. Yamashita, M. Yasuda, T. Isami, K. Tanabe, and K. Shima, *Tetrahedron*, 50, 9275 (1994).
49. T. Majima, C. Pac, and H. Sakurai, *J. Am. Chem. Soc.*, 102, 5265 (1980).
50. M. Yasuda, R. Kojima, H. Tsutsui, D. Utsunomiya, K. Ishii, K. Jinnouchi, T. Shiragami, and T. Yamashita, *J. Org. Chem.*, 68, 7618 (2003).
51. M. Yasuda, R. Kojima, H. Tsutsui, L. A. Watanabe, J. Hobo, T. Yamashita, T. Shiragami, and K. Shima, *Chem. Lett.*, 1999, 1269.
52. Unpublished results.
53. H. Weng and H. D. Roth, *J. Org. Chem.*, 60, 4136 (1995).
54. M. Yasuda, J. Kubo, and K. Shima, *Heterocycles*, 31, 1007 (1990); M. Yasuda, S. Hamasuna, K. Yamano, J. Kubo, and K. Shima, *Heterocycles*, 34, 965 (1992).
55. F. D. Lewis and G. D. Reddy, *Tetrahedron Lett.*, 33, 4249 (1992).
56. F. D. Lewis, G. D. Reddy, and B. E. Cohen, *Tetrahedron Lett.*, 35, 535 (1994).
57. K. Nieminen, J. Niiranen, H. Lemmetyinen, and I. Sychtchikova, *J. Photochem. Photobiol. A: Chem.*, 61, 235 (1991).
58. J. P. Dinnocenzo, T. R. Simpson, H. Zuilhof, W. P. Todd, and T. Heinrich, *J. Am. Chem. Soc.*, 119, 987 (1997); J. P. Dinnocenzo, H. Zuilhof, D. R. Liberman, T. R. Simpson, and M. W. McKechney, *J. Am. Chem. Soc.*, 119, 994 (1997).
59. V. R. Rao and S. S. Hixson, *J. Am. Chem. Soc.*, 101, 6458 (1979); P. G. Gassman, K. D. Olsen, L. Walter, and R. Yamaguchi, *J. Am. Chem. Soc.*, 103, 4977 (1981); P. G. Gassman and K. D. Olsen, *J. Am. Chem. Soc.*, 104, 3740 (1982); T. Gotoh, M. Kato, M. Yamamoto, and Y. Nishijima, *J. Chem. Soc., Chem. Commun.*, 1981, 90; K. Mizuno, J. Ogawa, H. Kagano, and Y. Otsuji, *Chem. Lett.*, 1981, 437; K. Mizuno, H. Ogawa, and Y. Otsuji, *Chem. Lett.*, 1981, 741; K. Mizuno, K. Yoshioka, and Y. Otsuji, *Chem. Lett.*, 1983, 941; P. G. Gassman and K. D. Olson, *Tetrahedron Lett.*, 24, 19 (1983); K. Mizuno, K. Yoshioka, and Y. Otsuji, *J. Chem. Soc., Chem. Commun.*, 1984, 1665; J. P. Dinnocenzo, W. P. Todd, T. R. Simpson, and I. R. Gould, *J. Am. Chem. Soc.*, 122, 2462 (1990); S. S. Hixson and Y. Xing, *Tetrahedron Lett.*, 32, 173 (1991).
60. C. Reichardt, *Solvent Effects in Organic Chemistry*, Verlag Chemie, Weinheim, p. 270 (1979).
61. M. Yasuda, Y. Matsuzaki, T. Yamashita, and K. Shima, *Chem. Lett.*, 1989, 551.
62. C. Pac and O. Ishitani, *Photochem. Photobiol.*, 48, 765 (1988).
63. M. Yasuda, T. Harada, Y. Ansho, and K. Shima, *Bull. Chem. Soc. Jpn.*, 66, 1451 (1993).

7 DNA-Templated Assembly of Helical Multichromophore Aggregates

Bruce A. Armitage

CONTENTS

7.1 INTRODUCTION

Supramolecular arrays of chromophores often exhibit fascinating photochemical and photophysical properties which are distinct from those of the individual molecular components, as described in earlier volumes of this series. For example, photopolymerization of diacetylenes [1,2], photoinduced charge transfer through DNA [3–6], and photoresolution of racemic materials [7] all rely on the assembly of organized, supramolecular structures. As is often the case, Nature figured out the benefits of chromophore assemblies long before chemists ever did, as demonstrated by light harvesting antenna complexes that absorb solar light then transfer the excitation energy to reaction center proteins where photoinduced charge separation occurs at the start of photosynthesis [8]. The aggregated chromophores in the light harvesting complexes absorb light in a different region of the solar spectrum than do the isolated chromophores and extend the action spectrum for photosynthesis to wavelengths that are not utilized by the reaction center protein alone.

As our appreciation for the properties of supramolecular chromophore assemblies grows, we are naturally motivated to develop methods for the controlled synthesis of such structures. While a great deal can be learned from comparing a monomer to an aggregate containing n chromophores, much more information is available if the aggregate can be assembled one unit at a time. Supramolecular synthesis is an adolescent field at this time, mainly because our ability to design structures held together by collections of relatively weak interactions such as hydrogen bonds and van der Waals forces is nowhere near as sophisticated as our intuition about covalent structures. Once again, Nature provides a hint on how to proceed: rather than following the Quixotic strategy of relying on chlorophyll molecules to self-assemble in a membrane to produce an aggregate with an appropriate structure, protein hosts, or "templates" that precisely control the number of chromophores in the aggregate as well as their positions and orientations were developed [8].

This chapter describes the use of a different type of polymer as a template for the controlled assembly of chromophore aggregates: DNA. The fiftieth anniversary of the double-helical model for DNA [9] brought forth numerous perspectives on its role in storing, reading, and copying genetic information [10]. While the biological implications of the double helix are profound, from a supramolecular chemist's perspective, the DNA structure provides both an inspiration and a target for designing complex structures held together by noncovalent forces. Indeed, the success of organic and inorganic chemists in rationally designing small molecules that bind to DNA is unparalleled in the field of biomolecular recognition [11,12]. DNA-binding small molecules are usually of interest for DNA detection or interference with biological DNA-protein interactions related to gene expression. However, the systems described below exemplify a new role for DNA in supramolecular chemistry: as a template for the controlled growth of multichromophore assemblies. The DNA structure comes into play at two levels. In one class of assemblies, the DNA template is little more than a helical polyanion that attracts cationic dyes. While the resulting aggregates have a number of interesting properties, similar structures can be assembled on other polyanionic templates such as helical polypeptides. The second class of assemblies involves a more intimate association with the DNA template. These dyes recognize the minor groove, the narrower of the two concave surfaces present in double-helical DNA. At this level, the fine structure of the DNA becomes important and affords precise control over the assembly. In fact, dye aggregates of this type can be grown, literally, one unit at a time, affording an unprecedented level of control over the synthesis and understanding of how the photophysical properties depend on the aggregate size. The chapter concludes with a description of DNA-based multichromophore "foldamers" and an example of an application that arose from DNA-templated aggregates.

7.2 DNA-TEMPLATED AGGREGATION OF CYANINE DYES

7.2.1 Cyanine Dye Properties

The cyanine dyes (Figure 7.1) were first reported more than 150 years ago [13] and have found applications in diverse fields ranging from color photography to fluorescence microscopy [14]. The benefits of cyanines include large molar extinction coefficients, relatively facile synthesis, and a broad range of structures with varying excitation and emission wavelengths. Cyanine dyes also have several favorable properties for binding to DNA:

- The positive charge on the chromophore promotes electrostatic attraction to the polyanionic DNA.
- The dyes are typically hydrophobic.
- The chromophore can be planar to promote intercalation, yet is sufficiently flexible to allow twisting to follow the minor groove [15].

The ability of cyanines to bind to DNA has led to the development of numerous fluorescence detection methods, mostly involving dyes that exhibit significant enhancements in fluorescence quantum yield upon binding. Covalent attachment of cyanines to DNA results in fluorescent labeling of the DNA, useful for a variety of applications including gene chip analysis and single molecule spectroscopy.

7.2.2 Discovery of Cyanine Dye Aggregation on DNA

While the impact of cyanine dyes on the biological sciences has been substantial, recent work has led to investigation of cyanine–DNA interactions from a materials science perspective. This direction of research was motivated by a surprising

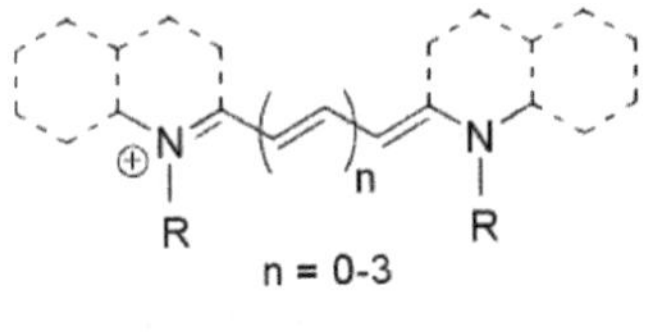

$$n = 0-3$$

FIGURE 7.1 Generic cyanine dye structure. Two nitrogen containing heterocycles are connected by a polymethine bridge. The odd number of carbons in the bridge permits resonance delocalization of the positive charge.

discovery involving the thiadicarbocyanine dye **DiSC₂(5)**. When this dye was mixed with double-helical DNA in which the sequence consisted of approximately 200 alternating adenine (A) and thymine (T) bases on each strand, [Poly(dA-dT)]₂, the dye absorption maximum was found to shift from 650 to 590 nm (Figure 7.2A) [16]. The same dye exhibits similar behavior in aqueous solution if the temperature is decreased or if the dye concentration is raised [17]. The spectral shift is due to noncovalent dimerization ($K = 6.7 \times 10^4$ M^{-1}) and further aggregation of the hydrophobic, polarizable dye has been observed for a wide variety of cyanines and other dyes [18].

In the case of the DNA solutions, neither the temperature nor the dye concentration was changed, indicating that the DNA was promoting aggregation of the dye. This was verified using circular dichroism (CD) spectropolarimetry, which takes advantage of the fact that achiral molecules exhibit no CD in solution, but when bound to a chiral molecule such as DNA can exhibit *induced* CD [19]. A strong, exciton coupled CD

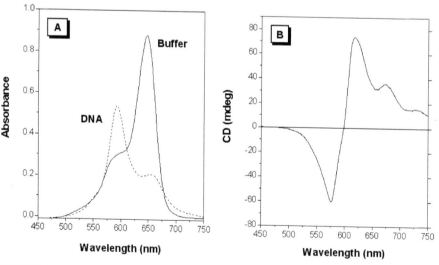

DiSC₂(5)

FIGURE 7.2 UV-vis (A) and CD (B) spectra recorded for cyanine dye DiSC₂(5) in the presence of [Poly(dA-dT)]₂. The blue-shifted absorption band and right-handed splitting pattern in the CD led to the conclusion that the dye assembled into a helical H-aggregate using the DNA as a template.

band was in fact observed for **DiSC$_2$(5)** in the presence of [Poly(dA-dT)]$_2$, but not in its absence (Figure 7.2B). This band was centered at 590 nm, matching the maximum absorbance of the dye when bound to DNA and connecting the UV-vis spectral shift with DNA binding.

In both the solution and DNA-templated **DiSC$_2$(5)** aggregates, the chromophores stack with little or no offset, leading to the hypsochromism observed in the absorption spectrum. The relationship between transition dipole moment alignment in dye aggregates and spectral perturbations was formalized by Kasha [20,21] based on work by Davydov [22] and is summarized in Figure 7.3. Dimerization leads to splitting of the excited state, with selection rules relating to the dipole alignment governing which transition will be active. When the dipoles are aligned in a head-to-head fashion, electronic transition to the upper state is allowed, resulting in a hypsochromic shift in the absorption spectrum. Dimers and aggregates that exhibit this property are known as H-dimers and H-aggregates, respectively. In contrast, when the dipoles are aligned head-to-tail, transition to the lower state is allowed and a bathochromic shift is observed. Aggregates of this type are known as J-aggregates, named after Jelley who first observed this phenomenon [23]. (Scheibe independently reported similar findings [24].)

Returning to the influence of DNA in promoting aggregation of **DiSC$_2$(5)**, the question arose as to the specific interactions between the dye and nucleic acid. The spectral features observed in UV-vis and circular dichroism could be explained by stacking of the dyes along the exterior of the duplex to form columnar structures,

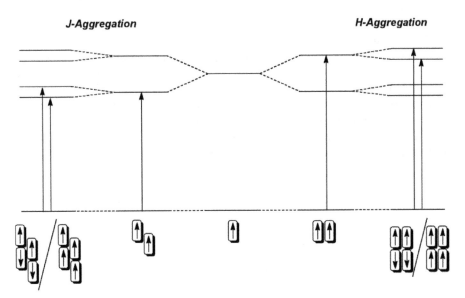

FIGURE 7.3 Exciton coupling model and allowed optical transitions for J- and H-aggregated dyes.

as originally reported for acridine orange and later for certain cationic porphyrins (Section 7.3). Alternatively, the dyes might aggregate using either the minor or major groove. (Intercalation of the dye between adjacent base pairs was ruled out by viscosity measurements, which showed that binding of the dye to the DNA did not cause an increase in the solution viscosity. Intercalators induce lengthening of DNA, leading to increases in viscosity [25].) Whether stacked on the DNA exterior or associated with one of the grooves, the dye aggregate clearly assembles by using the DNA double helix as a template.

7.2.3 DNA Sequence Dependence

Deeper insight into the nature of the **DiSC$_2$(5)**-DNA complex came from studies on the sequence dependence. Minor groove binding small molecules often show a preference for A/T-rich sequences of DNA [26]. We were particularly interested in the natural product distamycin, which not only binds in the minor groove, but does so at certain sequences as a noncovalent dimer [27,28]. A similar binding mode for **DiSC$_2$(5)** could explain the blue-shifted absorption band. Distamycin dimerizes in the minor groove at alternating A-T sequences, but not on G-C sequences. This is due in part to the presence of the exocyclic amino group on guanine, which projects into the minor groove and inhibits binding of small molecules [29]. However, substituting hypoxanthine (known as *deoxyinosine* or "dI" when attached to deoxyribose) for guanine opens the floor of the minor groove and restores dimerization by distamycin [30,31].

When the dye was mixed with the two alternating sequence duplexes, [Poly(dG-dC)]$_2$ and [Poly(dI-dC)]$_2$, the blue-shifted absorption band and induced CD band were observed only for the latter. This is consistent with the structures of the G-C and I-C base pairs: the two duplexes present the same array of functional groups in the major groove, but [Poly(dI-dC)]$_2$ lacks the exocyclic amino group present on guanine, making the minor groove more open and accessible to **DiSC$_2$(5)** (Figure 7.4). In contrast, the minor groove functional groups of [Poly(dI-dC)]$_2$ match those of [Poly(dA-dT)]$_2$. The similar behavior of **DiSC$_2$(5)** in the presence of [Poly(dI-dC)]$_2$ and [Poly(dA-dT)]$_2$ supports a minor-groove templated aggregate.

The minor groove was implicated further by studies using short DNA duplexes. Molecular modeling indicated that a single **DiSC$_2$(5)** H-dimer should span

Distamycin

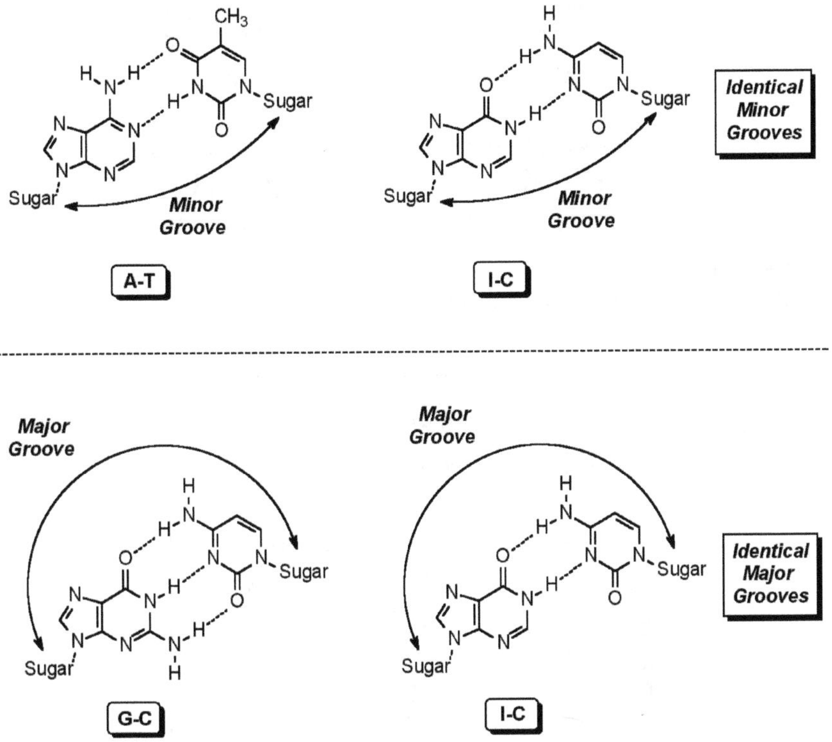

FIGURE 7.4 Comparison of functional groups in DNA major and minor grooves for A-T, G-C, and I-C base pairs.

approximately five base pairs within the minor groove [16]. The synthetic duplex **AT-5** was designed based on this, with the central five base pairs providing a dimerization site and the peripheral G-C pairs included to promote formation of a stable duplex (Figure 7.5). When the dye was added to this duplex, the characteristic blue-shifted absorption band was observed and a continuous variations experiment revealed formation of a specific 2:1 (dye:DNA) complex. When the length of the dimerization region was doubled (**AT-10**, Figure 7.5), the same shift was observed in the UV-vis but the stoichiometry increased to 4 dyes:duplex. These results indicate that the fundamental unit of the dye aggregate is not a monomer but rather a dimer and that the aggregate templated by the polymeric DNA duplexes propagates not by progressive stacking of additional dyes onto the original dimer but rather by end-to-end alignment of individual dimers within the minor groove.

The model shown in Figure 7.5 is also supported by CD spectra acquired on **AT-5** and **AT-10**. On the longer template, the CD spectrum for the dye showed the same exciton coupled profile observed for the polymeric DNA (Figure 7.6). The splitting pattern, namely positive at longer wavelength and negative at shorter, is indicative

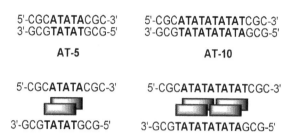

FIGURE 7.5 Dimerization and aggregation of a cyanine dye on DNA templates. AT-5 and AT-10 promote assembly of one and two dimers, respectively.

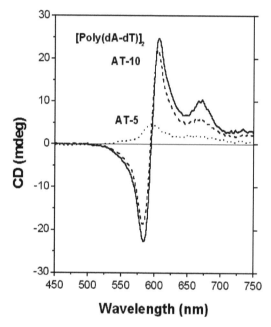

FIGURE 7.6 Induced CD spectra for DiSC₂(5) recorded in the presence of AT-5, AT-10, and [Poly(dA-dT)]₂. The splitting observed for the latter two cases indicates assembly of multiple adjacent dimers on the DNA template.

of chromophores that are electronically coupled and have a right-handed helical relationship [32], as expected based on the right-handed helical morphology of the DNA template. Interestingly, when the shorter **AT-5** template was used, the CD band was again observed at 590 nm, but no splitting was present. This shows that a second-order electronic coupling is responsible for the splitting of the CD band. The face-to-face overlap of two dyes in forming a dimer results in the shift of the absorption band from 650 to 590 nm, but it must be the end-to-end interaction between dimers that causes the splitting observed in the CD spectrum. There is no reason for the CD

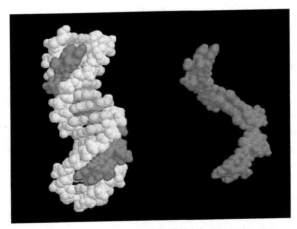

FIGURE 7.7   Molecular model of a DNA-templated $DiSC_2(5)$ aggregate consisting of three dimers aligned end-to-end in the minor groove. DNA is removed on the right to illustrate the right-handed helical morphology of the dye aggregate. (Reproduced from Reference 33 with permission from the American Chemical Society.)

spectrum of an aggregate that propagates in the face-to-face direction to abruptly change from a weak, positive band to an intense, split band. Therefore, these results are much easier to rationalize in terms of a minor groove templated aggregate than an exterior stacked structure reminiscent of acridine orange. A molecular model for an aggregate consisting of three $DiSC_2(5)$ dimers assembled in the DNA minor groove is shown in Figure 7.7 [33]. Efforts to obtain a high-resolution NMR structure of a minor-groove bound dimer have been unsuccessful so far, but INDO calculations reported by Yaron and coworkers indicate that the proposed model is consistent with the observed shift in the UV-vis spectrum [34]. It is important to emphasize that the aggregate would not be stable in the absence of the DNA template. The end-to-end interactions between adjacent H-dimers simply offer too little enthalpic stabilization to compensate for the substantial entropic penalty that would be incurred for associating two dimers in this fashion.

7.2.4   Cooperative Assembly of the Aggregate

The mechanism for growth of the DNA-templated aggregates was revealed by studying the concentration dependence of $DiSC_2(5)$ aggregation. At low micromolar concentrations, the dye is dissolved as a monomer in an aqueous buffer/methanol mixture. Addition of an appropriate DNA template under these conditions leads to immediate dimerization and aggregation of the dye. This indicates that there is no need for the dye to dimerize prior to binding to the DNA. Dimerization is also observed at concentrations where there is less than two dyes per dimerization site, demonstrating that assembly of a $DiSC_2(5)$ dimer in the DNA minor groove occurs cooperatively rather than binding monomeric dye until a 1:1 ratio is exceeded. This cooperativity likely arises due to van der Waals interactions: in the dimer, the two polarizable dye molecules can stack with one another and only contact the walls of the minor groove on either side. (The nonplanarity and low polarizability of the deoxyribose sugars should lead to weaker van der Waals interactions with a dye molecule.) For the two dyes to bind as separate monomers, both faces of both dye molecules will be forced to contact the groove walls.

Circular dichroism experiments involving **AT-10** revealed a second level of cooperativity involved in assembling the dye aggregate. As noted above, splitting of the CD band into positive and negative components is indicative of end-to-end interaction between nearby **DiSC$_2$(5)** dimers whereas a single positive band is characteristic of an isolated dimer. If binding is noncooperative, then splitting of the CD band should not be observed until the amount of dye added to **AT-10** exceeds a ratio of 2 dyes (i.e., one dimer) per duplex. However, splitting of the band was observed for as little as 0.5 dyes per duplex, indicating that assembly of one dimer in the groove greatly enhances assembly of a second dimer on the same template.

The origin of this cooperativity might be the impact of the bound dimer on the DNA structure: in order to accommodate two ligands bound as a dimer, some studies have shown that the minor groove has to approximately double its width [30,31].

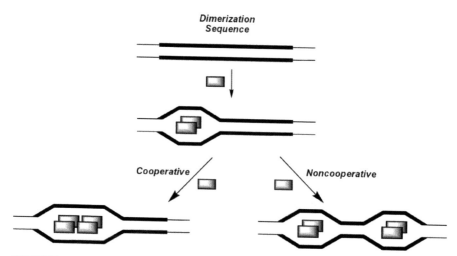

FIGURE 7.8   Assembly of a DNA-templated cyanine dye aggregate within the minor groove is cooperative. Widening of the groove to accommodate one dimer facilitates assembly of additional dimers at adjacent sites rather than at remote sites where the groove is narrower.

This perturbation likely extends beyond the bound dimer by some distance, meaning it will be easier to assemble a second dimer directly adjacent to the first dimer since the minor groove is already partially widened (Figure 7.8). However, a recent report from Wilson and coworkers illustrates dimeric binding of a dicationic ligand without significant widening of the groove [35]. Whether assembly of a dimer induces widening of the minor groove will likely depend on the DNA sequence and could impact the degree of cooperativity involved in propagation of the aggregate.

7.2.5 DYE STRUCTURAL REQUIREMENTS

DNA-templated aggregation has been investigated for each of the dyes shown in Table 7.1 [33]. For the benzothiazole series of dyes, aggregation in water increases with the length of the polymethine bridge [17]. However, aggregation on DNA followed the order **DiSC$_2$(5) > DiSC$_2$(7) > DiSC$_2$(3)**, meaning the order is inverted for the penta- and heptamethine dyes. This could be due to the induced fit nature of the binding between the dye and DNA: optimal binding in the groove should require the dye to twist to follow the helical pitch of the DNA. This structural distortion might be sufficiently costly for the longer dye as to weaken dimerization in the groove.

TABLE 7.1

Cyanine Dyes Used for Aggregation on DNA

Dye	X	n
DiSC$_2$(3)	S	1
DiSC$_2$(5)	S	2
DiSC$_2$(7)	S	3
DiQC$_2$(3)	(CH=CH)	1
DiQC$_2$(5)	(CH=CH)	2
DiOC$_2$(5)	O	2
DiIC$_1$(5)[a]	C(CH$_3$)$_2$	2

[a] The N-substituents on this dye are methyl rather than ethyl.

DiSC$_{3+}$(5) : R = (CH$_2$)$_3$N(CH$_3$)$_3$

DiSC$_{3-}$(5) : R = (CH$_2$)$_3$SO$_3$

In comparing dyes with the same bridge length (pentamethine) but different heterocycles, the order of aggregation was quinoline > benzothiazole > benzoxazole > dimethylindole [33]. This order follows the tendency of these dyes to aggregate in pure water. Aggregation on DNA is actually undetectable for the dimethylindole derivative, most likely because of steric clashes between the methyl groups from separate dyes when trying to assemble the dimer. In all other cases, blue-shifted absorption maxima were observed, indicating preferential assembly of H-aggregates.

Electrostatic factors also contribute significantly to the aggregation of cyanines on DNA, from the perspective of modulating both the dye-DNA attractions and the dye–dye repulsions. The N-substituents on the dye were varied from neutral (**DiSC$_2$(5)** : ethyl) to cationic (**DiSC$_{3+}$(5)** : trimethylammonium propyl) and anionic (**DiSC$_{3-}$(5)** : propylsulfonate). The anionic substituents completely abolished aggregation on DNA [36], while the cationic substituents still permitted aggregation, albeit with lower cooperativity [37]. The latter effect can be explained by the strong interdye repulsions, leading to preferential binding of the dye as a monomer. Nevertheless, both H- and J-aggregates can be formed from **DiSC$_{3+}$(5)** at higher dye:DNA ratios (*vide infra*).

7.2.6 ELECTRONIC PROPERTIES

H-dimerization of cyanine dyes in aqueous solution usually leads to broadening of the main absorption band relative to the monomer dye [17]. This occurs because the geometric constraints on the dimer are not rigid: there are a fairly large number of translational and rotational degrees of freedom for the two monomers within the dimer complex and each structure can have a slightly different absorption maximum, leading to the observed broadening. However, when **DiSC$_2$(5)** dimerizes in the minor groove of DNA, the absorption band becomes quite sharp and is not only narrower than a dimer in solution, but also narrower than the monomer in solution. For example, the full-width half-maximum (FWHM) of **DiSC$_2$(5)** in methanol is 925 cm^{-1}, while the FWHM for the dimer band on [Poly(dI-dC)]$_2$ decreases to 772 cm^{-1} [33]. This illustrates how the DNA-templated dimer is constrained by the minor groove into a rigid, well-defined supramolecular structure.

The structural model for the DNA-templated aggregates indicates that two types of electronic coupling should be present. First, the face-to-face stacking of dye monomers to form a dimer will lead to splitting of the excited state (primary splitting). Then, for extended aggregates, end-to-end interactions between adjacent dimers will lead to further splittings (secondary splittings). The magnitude of these splittings can be estimated from the spectral shifts, as investigated in detail by Peteanu and coworkers [38]. DNA templates for assembly of **DiSC$_2$(5)** aggregates consisting of one, two, three or ca. 35 dimers were used and UV-vis spectra were recorded in frozen glasses, which allows resolution of the two bands that arise from end-to-end splitting. Table 7.2 shows how both types of splitting vary with aggregate length. The primary splittings are almost invariant with aggregate length, although a slight (ca. 5%) decrease is observed for the polymeric template. In contrast, the secondary splitting doubles in going from two dimers to the polymeric aggregate, reflecting considerable electronic delocalization in the longer aggregates. This can be understood by realizing that in an aggregate consisting of N dimers, all dimers except those at the very ends will be electronically coupled to two additional dimers, one on either side. However, in an aggregate consisting of only two dimers, each dimer will be coupled only to one other dimer. Hence, the secondary splitting should approximately double as N gets large and dimers at the ends of the aggregate contribute less to the overall splitting (Figure 7.9). Table 7.2 also shows that the primary splitting is

TABLE 7.2
Primary and Secondary Energy Splittings for H-Aggregates of Different Lengths

H-Dimers (#)	Primary Splitting (cm^{-1})	Secondary Splitting (cm^{-1})
1	3480	—
2	3500	330
5	3470	420
35 ± 5	3360	640

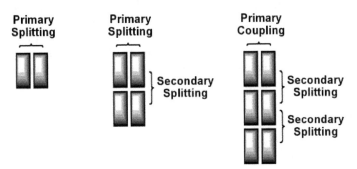

FIGURE 7.9 Representation of electronic couplings associated with cyanine dye aggregation on DNA. Face-to-face stacking within a dimer leads to primary splitting, observed as a blue-shift in the absorption spectrum. End-to-end interactions between dimers yield secondary splittings of the blue-shifted band.

TABLE 7.3
Effect of Cyanine Dye Structure on Primary and Secondary
Splittings in H-Aggregates

Dye	Primary Splitting (cm^{-1})	Secondary Splitting (cm^{-1})
DiSC$_2$(3)	2030	—
DiSC$_2$(5)	3118	830
DiSC$_2$(7)	3540	721
DiQC$_2$(3)	2968	774
DiQC$_2$(5)	3118	981
DiOC$_2$(5)	3122	—

tenfold greater than the secondary splitting for the two-dimer aggregate, but this difference decreases to fivefold for the polymeric aggregate. These results illustrate the need for a further refinement in the model shown in Figure 7.3. Thus, propagation of the aggregate beyond two dimers causes further splittings of the exciton states and greater secondary splittings. This model has also been discussed by Yarmoluk and coworkers in the context of DNA-templated J aggregates (*vide infra*).

Table 7.3 lists the effect of dye structure on the primary and secondary splittings [33]. The primary splitting increased with the polymethine bridge length, as expected since longer dyes should have larger stacking interactions [17]. There was no difference in the primary splitting among benzoxazole, benzothiazole, and quinoline-derived dyes having a pentamethine bridge. It is difficult to draw conclusions concerning the secondary splitting due to the limited data set.

Peteanu and coworkers also used Stark effect spectroscopy to analyze the changes in dipole moment, $|\Delta\mu|$, and polarizability, $<\Delta\alpha>$, that occur upon excitation of **DiSC$_2$(5)** in DNA-templated aggregates of varying length [38]. In contrast to the secondary splitting, $<\Delta\alpha>$ was essentially independent of the aggregate length. This indicates that the change in polarizability is governed primarily by intradimer interactions, even though the excitation can be delocalized over several dimers. Meanwhile, $|\Delta\mu|$ decreased monotonically from a value of 1.1D for a **DiSC$_2$(5)** monomer to 0.45D for a polymeric aggregate, which is consistent with a theoretical treatment of aggregated chromophores [39].

7.2.7 HETEROAGGREGATION

The fact that a variety of different cyanine dyes were capable of aggregating on DNA suggested that it should be possible to form heteroaggregates from different combinations of dyes, a phenomenon that was previously investigated in aqueous solution [17]. Various pairs of the dyes shown in Table 7.1 did, in fact, form heteroaggegates on DNA, although never as a pure supramolecular species [33]. For example, the heteroaggregate absorption band can be observed in between the two

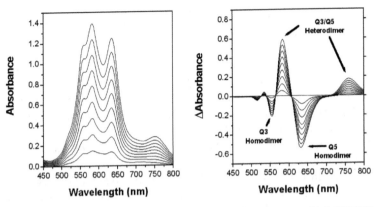

FIGURE 7.10 Heteroaggregation of quinoline-based cyanine dyes on [Poly(dI-dC)]$_2$. Left: UV-vis spectra. Right: Difference spectra calculated by subtracting spectra for individual dyes from dye mixture spectra shown on left. Q3 = DiQC$_2$(3), Q5 = DiQC$_2$(5) (see Table 7.1 for structure).

homoaggregate bands in the UV-vis spectrum for **DiQC$_2$(3)** and **DiQC$_2$(5)** (Figure 7.10). Heteroaggregation was most favored for dyes with similar bridge lengths, as observed previously in water [17].

The inability to drive heteroaggregation to completion can be attributed to weaker van der Waals' interactions between two different dyes relative to two identical dyes. In order to compensate for this, complementary interactions between two different dyes can be introduced. The first example of this involved the dyes with anionic and cationic substituents, **DiSC$_3$-(5)** and **DiSC$_{3+}$(5)**, respectively [36]. As noted above, **DiSC$_3$-(5)** does not aggregate on DNA while **DiSC$_{3+}$(5)** does aggregate, but only at elevated concentrations. Mixing the two dyes together at low concentrations, where neither dye aggregates alone, yields blue-shifted UV-vis and exciton coupled CD spectra similar to those observed for the homoaggregate formed by **DiSC$_2$(5)**. Thus, electrostatic complementarity between the N-substituents of different dyes can be used to promote heteroaggregation. Alternative strategies involving complementary substituents on the heterocycles should also favor heteroaggregation.

7.2.8 J-Aggregates

The cyanine dyes described above preferentially assemble into H-dimers on DNA. While this leads to interesting supramolecular structures, preparation of DNA-templated J-aggregates could have wider ranging impact. One key difference between H- and J-aggregates is that fluorescence in the former is quenched. This is due to rapid relaxation from the upper to the lower exciton state. Fluorescence from this state is forbidden, leading to virtually no emission from the H-aggregated dyes [16]. In contrast, transition to the lower state is allowed for J-aggregates, meaning that intense fluorescence is often observed [40]. DNA-templated J-aggregates are also desirable because chiral J-aggregates are predicted to have interesting nonlinear optical properties [41,42].

The first example of a DNA-templated J-aggregate was reported by Nordén and Tjerneld [43]. The quinoline-based dye **DiQC$_2$(1)** (also referred to as *pseudoisocyanine* or *PIC*) exhibited a red-shifted, exciton-coupled CD spectrum in the presence of calf thymus DNA. The aggregate required high dye:DNA ratios and low ionic strengths in order to be observed. Interestingly, the same dye forms J-aggregates in the absence of the DNA template, but exhibits the opposite dependence on ionic strength: the nontemplated aggregate is stabilized at high ionic strength due to decreased dye–dye repulsions. In the presence of DNA, high ionic strength decreases dye-DNA attractions, resulting in dissociation of the aggregate from the template. The authors proposed assembly of a right-handed, helical J-aggregate wrapping around the DNA template and held in place by relatively weak electrostatic attractions. While clear evidence for this structure was obtained from the CD and linear dichroism (LD) spectra, no distinct red-shifted band was observed in the UV-vis spectrum, indicating that the DNA-templated aggregate constituted a minor species. This might be due to the use of a random sequence DNA template rather than one of the homopolymer or alternating copolymers used in later work.

More recently, we reported formation of a DNA-templated J-aggregate by the tricationic cyanine dye **DiSC$_{3+}$(5)** [44]. Adding this dye to [Poly(dI-dC)]$_2$ led to the growth of a narrow, red-shifted absorption band that was split into two components (Figure 7.11A). The CD spectrum showed an intense, right-handed exciton couplet, indicating that the aggregate was bound to the DNA (Figure 7.11B). While the reason

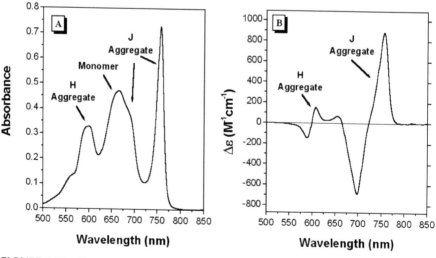

FIGURE 7.11   H- and J-aggregation of DiSC$_{3+}$(5) on [Poly(dI-dC)]$_2$ revealed by (A) UV-vis and (B) CD spectra.

J-aggregation is observed for this dye but not the monocationic **DiSC$_2$(5)** is unclear, one possibility involves repulsions between adjacent dye molecules. (This would also explain why relatively high dye:DNA ratios were required to assemble the J-aggregate.) Slipping two dyes into the offset geometry required to allow transition to the lower exciton state can increase the distance between their cationic substituents, decreasing Coulombic repulsions.

Two lines of evidence support the assembly of an aggregate in which the dye molecules are cofacially offset, as shown schematically in Figure 7.3. First, the primary and secondary energy splittings were determined by Peteanu and coworkers for H- and J-aggregated **DiSC$_{3+}$(5)**. As shown in Table 7.4, the primary splitting was 30% larger for the H-aggregate, consistent with the better stacking of the dyes in this arrangement [45]. However, the secondary splitting was ca. 333% larger for the J-aggregate, which reflects the better interdimer coupling that is possible when the two chromophores within a given dimer are offset from one another (Figure 7.3). This leads to the well-known "brickwork" structure for J-aggregates, although this term is perhaps less descriptive when considering the quasi one-dimensional aggregates assembled on the DNA template.

The second experiment that supports an offset geometry for the dye molecules involved short oligonucleotide duplexes. In contrast to the H-aggregates formed by **DiSC$_2$(5)**, there was a strong dependence of the **DiSC$_{3+}$(5)** J-aggregate stability on the length of the DNA template [44]. Specifically, the J-aggregate was much more stable on a polymeric DNA (*ca.* 200 base pairs) template where 30 to 40 dimers could bind than on shorter duplexes that would permit binding of only one or two dimers (Figure 7.12). H-aggregation was preferred on the short duplexes, an observation that is consistent with the model shown in Figure 7.13; for assembly of a J-aggregate, the offset in the chromophores comes at the expense of van der Waals interactions between any two dyes. The contribution of the unstacked dyes at either end of the aggregate contributes substantially to the total free energy change for short aggregates, thereby leading to shifting of the dyes into an H-aggregate morphology to maximize stacking interactions. For a larger aggregate (i.e., assembled on a longer duplex), the contribution of these end effects is minimized and allows assembly of the J-aggregate.

TABLE 7.4
Comparison of Primary and Secondary Splittings for H- vs. J-Aggregated DiSC$_{3+}$

Aggregate	Primary Splitting (cm^{-1})	Secondary Splitting (cm^{-1})
H	3300	360
J	2550	1200

From Ref. 5.

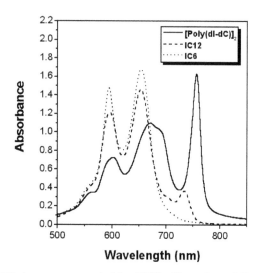

FIGURE 7.12 UV-vis spectra recorded for $DiSC_{3+}(5)$ on three different DNA templates: [Poly(dI-dC)]$_2$, IC12, and IC6 favor assembly of ca. 35, 2, and 1 dimer, respectively. J-aggregation, indicated by absorption at 750 nm, decreases with the template length.

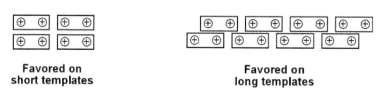

Favored on **short templates**   **Favored on** **long templates**

FIGURE 7.13 Formation of H- vs. J-aggregates by $DiSC_{3+}(5)$ depends on the length of the DNA template. Shorter templates favor H-aggregation to maximize van der Waals' interactions while longer templates favor J-aggregation to minimize electrostatic repulsions between cationic substituents (represented as plus signs).

Under all conditions tested, the UV-vis and CD spectra of J-aggregated $DiSC_{3+}(5)$ on DNA exhibited absorption bands at both longer and shorter wavelengths than the monomer band [44]. This could be attributed either to the presence of both H- and J-aggregates in the sample (although not necessarily on the same DNA duplex), or to allowed transitions to both of the exciton states. Peteanu and coworkers used fluorescence spectroscopy to address this question [45]. Excitation of the J-band led to near-resonant fluorescence, as is typical for J-aggregates. However, no fluorescence was observed when the H-band was excited. This indicated that H- and J-bands were not arising from the same species, since excitation in the H-band should lead to relaxation to the lower state and subsequent fluorescence. Rather, these results indicated that two distinct aggregates, H and J, are formed under these conditions.

Further insight into the structural organization of the J-aggregate was obtained by Peteanu and coworkers, who used electroabsorption (Stark effect) spectroscopy to characterize the $DiSC_{3+}(5)$ aggregates [45]. The average change in polarizability upon excitation of the H- and J-aggregates, $<\Delta\alpha>$, was found to be approximately twice as large for the J-aggregate. This was interpreted as being due to larger delocalization of the excitation due to better electronic coupling in the J-aggregate and is consistent with the larger interdimer coupling for this structure. The $<\Delta\alpha>$ was also found to be 5 to 6 times larger for the J-aggregate than for an isolated $DiSC_{3+}(5)$ monomer, indicating that the aggregate "size" is roughly 5 to 6 cyanine dye molecules. Thus, while the aggregates were assembled on a DNA template that was 150 to 200 base pairs long, the actual J-aggregates are relatively small. It should be noted that these experiments were done in frozen glass media containing up to 40% organic solvent. While the low temperature achievable in the glass (77K) leads to excellent resolution of different components in the spectra, it might alter the supramolecular structure of the DNA and dye aggregate. Nevertheless, these experiments provided the clearest information regarding the electronic coupling and co-existence of H- and J-aggregates. The contributions of dye structure (charged substituents, curvature, flexibility) and DNA template properties will need to be explored in greater detail in order to synthesize much larger J-aggregates that exhibit larger exciton delocalization.

A second example of a DNA-templated cyanine dye J-aggregate was reported for **Cyan βiPr**, an isopropyl-substituted analogue of $DiSC_1(3)$ [46]. This dye exhibited a narrow, red-shifted absorption band in the presence of the nonalternating sequence Poly(dA)-Poly(dT) and the band was split into two components. No J-aggregation occurred when the alternating sequence [Poly(dG-dC)]₂ was used. The J-band only appeared at high dye:DNA ratios, perhaps because the nonalternating A-T sequence is not optimal for aggregation of other cyanine dyes. Similar experiments with the alternating [Poly(dA-dT)]₂ might lead to more favorable aggregation of this dye. While J-aggregation by $DiSC_{3+}(5)$ can be attributed to electrostatic repulsion,

Cyan βiPr

Cyan βiPr J-aggregates are likely favored by the nonplanar isopropyl group in the center of the bridge. This precludes effective stacking into an H-dimer due to steric repulsions, forcing the dyes to slip into an offset geometry.

The **Cyan βiPr** J-aggregates exhibit intriguing fluorescence properties [47]. First, the fluorescence quantum yield (Φ_f) increased with dye concentration, correlating with the growth of the J-bands in the UV-vis spectrum. Second, the value of the quantum yield depended on which of the two J-bands was excited. Excitation into the lower energy band consistently produced higher quantum yields than excitation into the higher energy band. The fluorescence spectrum was the same in either case, indicating that excitation to the upper J-state resulted in relaxation to the lower J-state, from which emission occurred. Evidently the transition from the upper to the lower state was less than 100% efficient, leading to the wavelength dependence for the quantum yield. This model was supported by the finding that the difference in quantum yield decreased with increasing dye concentration. As the dye concentration increases and the J-aggregate gets larger, additional splittings within the upper and lower J-states should result in a net decrease in the energy separating them (Figure 7.14). This

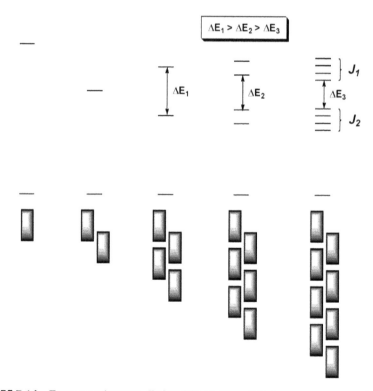

FIGURE 7.14   Energy gap between distinct J-bands (J_1 and J_2) decreases as aggregate length increases. Excitation to J_1 leads to nonradiative decay to J_2 and subsequent fluorescence.

should facilitate relaxation from the upper to lower state and, therefore, a weaker wavelength dependence for the fluorescence quantum yield.

7.3 DNA-TEMPLATED ASSEMBLY OF STACKED PORPHYRIN AGGREGATES

The ability of double-helical DNA to act as a template for the assembly of multichromophore arrays dates back to 1959, when Bradley and coworkers reported on the formation of acridine orange aggregates in the presence of DNA [48–51]. A model was proposed in which the cationic dye molecules were stacked cofacially and associated electrostatically with the polyanionic DNA backbone. This aggregation mode is quite different from the minor-groove templated motif exhibited by the cyanine dyes, where subtle variations in the DNA sequence and structure can have significant effects on aggregation. For example, aggregation by $DiSC_2(5)$ on natural DNA, which is effectively a random sequence of A-T and G-C pairs, is nearly undetectable, while acridine orange readily aggregates on mixed sequence DNAs.

7.3.1 PORPHYRIN PROPERTIES

While the acridine orange studies provided the first example of a DNA exterior-templated aggregate, more recent work has focused on cationic porphyrins as the aggregating chromophore [52]. Porphyrins are of interest not only for the important place they hold as light harvesting components in photosynthetic systems, but also because they have a diverse range of DNA-recognition modes [53]. Intercalation and minor groove binding have been reported for monomeric porphyrins, in addition to the exterior stacking mode described in this section. Porphyrins also have favorable optical properties, most notably the intense Soret band, which permit characterization by a variety of spectroscopic techniques.

7.3.2 PORPHYRIN AGGREGATION ON DNA

In 1982, Fiel and coworkers described the appearance of an exciton-split CD band when the tetracationic porphyrin **TMAP** was added to calf thymus (ct) DNA [54]. As was the case for the cyanine dyes, splitting of the CD band was interpreted as being due to electronic coupling between nearby porphyrin chromophores. This CD profile was observed only at relatively high porphyrin-DNA ratios, most likely because of electrostatic repulsions between the porphyrins that disfavor close approach at lower binding ratios. The authors proposed that the porphyrins formed a right-handed helical stacked aggregate surrounding the DNA core (Figure 7.15). The CD spectrum was more consistent with a left-handed helix for the porphyrin aggregate, though, with the positive component being at shorter wavelength than the negative component.

In 1988, Pasternack and coworkers reported the appearance of an unusually large and exciton split CD band for the meso-substituted porphyrin *trans*-H₂P [55]. The

TMAP

*trans-*H$_2$P

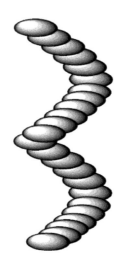

FIGURE 7.15   Illustration of helical aggregate of stacked chromophores. Note the distinction between stacked aggregate and cyanine dye aggregates described above: the latter only stack to form dimers with aggregate propagating by end-to-end alignment of dimers. Stacked aggregates propagate in the face-to-face direction.

intense CD signal was only observed for relatively high ionic strengths, but was present for random sequence ct DNA as well as the alternating copolymer duplexes [Poly(dA-dT)]$_2$ and [Poly(dG-dC)]$_2$. The corresponding *cis* isomer also gave rise to a large, split CD band, but only on the random sequence DNA and at significantly higher ionic strengths. Meanwhile, no exciton coupling or large intensities were

H₂-P4

observed in the CD spectra of the tetrakis derivative **H₂-P4** for any of the DNAs over a wide range of ionic strengths.

Similar to the model proposed by Fiel and coworkers for **TMAP**, the authors proposed the formation of stacked porphyrin aggregates using the DNA as a template. The ionic strength dependence arises from the need to minimize electrostatic repulsions within the porphyrin aggregate, although if the salt concentration is too high, the attraction between the porphyrin and the DNA is reduced leading to a loss of the CD signal. The fact that the aggregates assemble on all three DNA templates is in stark contrast to the minor groove-templated assemblies formed by the cyanine dyes, where alternating A-T or I-C sequences are strongly preferred. This illustrates that the porphyrins do not make extensive contacts with the DNA template but rather simply sense a helical polyanionic template. To further emphasize this point, Pasternack and coworkers demonstrated that *trans*-**H₂P** assembles into similar aggregates on helical polypeptide templates, with induced chirality following that of the underlying polymer [56].

The aggregation behavior described above for *trans*-**H₂P** also extends to the copper-metallated analogue *trans*-**CuP**, although higher porphyrin concentrations and/or ionic strengths are needed to drive assembly of the aggregate [57]. In contrast, the gold analogue *trans*-**AuP** does not show any evidence for aggregation on DNA, but rather prefers to bind to ct DNA as an intercalated monomer based on CD spectra.

7.3.3 CHARACTERIZATION OF THE PORPHYRIN AGGREGATE SIZE

While the CD spectra provide important information regarding how the porphyrin molecules are organized within the aggregate, little is learned about the dimensions of the aggregates. However, resonant light scattering (RLS) experiments indicated that the aggregates can be extremely large and that the supramolecular structure is considerably more complex than a simple stack of porphyrin molecules twisting around a single DNA helix [57]. RLS uses light having a wavelength that is normally absorbed by the chromophores. When the chromophores are aggregated, significantly increased light scattering can be observed and used to estimate the number of chromophores involved in a discrete aggregate [58,59]. For the case of **trans-H₂P**, DNA-templated aggregates containing 10^5 to 10^6 porphyrin molecules were detected, independent of the length or sequence of the DNA template [60]. It is impossible to stack this many porphyrin molecules along the exterior of a single DNA duplex, itself only a few hundred base pairs in length. RLS experiments in the UV region where the DNA absorbs indicated that the DNA is not aggregated. The only explanation for these observations is that the porphyrins form multilayered helical superstructures arranged around a core containing a single DNA duplex. Thus, the actual supramolecular structure of these aggregates is more complex than originally imagined.

7.3.4 TENTACLE PORPHYRINS

An intriguing addition to the porphyrin story are the "tentacle" porphyrins **TθOPP** and **TθF4TAP**, which are tetracationic but have the positive charges decoupled from the porphyrin by a flexible spacer [61–64]. Both tentacle porphyrins readily assemble into stacked aggregates in the presence of DNA. While the CD spectra for the aggregates are similar to those observed for **trans-H₂P**, significant differences are observed. First, the aggregates assemble both in low and high ionic strength conditions. This could reflect the ability of the positively charged groups to avoid one

TθOPP

TΘF4TAP

another better than in the more rigid *trans*-H_2P. Second, there is a strong preference for stacking on A-T vs. G-C sequences. One possible explanation for this observation is that one of the ammonium groups projects into the minor groove, which is known to be electronegative and contributes to the attraction of cationic minor groove binders [65,66]. Computational studies indicate that the minor groove at G-C sequences is considerably less electronegative. Differences in hydration might also play a role. Finally, the aggregation occurs at very low porphyrin-DNA base pair ratios (1:100), indicative of considerable cooperativity in assembly. Light scattering experiments on these aggregates have not been reported, so it is impossible to compare the overall dimensions with those formed by *trans*-H_2P.

Another interesting feature of the tentacle porphyrin aggregates is the appearance of two distinct species, characterized by differences in their CD intensities. At high porphyrin:DNA ratios, the CD intensities are actually smaller than at lower ratios. A model has been proposed in which the lower ratios feature "moderately stacked" aggregates, while higher ratios cause the porphyrins to stack together more closely [62]; why the more extensively stacked aggregate would lead to a weaker induced CD signal is not clear, though.

7.3.5 HETEROAGGREGATION

The stacked porphyrin arrays assembled on DNA templates are reminiscent of the columnar liquid crystalline (LC) phases produced by discotic mesogens, including several porphyrin derivatives [67]. Different discotic compounds with complementary interactions have been shown to form mixed LC phases [68], a concept that has been extended to the DNA-templated porphyrin stacks [69]. The two porphyrins H_2-**P4**

P^{4-}

and P^{4-} use electrostatic complementarity to form a mixed aggregate in the presence of duplex DNA under conditions where neither porphyrin forms a homoaggregate on its own. The induced CD signal for the heteroaggregate exhibited exciton coupling, although the intensity was considerably weaker than for the large aggregates obtained with *trans*-H_2P. This is similar to the example described above for heteroaggregation of cationic and anionic cyanine dyes in the minor groove of DNA.

The stacked aggregates formed by cationic porphyrins in the presence of DNA templates provide a fascinating example of supramolecular assembly. Recent work with triphenylphosphonium-substituted porphyrins indicates that a variety of meso-substituents can be tolerated in forming stacked aggregates [70], while an expanded "sapphyrin" macrocycle also showed DNA-templated stacking [71]. In contrast to the cyanine dye aggregates, less control is available over the dimensions of the porphyrin aggregates due to the apparent growth of the structure beyond a single shell around the DNA template, at least for some porphyrins. Nevertheless, experiments using short, oligonucleotide templates may provide the opportunity to prepare more well defined aggregates.

7.4 APPLICATIONS

No applications for DNA-templated dye aggregates have been reported at this time; rather, the studies described above have focused on structural and optical properties of these fascinating supramolecular structures. However, there is one related aggregation phenomenon that exhibits properties of both the cyanine dye and porphyrin aggregates and has found utility in the laboratory. It involves $DiSC_2(5)$, previously shown to aggregate in the minor groove of DNA, and peptide nucleic acid (PNA), which has the natural nucleobases from DNA attached to a peptide-like backbone (Figure 7.16) [72–74]. PNA oligomers recognize their complementary DNA and RNA strands by following the Watson–Crick rules for base pairing and form high affinity double- and triple-helical complexes. These properties have generated considerable interest in developing PNA for clinical diagnostic and therapeutic applications.

Standard DNA binding compounds such as ethidium bromide, which intercalates, and distamycin, which binds in the minor groove, exhibit significantly lower affinity for PNA-DNA duplexes [75]. However, addition of **DiSC$_2$(5)** to a ten base pair PNA-DNA duplex leads to an immediate and striking color change from blue to purple, which is due to a shift in the absorption maximum from 650 to 534 nm (Figure 7.17A) [76]. CD spectroscopy revealed a strong, exciton coupled band indicative of a right-handed helical aggregate formed by the dye in the presence of the PNA-DNA template (Figure 7.17B). Over time, the aggregates cluster to form purple thread-like structures that precipitate from solution. Thus, the helical aggregate resembles initially those formed by this dye on DNA-DNA templates, but later grows to form very large supramolecular structures, similar to the porphyrins. Other monocationic dyes such as **DiSC$_2$(3)** and **DiSC$_2$(7)** also aggregate, but the tricationic analogue **DiSC$_{3+}$(5)** does not.

The visible color change and strong induced CD provide instantaneous indicators of whether a PNA strand has hybridized to its target [77,78]. When **DiSC$_2$(5)** is added to a mixture of PNA and a noncomplementary DNA strand, the color of the solution remains blue. Furthermore, aggregation of the dye occurs on a wide variety of base sequences, ranging from random mixtures of purines and pyrimidines to repeating homopurine-homopyrimidine sequences. Thus, while there might be slight differences in affinity depending on the sequence composition, a colorimetric response will likely be observed in most, if not all cases.

If the PNA-DNA hybrid forms but contains a single mismatch, the dye can discriminate this based on temperature-dependent UV-vis measurements [76]. The dye aggregate is readily dissociated by heating the sample; the temperature at which the aggregate dissociates is higher for the fully matched duplex than for the singly mismatched duplex. Alternatively, enzymatic nucleases that degrade single stranded

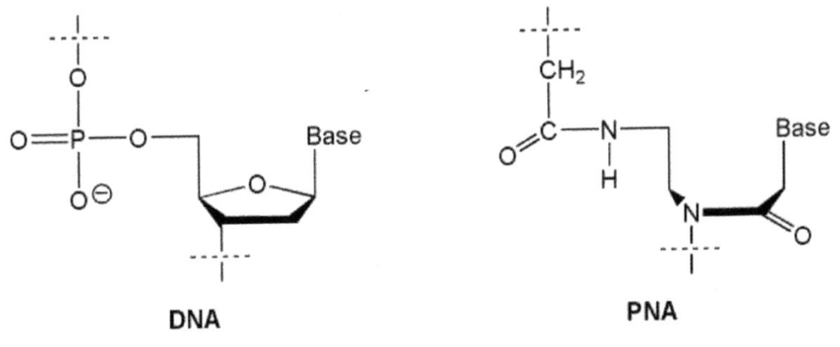

FIGURE 7.16 Comparison of DNA and PNA chemical structures.

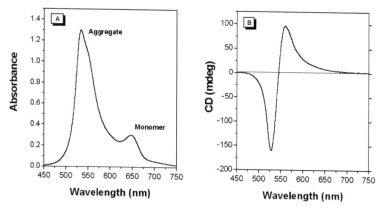

FIGURE 7.17 UV-vis (A) and CD (B) spectra recorded for DiSC$_2$(5) in the presence of a 10 base pair PNA-DNA duplex.

DNA can be used to destroy mismatched duplexes and provide room temperature discrimination of mismatches [79]. Two separate reports of using **DiSC$_2$(5)** and PNA for genetic screening have appeared [79,80]. While the sensitivity of these methods is limited by the extinction coefficient of the dye aggregate and the affinity of the dye for the duplexes, the cyanine dye colorimetric indicator remains a useful laboratory reagent for assessing PNA hybridization.

7.5 COVALENTLY LINKED DYE AGGREGATES: DNA-BASED FOLDAMERS

"Foldamers" are oligomeric or polymeric compounds that fold into well-defined secondary structures [81,82]. Proteins and RNA are examples of naturally occurring foldamers and numerous synthetic examples have been reported in the past ten years. The foldamer concept has been applied to the assembly of DNA-templated dye aggregates by Asamuna and Komiyama, who synthesized a series of DNA-dye conjugates and studied their spectroscopic properties [83]. The azo dye methyl red was covalently linked to either D- or L-threonine (Figure 7.18) and converted into a phosphoramidite, which was then incorporated into a DNA oligomer using standard solid phase synthesis methods. A number of oligomeric and polymeric dyes have been reported in the past, but the solid phase approach allows preparation of a homogenous compound, where the number and position of chromophores is completely defined.

A series of DNA oligomers having from 0 to 6 methyl red dyes sandwiched between two hexanucleotide domains was prepared. As the number of consecutive dyes increased, the absorption spectrum of the dye shifted progressively to shorter wavelengths and became significantly narrower. The authors attributed this to

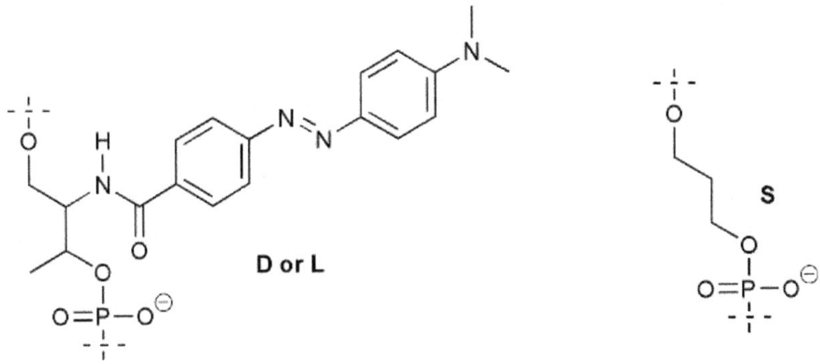

FIGURE 7.18 Methyl red (left) and spacer (right) units used to study assembly of dye-DNA foldamers. Methyl red units attached to D- or L-threonine.

formation of an H*-aggregate, which is distinguished from a standard H-aggregate by the narrowed absorption band [18]. If the dye units alternated with spacers, weakening the stacking of the chromophores, the H*-aggregate was disfavored. Exciton splitting was detected by CD spectroscopy, and the pattern was indicative of a right-handed helical conformation for the aggregate. In contrast, when the dye was connected to the backbone by the L-threonine, the H*-aggregate exhibited a left-handed splitting in the CD. This demonstrated that the chirality of the dye helix was determined by the threonine linker rather than by the flanking DNA which was constant in both series of oligomers.

Hybridization of the dye-conjugated oligomers with complementary DNA strands bearing 0 to 6 spacer residues had varying effects on the H*-aggregate. When no spacer was present in the complement, the UV-vis spectrum showed a red shift and considerable broadening while the CD spectrum lost intensity. As the number of spacers increased from 1 to 6, the UV-vis spectrum shifted further to the red but became as narrow as the single stranded oligomer while the CD spectrum recovered most of the original intensity. These results indicate that the lack of a spacer group severely distorts the dye aggregate while the spacer elements allow the dye-containing strand to relax and reassemble the aggregate. In all cases, the CD spectrum for the D-threonine linked aggregates exhibited right-handed splitting. Interestingly, the L-threonine linked aggregate exhibited a weak, broadened CD spectrum with multiple components. It is possible that the preferred chirality of the aggregate, namely left-handed, is opposed by the right-handed DNA duplex regions flanking the dyes, leading to the distorted CD spectrum.

This example illustrates how covalent and noncovalent forces can be used for the assembly of novel dye aggregates with interesting optical properties. The stepwise synthesis process could be used not only for preparing homoaggregates, as in the case with methyl red, but also for making heteroaggregates with fine control over the placement of the different chromophores.

7.6 CONCLUSION

The point of this chapter was to describe how double-helical DNA has been used as a template on which helical dye aggregates with novel photophysical properties can be assembled. While potential applications in nonlinear optics and DNA detection can be envisioned, an additional advantage of these systems lies in the ability to control the supramolecular structure and investigate how the photophysics change as the aggregate grows. The use of DNA in this context relates to the intense recent interest in incorporating DNA into various nanostructures and devices [84,85]. While these structures have been investigated primarily with electronic or mechanical functions in mind, assembling dye aggregates along the DNA helix could allow integration of optical functions into DNA-based devices. Improvements in our understanding of how dye structure affects the sequence-dependence of DNA binding will provide a deeper level of control over the system, since discrete dye aggregates could be assembled in different regions on a single DNA template.

ACKNOWLEDGMENTS

Work performed in the Armitage lab has been generously supported by the National Science Foundation, Howard Hughes Medical Institute, and Carnegie Mellon University.

REFERENCES

1. G Wegner. *Naturforsch* 24B: 824–832, 1969.
2. G Wegner. *Makromol Chem* 154: 35–48, 1972.
3. GB Schuster. *Acc Chem Res* 33: 253–260, 2000.
4. FD Lewis, Y Wu. *J Photochem Photobiol C: Photochem Rev* 2: 1–16, 2001.
5. B Giese. *Annu Rev Biochem* 71: 51–70, 2002.
6. S Delaney, JK Barton. *J Org Chem* 68: 6475–6483, 2003.
7. J Li, GB Schuster, K-S Cheon, MM Green, JV Selinger. *J Am Chem Soc* 122: 2603–2612, 2000.
8. AN Melkozernov, RE Blankenship, *Photosynthesis Res.* 85: 35–50, 2005.
9. JD Watson, FHC Crick. *Nature* 171: 737–738, 1953.
10. Special issue of *Nature,* April 24, 2003 (Volume 422, Issue 6934).
11. JA Mountzouris, LH Hurley. In: SM Hecht, Ed. *Bioorganic Chemistry: Nucleic Acids.* New York: Oxford University Press, 1996, pp 288–323.
12. WD Wilson. In: GM Blackburn, MJ Gait, Eds. *Nucleic Acids in Chemistry and Biology.* Oxford: Oxford University Press, 1996, pp 329–374.
13. CHG Williams. *Trans R Soc Edinburg* 21: 377, 1856.
14. A Mishra, RK Behera, PK Behera, BK Mishra, GB Behera. *Chem Rev* 100: 1973–2011, 2000.
15. BA Armitage. *Top Curr Chem* 253: 55–76, 2005.

16. JL Seifert, RE Connor, SA Kushon, M Wang, BA Armitage. *J Am Chem Soc* 121: 2987–2995, 1999.
17. W West, S Pearce. *J Phys Chem* 69: 1894–1903, 1965.
18. AH Herz. *Photogr Sci Eng* 18: 323–335, 1974.
19. L Stryer, ER Blout. *J Am Chem Soc* 83: 1411–1418, 1961.
20. M Kasha. *Physical Processes in Radiation Biology*. Academic Press, New York, 1964.
21. M Kasha, HR Rawls, M Ashraf El-Bayoumi. *Molecular Spectroscopy. Proc. VIII European Congr. Molec. Spectroscopy*. London: Butterworths, 1965, p 371–392.
22. AS Davydov. *Theory of Molecular Excitons*. New York: Plenum Press, 1971.
23. EE Jelley. *Nature* 138: 1009–1101, 1936.
24. G Scheibe. *Angew Chem* 50: 212–219, 1937.
25. D Suh, JB Chaires. *Bioorg Med Chem* 3: 723–728, 1995.
26. BH Geierstanger, DE Wemmer. *Annu Rev Biophys Biomol Struct* 24: 463–493, 1995.
27. JG Pelton, DE Wemmer. *Proc Natl Acad Sci USA* 86: 5723–5727, 1989.
28. JG Pelton, DE Wemmer. *Biochemistry* 27: 8088–8096, 1989.
29. U Sehlstedt, SK Kim, B Nordén. *J Am Chem Soc* 115: 12258–12263, 1993.
30. X Chen, B Ramakrishnan, ST Rao, M Sundaralingam. *Nature Struct Biol* 1: 169–175, 1994.
31. X Chen, B Ramakrishnan, M Sundaralingam. *J Mol Biol* 267: 1157–1170, 1997.
32. K Nakanishi, N Berova, RW Woody. *Circular Dichroism: Principles and Applications*. New York: VCH Publishers, 1994.
33. R Garoff, EA Litzinger, RE Connor, I Fishman, BA Armitage. *Langmuir* 6330–6337, 2002.
34. D Yaron, BA Armitage, I Raheem, S Kushon, JL Seifert. *Nonlinear Opt* 257–264, 2000.
35. FA Tanious, D Hamelberg, C Bailly, A Czarny, DW Boykin, WD Wilson. *J Am Chem Soc* 126: 143–153, 2004.
36. GL Silva, M Wang M, BA Armitage, unpublished data.
37. R Cao, CF Venezia, BA Armitage. *J Biomol Struct Dynam* 18: 844–856, 2001.
38. A Chowdhury, L Yu, I Raheem, L Peteanu, LA Liu, DJ Yaron. *J Phys Chem A* 107: 3351–3362, 2003.
39. NV Dubinin. *Opt Spectrosc (USSR)* 43: 49–51, 1977.
40. H Kuhn, C Kuhn. In: T Kobayashi, Ed. *J-Aggregates*. Singapore: World Scientific, 1996, pp 1–41.
41. T Kobayashi, K Misawa. In: T Kobayashi, Ed. *J-Aggregates*. Singapore: World Scientific, 1996, pp 161–180.
42. R Gadonas. In: T Kobayashi, Ed. *J-Aggregates*. Singapore: World Scientific, 1996, pp 181–197.
43. B Nordén, F Tjerneld. *Biophys Chem* 6: 31–45, 1977.
44. M Wang, GL Silva, BA Armitage. *J Am Chem Soc* 122: 9977–9986, 2000.
45. A Chowdhury, S Wachsmann-Hogiu, PR Bangal, I Raheem, LA Peteanu. *J Phys Chem B* 105: 12196–12201, 2001.
46. TY Ogul'chansky, MY Losytskyy, VB Kovalska, SS Lukashov, VM Yashchuk, SM Yarmoluk. *Spectrochim Acta A* 57: 2705–2715, 2001.
47. MY Losytskyy, VM Yaschuk, SM Yarmoluk. *Mol Cryst Liq Cryst* 385: [147]/(127)–[152]/(132), 2002.

48. DF Bradley, MK Wolf. *Proc Natl Acad Sci USA* 45: 944–952, 1959.
49. DF Bradley, G Felsenfeld. *Nature* 184: 1920–1922, 1959.
50. DMJ Neville, DF Bradley. *Biochim Biophys Acta* 50: 397–399, 1960.
51. AL Stone, DF Bradley. *J Am Chem Soc* 83: 3627–3634, 1961.
52. RF Pasternack. *Chirality* 15: 329–332, 2003.
53. RF Pasternack, EJ Gibbs. In: H Sigel, Ed. *Metal Ions in Biological Systems, Vol 33.* New York: Marcel Dekker, 1996, pp 367–377.
54. MJ Carvlin, N Datta-Gupta, RJ Fiel. *Biochem Biophys Res Commun* 108: 66–73, 1982.
55. EJ Gibbs, I Tinoco Jr., MF Maestre, PA Ellinas, RF Pasternack. *Biochem Biophys Res Commun* 157: 350–358, 1988.
56. RF Pasternack, A Giannetto, P Pagano, EJ Gibbs. *J Am Chem Soc* 113: 7799–7800, 1991.
57. RF Pasternack, C Bustamante, PJ Collings, A Giannetto, EJ Gibbs. *J Am Chem Soc* 115: 5393–5399, 1993.
58. PJ Collings, EJ Gibbs, TE Starr, O Vafek, C Yee, LA Pomerance, RF Pasternack. *J Phys Chem B* 103: 8474–8481, 1999.
59. RF Pasternack, PJ Collings. *Science* 269: 935–939, 1995.
60. RF Pasternack, S Ewen, A Rao, AS Meyer, MA Freedman, PJ Collings, SL Frey, MC Ranen, JC de Paula. *Inorg Chim Acta* 317: 59–71, 2001.
61. LG Marzilli, G Pethö, M Lin, MS Kim, DW Dixon. *J Am Chem Soc* 114: 7575–7577, 1992.
62. NE Mukundan, G Pethö, DW Dixon, MS Kim, LG Marzilli. *Inorg Chem* 33: 4676–4687, 1994.
63. NE Mukundan, G Pethö, DW Dixon, LG Marzilli. *Inorg Chem* 34: 3677–3687, 1995.
64. JE McClure, L Baudouin, D Mansuy, LG Marzilli. *Biopolymers* 42: 203–217, 1997.
65. A Pullman, B Pullman. *Quart Rev Biophys* 14: 289–380, 1981.
66. R Lavery, B Pullman, K Zakrzewska. *Biophys Chem* 15: 343–351, 1982.
67. RJ Bushby, OR Lozman. *Curr Opin Coll Interface Sci* 7: 343–354, 2002.
68. T Kreouzis, K Scott, KJ Donovan, N Boden, RJ Bushby, OR Lozman, Q Liu. *Chem Phys* 262: 489–497, 2000.
69. XD Che, MH Liu. *J Inorg Biochem* 94: 106–113, 2003.
70. K Lang, P Anzenbacher, P Kapusta, P Kral, P Kubat, DM Wagnerova. *J Photochem Photobiol B: Biology* 57: 51–59, 2000.
71. BL Iverson, K Shreder, V Kral, P Sansom, V Lynch, JL Sessler. *J Am Chem Soc* 118: 1608–1616, 1996.
72. PE Nielsen, M Egholm, RH Berg, O Buchardt. *Science* 254: 1498–1500, 1991.
73. PE Nielsen. In: PE Nielsen, Ed. *Peptide Nucleic Acids: Methods and Protocols (Methods in Molecular Biology).* Towana, NJ: Humana Press, 2002, pp 3–26.
74. PE Nielsen, G Haaima. *Chem Soc Rev* 26: 73–78, 1997.
75. P Wittung, SK Kim, O Buchardt, PE Nielsen, B Nordén. *Nucleic Acids Res* 22: 5371–5377, 1994.
76. JO Smith, DA Olson, BA Armitage. *J Am Chem Soc* 121: 2686–2695, 1999.
77. SA Kushon, JP Jordan, JL Seifert, PE Nielsen, H Nielsen, BA Armitage. *J Am Chem Soc* 123: 10805–10813, 2001.
78. B Datta, BA Armitage. *J Am Chem Soc* 123: 9612–9619, 2001.
79. M Komiyama, S Ye, X Liang, Y Yamamoto, T Tomita, J-M Zhou, H Aburatani. *J Am Chem Soc* 125: 3758–3762, 2003.

80. LM Wilhelmsson, B Nordén, K Mukherjee, MT Dulay, RN Zare. *Nucleic Acids Res* 30, e3, 2002.
81. SH Gellman. *Acc Chem Res* 31: 173–180, 1998.
82. DJ Hill, MJ Mio, RB Prince, TS Hughes, JS Moore. *Chem Rev* 101: 3893–4011, 2001.
83. H Asanuma, K Shirasuka, T Takarada, H Kashida, M Komiyama. *J Am Chem Soc* 125: 2217–2223, 2003.
84. NC Seeman. *Annu Rev Biophys Biomol Struct* 27: 225–248, 1998.
85. NC Seeman. *Chem Biol* 10: 1151–1159, 2003.

Index